Lecture Notes in Computer Science 2653

Edited by G. Goos, J. Hartmanis, and J. van Leeuwen

Springer
Berlin
Heidelberg
New York
Barcelona
Hong Kong
London
Milan
Paris
Tokyo

Rossella Petreschi Giuseppe Persiano
Riccardo Silvestri (Eds.)

Algorithms
and Complexity

5th Italian Conference, CIAC 2003
Rome, Italy, May 28-30, 2003
Proceedings

 Springer

Series Editors

Gerhard Goos, Karlsruhe University, Germany
Juris Hartmanis, Cornell University, NY, USA
Jan van Leeuwen, Utrecht University, The Netherlands

Volume Editors

Rossella Petreschi
Riccardo Silvestri
Università di Roma "La Sapienza"
Dipartimento di Infornatica
Via Salaria 113, 00198 Rome, Italy
E-mail: {petreschi/silvestri}@dsi.uniroma1.it

Giuseppe Persiano
Università di Salerno
Dipartimento di Infornatica ed Applicazioni "R.M. Capocelli"
Via S. Allende 2, 84081 Baronissi, SA, Italy
giuper@dia.unisa.it

Cataloging-in-Publication Data applied for

A catalog record for this book is available from the Library of Congress.

Bibliographic information published by Die Deutsche Bibliothek
Die Deutsche Bibliothek lists this publication in the Deutsche Nationalbibliografie;
detailed bibliographic data is available in the Internet at <http://dnb.ddb.de>.

CR Subject Classification (1998): F.2, F.1, E.1, I.3.5, G.2

ISSN 0302-9743
ISBN 3-540-40176-8 Springer-Verlag Berlin Heidelberg New York

Springer-Verlag Berlin Heidelberg New York
a member of BertelsmannSpringer Science+Business Media GmbH

http://www.springer.de

© Springer-Verlag Berlin Heidelberg 2003
Printed in Germany

Typesetting: Camera-ready by author, data conversion by Steingräber Satztechnik GmbH
Printed on acid-free paper   SPIN: 10932575   06/3142   5 4 3 2 1 0

Preface

The papers in this volume were presented at the 5th Italian Conference on Algorithms and Complexity (CIAC 2003). The conference took place during May 28–30, 2003, in Rome, Italy, at the Conference Centre of the University of Rome "La Sapienza."

CIAC started in 1990 as a national meeting to be held every three years for Italian researchers in algorithms, data structures, complexity theory, and parallel and distributed computing. Due to a significant participation of foreign researchers, starting from the second edition, CIAC evolved into an international conference. However, all the editions of CIAC have been held in Rome. The proceedings of CIAC were published by World Scientific for the first edition and by Springer-Verlag in the Lecture Notes in Computer Science series (volumes 778, 1203 and 1767) for the subsequent editions. A selection of the papers of the fourth edition was published in a special issue of Theoretical Computer Science Vol. 285(1), 2002. This year we expect to publish an extended version of selected papers presented at the conference in a special issue of the journal Theory of Computing Systems.

In response to the call for papers for CIAC 2003, 57 papers were submitted, from which the Program Committee selected 23 papers for presentation at the conference from 18 countries. Each paper was evaluated by at least three Program Committee members with the help of 63 external reviewers.

In addition to the selected papers, the Organizing Committee invited Charles E. Leiserson (Cambridge), David Peleg (Rehovot), Michael O. Rabin (Cambridge and Jerusalem), John E. Savage (Providence), and Luca Trevisan (Berkeley) to give plenary lectures at the conference. Moreover, three tutorials by David Peleg, Jayme L. Szwarcfiter (Rio de Janeiro) and Luca Trevisan were offered in the days preceding the conference.

We wish to express our appreciation to all the authors of the submitted papers, to the Program Committee members and the referees, to the Organizing Committee, and to the plenary and tutorial lecturers who accepted our invitation. We are grateful to the University of Rome "La Sapienza" for giving us the opportunity to use the Conference Centre, to the Department of Computer Science of Rome University for offering the use of their facilities, and to the Università di Salerno for partially funding CIAC 2003.

Rome, March 2003

Giuseppe Persiano
Rossella Petreschi
Riccardo Silvestri

Organization

Organizing Committee

R. Silvestri (Chair, Rome)
G. Bongiovanni (Rome)
I. Finocchi (Rome)

A. Monti (Rome)
R. Petreschi (Rome)

Program Committee

G. Persiano (Chair, Salerno)
A. Brandstadt (Rostock)
T. Calamoneri (Rome)
A. Clementi (Rome)
R. Giancarlo (Palermo)
R. Grossi (Pisa)
K. Iwama (Kyoto)
C. Kaklamanis (Patras)

K. Jansen (Kiel)
A. Marchetti-Spaccamela (Rome)
Y. Rabani (Haifa)
J.L. Szwarcfiter (Rio de Janeiro)
P. Toth (Bologna)
J. Urrutia (Mexico City)
G.J. Woeginger (Twente)

Additional Referees

C. Ambuehl
A. Barnoy
C.F. Bornstein
I. Caragiannis
Z. Chen
J. Chlebikova
P. Cintioli
R. Corra
C.M.H. de Figueiredo
C. Demetrescu
G. Di Crescenzo
M. Di Ianni
O. Dunkelman
L. Epstein
C.G. Fernandes
I. Finocchi
A.V. Fishkin
A. Frangioni
C. Galdi
O. Gerber
K.S. Guimares

T. Hagerup
S. Hirose
H. Ito
P. Kanellopoulos
S. Khanna
S. Klein
C. Konstantopoulos
E. Kushilevitz
R. Labahn
V.B. Le
M. Leoncini
F. Luccio
C. Malvenuto
M. Margraf
A. Massini
J. Meidanis
S. Miyazaki
A. Monti
M. Moscarini
A. Navarra
L. Nogueira

M. Ogihara
A. Panconesi
M. Parente
P. Penna
L.D. Penso
R. Petreschi
V. Raman
G. Rossi
R. Silvestri
J. Soares
H. Tamaki
A. Taraz
T. Tsukiji
A. Urpi
I. Visconti
P. Vocca
B. Voecking
I. Wegener
H. Zhang
G. Zhang
D. Wagner

Invited Presentations

Tutorials

Localized Network Representations
David Peleg

The Weizmann Institute,
Rehovot, Israel

Optimal Binary Search Trees
with Costs Depending
on the Access Paths
Jayme Luiz Szwarcfiter

Universidade Federal do RdJ,
Rio de Janeiro, Brasil

On the Generation of Extensions
of a Partially Ordered Set
Jayme Luiz Szwarcfiter

Universidade Federal do RdJ,
Rio de Janeiro, Brasil

Error-Correcting Codes
in Complexity Theory
Luca Trevisan

University of California,
Berkeley, USA

Invited Talks

Cache-Oblivious Algorithms
Charles E. Leiserson

MIT, Cambridge, USA

Spanning Trees with Low
Maximum/Average Stretch
David Peleg

The Weizmann Institute,
Rehovot, Israel

Hyper Encryption
and Everlasting Secrets, a Survey
Michael O. Rabin

Harvard University, Cambridge, USA

Computing
with Electronic Nanotechnologies
John E. Savage

Brown University, Providence, USA

Sub-linear Time Algorithms
Luca Trevisan

University of California, Berkeley, USA

Table of Contents

Localized Network Representations

David Peleg

The Weizmann Institute, Rehovot, Israel

Abstract. The talk will concern compact, localized and distributed network representation methods. Traditional approaches to network representation are based on global data structures, which require access to the entire structure even if the sought information involves only a small and local set of entities. In contrast, localized network representation schemes are based on breaking the information into small local pieces, or labels, selected in a way that allows one to infer information regarding a small set of entities directly from their labels, without using any additional (global) information. The talk will concentrate mainly on combinatorial and algorithmic techniques, such as adjacency and distance labeling schemes and interval schemes for routing, and will cover a variety of complexity results.

R. Petreschi et al. (Eds.): CIAC 2003, LNCS 2653, p. 1, 2003.
© Springer-Verlag Berlin Heidelberg 2003

Optimal Binary Search Trees
with Costs Depending on the Access Paths

Jayme L. Szwarcfiter

Universidade Federal do Rio de Janeiro, RJ, Brasil

Abstract. We describe algorithms for constructing optimal binary search trees, in which the access cost of a key depends on the k preceding keys, which were reached in the path to it. This problem has applications to searching on secondary memory and robotics. Two kinds of optimal trees are considered, namely optimal worst case trees and weighted average case trees. The time and space complexity of both algorithms are $O(n^{k+2})$ and $O(n^{k+1})$, respectively. The algorithms are based on a convenient decomposition and characterizations of sequences of keys, which are paths of special kinds in binary search trees. Finally, using generating functions, the exact number of steps performed by the algorithms has been calculated. The subject will be introduced by a general discussion on the construction of optimal binary search trees.

R. Petreschi et al. (Eds.): CIAC 2003, LNCS 2653, p. 2, 2003.
© Springer-Verlag Berlin Heidelberg 2003

On the Generation of Extensions
of a Partially Ordered Set

Jayme L. Szwarcfiter

Universidade Federal do Rio de Janeiro, RJ, Brasil

Abstract. A partially ordered set (or simply order) P is a set of elements E, together with a set R of relations of E, satisfying reflexivity, anti-symmetry and transitivity. The set E is called the ground set of P, while R is the relation set of it. There are many special orders. For example, when any two elements of E are related, the order is a chain. Similarly, we can define tree orders, forest orders and many others. An extension P' of P is an order P' having the same ground set as P, and such that its relation set contains R. When P' is a chain then P' is a linear extension of P. Similarly, when P' is a forest then it is a forest extension of P. We consider the algorithmic problem of generating all extensions of a given order and also extensions of a special kind. The subject will be introduced by a general discussion on partially ordered sets.

R. Petreschi et al. (Eds.): CIAC 2003, LNCS 2653, p. 3, 2003.

Error-Correcting Codes in Complexity Theory

Luca Trevisan

University of California, Berkeley, USA

Abstract. Error-correcting codes and related combinatorial constructs play an important role is several recent (and old) results in complexity theory. This course will give a brief overview of the theory, constructions, algorithms, and applications of error-correcting codes. We will begin with basic definitions and the constructions of Reed-Solomon, Reed-Muller, and low-weight parity-check codes, then see unique-decoding and list-decoding algorithms, and finally, as time allows, applications to secret-sharing, hashing, private information retrieval, average-case complexity and probabilistically checkable proofs.

R. Petreschi et al. (Eds.): CIAC 2003, LNCS 2653, p. 4, 2003.
© Springer-Verlag Berlin Heidelberg 2003

Cache-Oblivious Algorithms

Charles E. Leiserson

MIT Laboratory for Computer Science, Cambridge, MA 02139, USA

Abstract. Computers with multiple levels of caching have tradition-
ally required techniques such as data blocking in order for algorithms to
exploit the cache hierarchy effectively. These "cache-aware" algorithms
must be properly tuned to achieve good performance using so-called
"voodoo" parameters which depend on hardware properties, such as
cache size and cache-line length.

Surprisingly, however, for a variety of problems – including matrix mul-
tiplication, FFT, and sorting – asymptotically optimal "cache-oblivious"
algorithms do exist that contain no voodoo parameters. They perform
an optimal amount of work and move data optimally among multiple
levels of cache. Since they need not be tuned, cache-oblivious algorithms
are more portable than traditional cache-aware algorithms.

We employ an "ideal-cache" model to analyze these algorithms. We prove
that an optimal cache-oblivious algorithm designed for two levels of memory is also optimal across a multilevel cache hierarchy. We also show that
the assumption of optimal replacement made by the ideal-cache model
can be simulated efficiently by LRU replacement. We also provide some
empirical results on the effectiveness of cache-oblivious algorithms in
practice.

R. Petreschi et al. (Eds.): CIAC 2003, LNCS 2653, p. 5, 2003.
© Springer-Verlag Berlin Heidelberg 2003

Spanning Trees
with Low Maximum/Average Stretch

David Peleg

The Weizmann Institute, Rehovot, Israel

Abstract. The talk will provide an overview of problems and results concerning spanning trees with low maximum or average stretch and trees with low communication cost, in weighted or unweighted graphs and in metrics, and outline some techniques for dealing with these problems.

R. Petreschi et al. (Eds.): CIAC 2003, LNCS 2653, p. 6, 2003.

Hyper Encryption and Everlasting Secrets

A Survey

Michael O. Rabin

Harvard University, Cambridge, USA*

A fundamental problem in cryptography is that of secure communication over an insecure channel, where a sender Alice wishes to communicate with a receiver Bob, in the presence of a powerful Adversary $\mathcal{AD}$. The primary goal of encryption is to protect the privacy of the conversation between Alice and Bob against $\mathcal{AD}$. Modern cryptographic research has identified additional essentially important criteria for a secure encryption scheme. Namely that the encryption be non-malleable, be resistant to various chosen plaintext and ciphertext attacks, and if so desired, will allow the receiver to authenticate the received message and its sender. All these issues are now settled for the case that the Adversary $\mathcal{AD}$ is computationally unbounded.

If we wish to achieve unconditionally secure communication against an all-powerful adversary, we must use private-key cryptosystems.In a public-key cryptosystem, the encryption function using the public key, and the decryption function using the private key, are inverses of each other, and the public key is publicly known. Thus, a computationally unbounded adversary can, from the public key, compute the corresponding private key, using unlimited computing power. Private-key encryption, in which the sender and the receiver share a common secret key for the encryption and decryption of messages, were proposed already in antiquity, dating back to as early as Julius Caesar, and are widely used today. A provably secure example of private-key encryption is the simple *one-time pad* scheme of Vernam.

In the one-time pad scheme, for the encryption and transmission of each message M, the sender Alice and the receiver Bob establish a shared random secret key X called a *one-time pad*, with $|X| = |M|$, where $|\cdot|$ denotes the length (i.e. number of bits) of a binary string. Alice encrypts M by computing $C = M \oplus X$, where $\oplus$ denotes bit-wise XOR. Bob decrypts C by computing $C \oplus X = M \oplus X \oplus X = M$.

The one-time pad scheme achieves information-theoretic secrecy, provided that for each transmission of a message, a *new* independent, uniformly random one-time pad, whose size equals that of the message, is established between Alice and Bob. In fact, a single careless re-use of the same one-time pad to encrypt a second message can result in the decryption of both ciphertexts. The non-reusability of the shared key renders this method impractical except for special situations where a large code book can be securely shipped from a sender to a receiver ahead of the transmissions. Furthermore, if an adversary captures and

* Supported by NSF contract CCR-0205423. e-mail: rabin@deas.harvard.edu

R. Petreschi et al. (Eds.): CIAC 2003, LNCS 2653, pp. 7–10, 2003.
© Springer-Verlag Berlin Heidelberg 2003

stores an encrypted message $C = M \oplus X$, where X is the one-time pad, and later on gets X (steals the code book), then he can decode: $M = X \oplus C$. The same holds for any existing private or public key encryption scheme.

In a seminal work, Shannon proved that if the Adversary $\mathcal{AD}$ has complete access to the communication line and is unbounded in computational power, then information-theoretically secure private-key communication is only possible if the entropy of the space of secret keys is as large as that of the plaintext space. Essentially this implies that for achieving provable security, the shared one-time pad method, with its drawbacks, is optimal. Thus to overcome the limitations imposed by Shannon's result and have provable information theoretic security against a computationally unbounded Adversary, some other restriction has to be imposed.

Striving to produce a provably secure encryption scheme, which has the additional property that revealing the decryption key to an adversary after transmission of the ciphertext does not result in the decryption of the ciphertext, is important. Computational complexity based cryptosystems rely on unproven assumptions. If an adversary captures and stores ciphertexts, subsequent advances in algorithms or in computing power, say the introduction of quantum computers, may allow him to decrypt those ciphertexts.

A moment's thought shows that if the adversary can capture all the information visible to the sender and receiver, and later obtains the decryption key, then he will be able to decrypt the ciphertext just as the receiver does. Thus one is led to consider an adversary who is computationally unbounded but is limited in his storage capacity.

The Bounded Storage Model

We consider the bounded storage model, where the security guarantees are based on a *bound on the storage capacity* of the adversary. This model, introduced by Maurer in his influential work [Mau92], assumes that there is a known bound B, possibly very large but fixed at any given time, on the Adversary $\mathcal{AD}$'s storage capacity. It is important to differentiate between this model and the *bounded space model*.

The bounded space model considers situations where the space available to the code breaker for *computation* is limited, e.g. log space or linear space. Thus, the space bound is in effect a limitation on the computational power of the Adversary. The bounded *storage* model, however, stipulates no limitation on the computational power of the Adversary $\mathcal{AD}$ who tries to subvert the security of the protocol. Imposing the bound B on the Adversary's storage allows us to construct efficient encryption/decryption schemes that are *provably information-theoretically secure*, and require of the sender Alice and the receiver Bob very modest computations and storage space.

In the basic form, the encryption schemes of [AR99], [ADR02], [DR02], [Mau92] work as follows. Alice and Bob utilize a publicly accessible string α of n random bits, where for a fixed $\gamma < 1$, γn is larger than the bound B on the Adversary $\mathcal{AD}$'s storage capacity. In a possible implementation, the string α may be one in a stream of public random strings $\alpha_1, \alpha_2, \ldots$, each of length n, continually created in and beamed down from a satellite system, and available to all. Alice and Bob share a randomly chosen secret key s. For transmitting an encrypted message M, $|M| = m$, Alice and Bob listen or have listened to some $\alpha = \alpha_i$ (possibly a previously beamed α_i, see below), and by use of s create a common one-time pad $X(s, \alpha)$. In the schemes of [AR99], [ADR02], [Mau92], computing $X(s, \alpha)$ requires just fast XOR operations and very little memory space of Alice and Bob. Alice encrypts M as $C = M \oplus X(s, \alpha)$, and Bob decrypts C by $M = C \oplus X(s, \alpha)$. Every α is used only once. But the same α can be simultaneously used by multiple Sender/Receiver pairs without degrading security.

The Adversary $\mathcal{AD}$ listens to α, computes a *recording function* $A_1(\alpha) = \eta$, where $|\eta| = B \leq \gamma n$, $\gamma < 1$, and stores η. The Adversary also captures the ciphertext C. Later $\mathcal{AD}$ is given s, and applies a decoding algorithm $A_2(\eta, s, C)$ to gain information about M. There is no limitation on the space or work required for computing $A_1(\alpha)$ and $A_2(\eta, s, C)$. Theorem 1 of [AR99] and Theorem 2 of [ADR02], and the Main Theorem of [DR02], say that the Aumann-Rabin and Aumann-Ding-Rabin schemes are *absolutely semantically secure*. Namely, for $B = 0.3n$ (later we will use $\gamma = 1/6$ without loss of generality),[1] for every two messages $M_0, M_1 \in \{0, 1\}^m$, for an encryption $C_\delta = M_\delta \oplus X(s, \alpha)$, where $\delta \in \{0, 1\}$ is randomly chosen,

$$| \Pr [A_2(A_1(\alpha), s, C_\delta) = \delta] - 1/2| \leq m \cdot 2^{-k/3}. \tag{1}$$

Here, k is a security parameter, e.g. $k = 200$. The security obtained in the above results is the absolute version (i.e. in a model allowing a computationally unbounded adversary) of *semantic security* pioneered by Goldwasser and Micali. Note that the storage bound B is needed only at the time α is broadcast, and the recording function $\eta = A_1(\alpha)$ is computed. The decoding algorithm $A_2(\eta, s, C)$ can use unbounded computing power and storage space, and thus subsequent increase in storage does not help the Adversary $\mathcal{AD}$.

Extraction of Randomness

An important issue in Hyper Encryption is the size of the common secret key S used by the sender and receiver. Reducing the key size is intimately connected to the problem of extracting nearly perfect uniform randomness from a weaker source of randomness. However, the extractors required for implementations of

---

[1] The pedagogical choice of $\gamma = 0.3$ is for convenience only, and leads to the probability $2^{-k/3}$ in (1). Similar results are obtainable for any constant $\gamma < 1$.

Hyper Encryption must have specialized properties beyond those of general extractors. Important advances were made on this problem in the past two years in [DM02], [Lu02], [Vad02].

References

ADR02. Yonatan Aumann, Yan Zong Ding, and Michael O. Rabin. Everlasting security in the bounded storage model. *IEEE Transactions on Information Theory*, 48(6):1668–1680, June 2002.

AR99.   Yonatan Aumann and Michael O. Rabin. Information theoretically secure communication in the limited storage space model. In *Advances in Cryptology—CRYPTO '99*, Lecture Notes in Computer Science, pages 65–79. Springer-Verlag, 1999, 15–19 August 1999.

CM97.   Christian Cachin and Ueli Maurer. Unconditional security against memory-bounded adversaries. In Burton S. Kaliski Jr., editor, *Advances in Cryptology — CRYPTO '97*, volume 1294 of *Lecture Notes in Computer Science*, pages 292–306. Springer-Verlag, 1997.

DR02.   Yan Zong Ding and Michael O. Rabin. Hyper-encryption and everlasting security (extended abstract). In *STACS 2002 — 19th Annual Symposium on Theoretical Aspects of Computer Science*, Lecture Notes in Computer Science, pages 1–26. Springer-Verlag, 14–16 March 2002.

DM02.   Stefan Dziembowski and Ueli Maurer. Tight security proofs for the bounded-storage model. In *Proceedings of the Thirty-Fourth Annual ACM Symposium on Theory of Computing*, pages 341–350, Montreal, 19–21 May 2002.

Lu02.   Chi-Jen Lu. Hyper-encryption against space-bounded adversaries from online strong extractors. In *Advances in Cryptology—CRYPTO '02*, Lecture Notes in Computer Science. Springer-Verlag, 2002, 18–22 August 2002. To appear.

Mau92. Ueli Maurer. Conditionally-perfect secrecy and a provably-secure randomized cipher. *Journal of Cryptology*, 5(1):53–66, 1992.

Vad02. Salil P. Vadhan. On constructing locally computable extractors and cryptosystems in the bounded storage model. Cryptology ePrint Archive, Report 2002/162, 2002. `http://eprint.iacr.org/`.

Computing with Electronic Nanotechnologies*

John E. Savage

Department of Computer Science, Brown University, Providence, RI 02912-1910, USA

Abstract. Computing with electronic nanotechnologies is emerging as a real possibility. In this talk we explore the role theoretical computer scientists might play in understanding such technologies and give examples of completed research.

We consider two representative problems, whether nanowire (NW) address decoders can be self-assembled reliably and whether data can be stored efficiently in crossbar nanoarrays, a means for data storage and computation.

Recent research suggests that address decoders for a set of parallel NWs can be realized using modulation doping. This is a process in which NWs are grown with embedded electronic switches (field-effect transistors (FETs)) that can be controlled by microwires. h-hot addressing allows a small number of microwires to activate one NW by activating its h FETs. We examine the feasibility of stochastically self-assembling an address decoder using such technology.

The crossbar array, two orthogonal sets of parallel wires placed one above the other, is one of the most promising nanotechnology architectures under consideration. Small crossbars have been self-assembled from carbon nanotubes (CNTs) and semiconducting NWs. Two media for binary data storage in nanoarrays have been proposed, namely, changing the state of molecules layered between orthogonal sets of wires and making or breaking mechanical contacts between these wires.

Since nanoarrays are expected to be very large, we examine two key questions: (a) "What are the most efficient ways of entering data into such arrays?" and (b) "How difficult is it to find a minimal or near-minimal number of steps to program an array using h-hot addressing when either 1s or 0s can be written into subarrays on each step?"

A partial answer to (a) is that some commonly occurring arrays can be programmed much more rapidly when both 1s and 0s can be written than when only 1s can be written. The answer to (b) is that it is **NP**-hard unless the number of 1s in each row or column is bounded or h is large.

* This research was funded in part by NSF Grant CCR-0210225.

R. Petreschi et al. (Eds.): CIAC 2003, LNCS 2653, p. 11, 2003.

Efficient Update Strategies
for Geometric Computing with Uncertainty

Richard Bruce[1], Michael Hoffmann[1], Danny Krizanc[2], and Rajeev Raman[1]

[1] Department of Mathematics and Computer Science, University of Leicester,
Leicester LE1 7RH, UK
`r.bruce@mcs.le.ac.uk`, `m.hoffmann@mcs.le.ac.uk`, `r.raman@mcs.le.ac.uk`
[2] Department of Mathematics and Computer Science, Wesleyan University,
Middletown, CT 06459, USA
`dkrizanc@cs.wesleyan.edu`

Abstract. We consider the problems of computing *maximal points* and
the *convex hull* of a set of points in 2D, when the points are "in motion."
We assume that the point locations (or trajectories) are *not* known precisely and determining these values exactly is feasible, but expensive. In
our model, the algorithm only knows areas within which each of the input
points lie, and is required to identify the maximal points or points on the
convex hull correctly by *updating* some points (i.e. determining exactly
their location). We compare the number of points updated by the algorithm on a given instance to the minimum number of points that must
be updated by an omniscient adversary in order to provably compute
the answer correctly. We give algorithms for both of the above problems
that always update at most 3 times as many points as the adversary, and
show that this is the best possible. Our model is similar to that of [5,2].

1 Introduction

In many applications, an intrinsic property of the data one is dealing with is that
it is "in motion," i.e., changing value (within prescribed limits) over time. For
example, the data may be derived from a random process such as stock market
quotes or queue lengths in switches or it may be positional data for moving
objects such as planes in an air traffic control area or users in a mobile ad-hoc
network. A lot of work has gone into developing on-line algorithms [1] and kinetic
data structures [4] in order to compute efficiently in these situations.

Most of the previous approaches to data in motion assume that the actual
data values are known precisely at all times or that their is no cost in establishing
these values. This is not always the case. In reality finding the exact value of some
data item may involve costs in time, energy, money, bandwidth, etc. Accurate
and timely stock quotes cost money. Remote access to the state of network queues
costs time and bandwidth. Querying battery-powered units of sensor networks
unnecessarily uses up precious energy.

In order to study the costs associated with updating data in an uncertain
environment, we consider the *update complexity* of a problem. A problem instance

R. Petreschi et al. (Eds.): CIAC 2003, LNCS 2653, pp. 12–23, 2003.

consists of a function of n inputs to be computed (e.g., the maxima) and a specification of the possible values each of the inputs might obtain (e.g., a set of n real intervals). An update strategy is a adaptive algorithm for deciding which of the inputs should be updated (i.e., be determined exactly) in order to correctly compute the function. An omniscient strategy (with knowledge of the exact values of all inputs) updates a set of inputs of minimum size which provably computes the function correctly. An update strategy is said to be *c-update optimal* if it updates at most $c * OPT + O(1)$ inputs, where OPT is the minimum number of updates performed by the omniscient strategy. The notion of update complexity is implicit in Kahan's [5] model for data in motion. In the spirit of online competitive analysis he defined the *lucky* ratio of an update strategy on a sequence of queries whereby a strategy competes against a "lucky" strategy with knowledge of the future. Kahan provides update optimal strategies for the problems of finding the maximum, median and the minimum gap of a set of n real values constrained to fall in a given set of n real intervals.

Motivated by the situation where one maintains in a local cache, intervals containing the actual (changing) data values stored at a remote location, Olston and Widom [7] studied a similar notion. In their model, the costs associated with updating data items may vary with the item and the function need not be computed exactly but only to within a given tolerance. A series of papers [2,3,6] establish tradeoffs between the update costs and the error tolerance and/or give complexity results for computing optimal strategies for such problems as selection, sum, average and computing shortest paths.

For the most part, the above results assume the uncertainty of a data item is best described by a real interval that contains it. In a number of situations, the uncertainty is more naturally captured by regions in two- (or higher) dimensional space. This is especially the case for positional data of moving objects with known upper bounds to their speed and possible constraints on their trajectories. The functions to be computed in these situations are most often geometric in nature. For example, to establish the which planes to deal with first, an air traffic controller would be interested in computing the closest pair of points in three dimensions. To apply greedy directional routing in an mobile ad-hoc network, a node must establish which of its neighbors is currently closest to the destination of a packet it is forwarding. To determine the extent of the coverage area of a mobile sensor network, one would like to compute the convex hull of the sensor's current positions.

In this paper, we describe a general method, called the *witness algorithm*, for establishing upper bounds on the update complexity of geometric problems. The witness algorithm is used to derive update optimal strategies for the problems of finding all maximal points and reporting all points on the convex hull of a set of moving points whose uncertainty may be described by the closure of open connected regions on the plane. The restriction to connected regions is necessary in order to ensure the existence of strategies with bounded update complexity. For both of these problems we provide examples that show our update strategies are optimal.

The remainder of the paper is organized as follows. In section 2 we describe a general method for establishing upper bounds on the update complexity for geometric problems. This approach is then used in sections 3 and 4 where optimal strategies for finding maximal points and points on the convex hull are given, respectively.

2   Preliminaries

We begin by giving some definitions. An input instance is specified by a set P of points in $\Re^2$, and associated with each point $p \in P$, an area A_p which includes p. Let $S \subseteq P$ be a set of points with some property ϕ, such as the set of all maximal points in P, or the set of points on the convex hull of P. For convenience, we say that $p \in P$ has the property ϕ if $p \in S$, and that A_p has the property ϕ if p has. The algorithm is given only the set $\{A_p | p \in P\}$, and must return return all areas which have the property ϕ.

In order to determine the areas with property ϕ, the algorithm may *update* an area A_p and determine the exact location of the point p with which it is associated. This reduces A_p to a *trivial* area containing only one point, namely p. The performance of an algorithm is measured in terms of the number of updates it performs to compute the answer; in particular it should be expressible as a function of the minimal number of updates OPT required to verify the solution. An update strategy is said to *c-update optimal* if it updates at most $c*OPT+O(1)$ inputs, and *update-optimal* if it is c-update optimal for some constant $c \geq 1$. Note that the algorithm may choose to return areas that are not updated as part of the solution. Indeed, in some instances, an algorithm may not need to update *any* areas to solve the problem.

In section 3 and 4 we give two update-optimal algorithms for the maximal points and convex hull problems, respectively. These two algorithms are instances of the same generic algorithm, the *witness algorithm*. In this section we describe this algorithm. We begin by giving some definitions. For a given set of areas $\mathcal{F} = \{A_1, \ldots, A_n\}$ we call $C = \{p_1, \ldots, p_n\}$ a *configuration* for $\mathcal{F}$ if $p_i \in A_i$ for $i = 1, \ldots, n$.

Definition 1. *Let $\mathcal{F}$ be a set of areas, $A \in \mathcal{F}$ and $p \in A$. Then:*

- *The point p is always ϕ if for any configuration C of $\mathcal{F} - \{A\}$, the point p has the property ϕ in $C \cup \{p\}$.*
- *The point p is never ϕ if for any configuration C of $\mathcal{F} - \{A\}$, the point p does not have the property ϕ in $C \cup \{p\}$.*
- *The point p is dependent ϕ if for at least one configuration C of $\mathcal{F} - \{A\}$, the point p has the property ϕ in $C \cup \{p\}$ and p is not always ϕ.*
- *a dependent ϕ point p depends on an area B, if for at least one configuration C of $\mathcal{F} - \{A, B\}$, there exist points b_1 and b_2 in B such that the point p has the property ϕ in $C \cup \{p, b_1\}$ and does not have the property ϕ in $C \cup \{p, b_2\}$.*

We extend this notion from points to areas.

Definition 2. *Let $\mathcal{F}$ be a set of areas. Let A be an area in $\mathcal{F}$. We say*

- *the area A is always ϕ if every point in A is always ϕ.*
- *the area A is partly ϕ if A contains at least one always ϕ point and A is not always ϕ.*
- *the area A is dependent ϕ if A contains at least one dependent ϕ point and A is not partly ϕ.*
- *the area A is never ϕ if every point in A is never ϕ.*

Definition 3. *Let $\mathcal{F}$ be a set of areas. Let C be a set of configuration for $\mathcal{F}$. A set W of areas in $\mathcal{F}$ is called a* witness set *of $\mathcal{F}$ with respect to C if for any configuration in C at least one area in W must have been updated to verify the solution. We say W is a* witness set *of $\mathcal{F}$ if W is a witness set of F with respect to all possible configurations of $\mathcal{F}$.*

The witness algorithm is as follows:
step 1: **while** (there exists at least one partly ϕ or one dependent ϕ area)
step 2:   **if** there exist a partly ϕ area
step 3:   update a witness set.
step 4:   **else** // there must exists a dependent ϕ area //
step 5:   update a witness set
step 6: **end**

Note that the idea is to concentrate first on partly ϕ areas and witness sets concerning these areas. Only if there are no partly ϕ areas left in the given instance, witness sets based on the existence of dependent areas will get updated.

The split in these two cases helps to identify witness sets.

Note further that, by updating certain areas we change the set of areas $\mathcal{F}$ and work from then on with a new set of areas $\mathcal{F}'$ where the updated areas are trivial. However, a witness set in $\mathcal{F}'$ is also witness set in $\mathcal{F}$ with respect to all possible configurations of $\mathcal{F}'$.

Lemma 1. *If there exists a constant k such that every witness set that gets updated by the witness algorithm is of size at most k, then the witness algorithm is k-update optimal.*

Proof. (sketch): We can partition the set up into areas which will be updated by the algorithm in blocks of size at most k and each block contains at least one area which must be updated to verify the result.

3   Maximal Points

We now give an update-optimal algorithm for the maximal problem when the areas are either trivial or closures of connected, open areas. We then note (Fig. 2) that there is no update-optimal algorithm for arbitrary areas.

Given two points $p = (p_x, p_y)$ and $q = (q_x, q_y)$ we say that $p > q$ if $p \neq q$ and $p_x \geq q_x$ and $p_y \geq q_y$. We say a line l *intersects* an area A if they share a common point. We say a line l *splits* an area A if $A - l$ is not connected.

Lemma 2. *Let p be a dependent maximal point. Let Y be the set of areas p depends on. Then, all areas of Y must be updated to show p is maximal.*

Proof. Let B be an area on which p depends. Hence there exists $b_1, b_2 \in B$ with $b_1 \not> p$ and $b_2 > p$. In order to verify that p is maximal the area B must be updated.

Lemma 3. *Let A be a partly maximal area, then there exists a witness set of size at most 2.*

Proof. Since the area A is partly maximal it contains at least one always maximal point. It also contains either a never maximal point or a dependent maximal point. We look at these cases separately.
Case a) The area A contains a never maximal point. In order to verify that A is maximal or not, we have to update the area A. Therefore the set $\{A\}$ is a witness set.
Case b) The area A contains a dependent maximal point p. Since the area A is partly, by updating only areas other then A the area A can change its status to only an always maximal area. Let B be an area on which p depends. By Lemma 2, in order to verify that p is maximal, we have to at least update B. Hence $\{B, A\}$ is a witness set of size two.

Lemma 4. *Let l be a horizontal or vertical line. Further let l split two areas A and B. Then the areas A and B are not both always maximal areas.*

Proof. Since all areas are closures of connected open areas the intersection of l with A must contain an open interval and similarly with B. So there exists $a \in A \cap l$ and $b \in B \cap l$ with $a \neq b$. Therefore either $a < b$ or $b < a$. Hence not both A and B are always maximal areas.

Lemma 5. *If there are no partly maximal areas, but there exists a dependent area, then there exists a witness set of size at most 3.*

Proof. Let A be the area with a dependent maximal point $p \in A$, such that there is no dependent maximal point q in any area such that $q > p$. In other words p is maximal among the all dependent maximal points. Note that p must exist since all area are closed. Let l_1 be vertical line starting at p and going upwards. Let l_2 be the horizontal line starting at p and going to the right. Since p is maximal among the dependent points, every point that lies in $l_1 \cup l_2 - \{p\}$ and also in an area other than A is an always maximal point. Let Q be the top right quadrant of l_1 and l_2 including l_1 and l_2 but not $\{p\}$. Since p is dependent there exists at least one area B with a point in Q and a point not in Q. Since p is maximum among the dependent points every point in Q is always. By our assumption that there are no partly areas the area B must be always. Hence it does not contain p. Let l_1' be the line l_1 shifted ϵ to the left and let l_2' be the line l_2 shifted ϵ down, where $\epsilon > 0$ is arbitrarily small. Since all areas are the closure of connected, open areas B must be split either l'_1 or l'_2. By Lemma 4 there are at most two areas that contain a point in Q as well as a point not in Q. Lets call these areas B and C and let $W = \{B, C, A\}$. Note, it is possible that only one of the areas B or C exists. To verify whether A is an always maximal area or not we have at least to update A or we have to verify that p is or is not maximal. For this we have to either update B or C or both. However W is a witness set.

We now show:

Theorem 1. *Under the restriction to the closure of open, connected areas or trivial areas, the witness algorithm for the maximal point problem is 3-update optimal. Furthermore, this is the best possible.*

Proof. The above lemmas show that there is always a witness set of size less than 3. By Lemma 1 we have that the witness algorithm is 3-update optimal. We now argue that there is no c-update optimal algorithm for any $c < 3$.

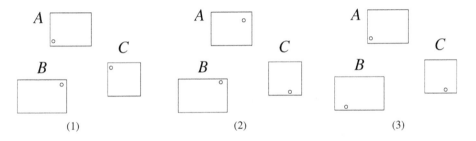

Fig. 1. The three configurations.

Consider the three areas A, B and C in Figure 1. The areas A and C are always maximal areas, but updates are needed to determine whether the point of area B is maximal. For any strategy S in updating these areas there exists a configuration of points for these three areas such that S requires three update where actually only one was needed. For example, consider that S updates A first. We choose one of (1) and (3) to be the input. If the algorithm updates B next, we say that (1) is the input, and force the algorithm to update C as well. However, both (1) and (3) can be "solved" by updating C and B respectively. To show the lower bound, we simply repeat this configuration arbitrarily often, with each configuration lying below and to the right of the previous one.

Remark 1. Finally, we note that there is no update-optimal algorithm for the maximal problem with arbitrary areas. Consider the situation shown in Figure 2. Assume there are n intervals on the line segment l such that after a horizontal projection each interval contains the projection of A. Hence each interval is maximal and for any strategy of determine whether A is maximal we have to update some of these intervals. For any given strategy there exists a configuration

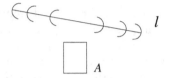

Fig. 2. Example showing that update-optimality is impossible for unrestricted areas.

such that the first $n - 1$ updated intervals correspond to points that lie in a horizontal projection to the left of A and the n's updated interval corresponds to a point such that no p oint in A can be maximal. If this interval would have been updated first the status of A could have been determined by only one update. Since n was arbitrary there exists no c-update optimal algorithm for arbitrary areas.

4   Convex Hull

Based on the witness algorithm, described in Section 2, we now give an algorithm for computing the convex hull. Again, we restrict the areas to either trivial areas or closures of connected open areas. We also require that every non-empty intersection of two non-trivial areas contains an ϵ-ball. This algorithm will again be 3-update optimal, and we will also show that this is the best possible. Let $\mathcal{F}$ be the set of areas.

Definition 4. *Let $\mathcal{F}$ be a set of areas in $\Re^2$ and let l be a line. Then l splits $\Re^2$ in three regions: two half planes (H_1, H_2) and l. We say l has an empty half if H_1 or H_2 does not intersect with any area in $\mathcal{F}$.*

Lemma 6. *Let p be a point in $A \in \mathcal{F}$. The point p is always on the convex hull if, and only if, for every configuration with p as the point of the area A, there exists a line l through p such that l has an empty half.*

Lemma 7. *Let A and B be two non-trivial areas in $\mathcal{F}$ with a non-empty inter-section. Further let $\mathcal{F} - \{A, B\}$ contain at least two areas C and D such that there exists $c \in C$ and $d \in D$ with $c \neq d$. Then A and B are not always convex hull areas.*

Proof. By our general condition on all areas in $\mathcal{F}$ the intersection of A and B must contain a non-empty open area E. Since $c \neq d$ and E is open there exist $p, q \in E$ such that p lies inside the triangle with vertices q, c and d, see Figure 3.

Lemma 8. *Let p be a point in an area $A \in \mathcal{F}$ and l be a line through p such that to one direction of p the line l splits an area $B \in \mathcal{F} - \{A\}$ and to the other direction of p the line l intersect with an area $C \in \mathcal{F} - \{A, B\}$. Further let $D \in \mathcal{F} - \{A, B, C\}$ such that either*

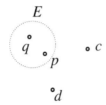

Fig. 3. Choice of p and q

  i) $D - \{p\}$ is not empty, if the area C is non-trivial,
ii) $D - l$ is not empty, if the area C is trivial.

Then p is not always on the convex hull.

Proof. Case i) By our assumption there exists a point $d \in D$ with $d \neq p$. Since C is non-trivial it must be the closure of an open area. Hence there exists a $c \in C$ such that p, c and d are not in a line and there exists an point $b \in B$ such that p lies in the inner of the triangle with vertices b, c and d.

    Case ii) The area C is trivial. So let c be the point in C. Let $d \in D - l$. Since l splits B there exists a point $b \in B$ such that p lies in the inner of the triangle with vertices b, c and d, see Figure 4.

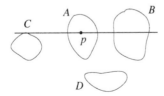

Fig. 4. A line through p that splits on area to one sides and touches another one to the other side

Lemma 9. *Let p be a dependent convex hull point in an area $A \in \mathcal{F}$. Then there exists a line l through p such that in one direction of p the line l splits an area $B \in \mathcal{F} - \{A\}$ and to the other direction of p the line l intersect with an area $C \in \mathcal{F} - \{A, B\}$ and there exists an area $D \in \mathcal{F} - \{A, B, C\}$ such that*

  i) $D - \{p\}$ is not empty, if the area C is non-trivial,
ii) $D - l$ is not empty, if the area C is trivial.

Proof. Since p is dependent convex hull there exists a configuration G and a line l such that $p \in l$ and l has an empty half H with respect to G. If all areas except A do not intersect with H then l has an empty half for any configuration and p would be therefore always convex hull. Since all areas are connected l splits at least one area other than A. By rotating l at the point p we can assume that l in one direction of p splits an area $B \in \mathcal{F} - \{A\}$ and intersects in the other direction with another area $C \in \mathcal{F} - \{A, B\}$.

Case i) Since p is dependent convex hull there exist three points in three different areas other then A, such that p lies inside the triangle created by these three points. Hence there exists exists at least three areas which are not identical to $\{p\}$. Hence, if C is not trivial there exists $D \in \mathcal{F} - \{A, B, C\}$ with $D - l$ is not empty.

Case ii) If C is trivial and all areas except A, B are lying completely on l then in any configuration G_1 with p as the point of the area A, all points in G_1 lie on the convex hull, since all points in G_1 except the point of the area B lie on one line. This contradicts that p is dependent convex hull. Hence if C is trivial there exists at least one other area in $\mathcal{F} - \{A, B\}$ which does not lie completely on l.

Lemma 10. *If there exists a partly convex hull area, then there exists a witness set of size at most* 3.

Proof. Let A be a partly convex hull area. If we assume there exists a never convex hull point in A. Then in order to verify whether A is on the convex-hull we have to update A. So $\{A\}$ is a witness set. For the rest of this proof let A contain a dependent convex hull point p.

If a verification does not update the area A, the point p must change its status through updating other areas to an always convex hull point.

By Lemma 9 there exists a line l through p such that in one direction of p the line l splits an area $B \in \mathcal{F} - \{A\}$ and to the other direction of p the line l intersect with an area $C \in \mathcal{F} - \{A, B\}$ and there exists an area $D \in \mathcal{F} - \{A, B, C\}$ such that

i) $D - \{p\}$ is not empty, if the area C is non-trivial,
ii) $D - l$ is not empty, if the area C is trivial.

We look at these cases separately.

Case i) C is non-trivial. Since A is partly convex hull and connected, there exists a point $q \in A$ and a line l' through q such that in one direction of q the line l splits the area B and to the other direction of q the line l intersects with C. By Lemma 8 without updating B or C, not both point p and q can change their status to always convex hull points. Hence $\{A, B, C\}$ is a witness set.

Case ii) C is trivial. By Lemma 8 without updating B or D the point p can not change its status to always convex hull point. Hence $\{A, B, D\}$ is a witness set.

The following definitions and lemmas will lead to the Theorem 2, which shows that, we can also a witness of size at most 3, if there are no partly convex hull areas but at least one dependent area.

Definition 5. *Let $\mathcal{F}$ consist only of always convex hull areas. Let A and B be two areas in $\mathcal{F}$. We call A and B neighbors if in any configuration the points of A and B are adjacent on the convex hull.*

Lemma 11. *Let $\mathcal{F}$ consist of only always convex hull areas and let $\mathcal{F}$ not contain multiple identical trivial areas. Further let $\mathcal{F}$ contain three or more areas. Then every area in $\mathcal{F}$ has exactly two neighbors.*

Proof. Since $\mathcal{F}$ has no multiple trivial areas and every non-trivial area is a closure of an open area there exists a configuration C_1 of $\mathcal{F}$ such that all points in C_1 are distinct. Let A be an area in $\mathcal{F}$ and let a be the point in C_1 that corresponds to the area A. Since all points in C_1 are distinct the point a has exactly two neighbors ($b \in B \in \mathcal{F}$ and $c \in C \in \mathcal{F}$) on the convex hull with respect to C_1. We will show that the areas A and B as well as A and C are neighboring areas.

Let us assume that A and B are not neighboring areas. So, there exists an area $D \in \mathcal{F}$ and a configuration C_2 with $a' \in A$, $b' \in B - \{a'\}$ and $d' \in D - \{a', b'\}$ such that a' and d' are neighbors and a' and b' are not. Since all of these points are disjoint we can change our set of areas as follows.

$$\mathcal{F}' = \{S - \bigcup_{T \in \{A,B,C,D\}, T \text{is trivial}, S \neq T} T : S \in \{A, B, C, D\}\}$$

Since all areas in $\mathcal{F}$ are either closures of open areas or trivial, all areas in $\mathcal{F}'$ are connected. Note also that by Lemma 7 none of the areas in $\mathcal{F}'$ intersect with another area in $\mathcal{F}'$. Let d be the corresponding point of the area D in the configuration C_1 and let c' be the corresponding point of the area C in the configuration C_2. Note that C_1 and C_2 when restricted to points corresponding to the areas A, B, C and D are configuration for $\mathcal{F}'$. Since the areas A, B, C and D are connected we can continuously transform the configuration C_1 to the configuration C_2 (see Figure 5). Since all areas in $\mathcal{F}'$ are disjoint and always convex hull, this is not possible; a contradiction. Hence A and B are neighboring areas. Similarly for the areas A and C.

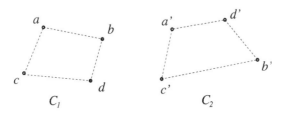

Fig. 5. Different configurations

Definition 6. *Let $\mathcal{F}$ be a collection of areas. Let A and B be in $\mathcal{F}$. Let $\mathcal{F}' = \{C \in \mathcal{F} : C \text{ is always convex hull}\}$. We call A and B always neighbors if A and B are always convex hull areas and A and B are neighbors in $\mathcal{F}'$.*

Definition 7. *Let A and B be always neighbors. We call the set*

$$N_{A,B} = \{p : \exists q \in A, r \in B \text{ with } p \in edge_{q,r}\}$$

the neighboring band of A and B

Lemma 12. *If there are no partly convex hull areas in $\mathcal{F}$, but $\mathcal{F}$ contains at least three always convex hull areas, then any dependent convex hull area $A \in \mathcal{F}$ must intersect with at least one neighboring band.*

Proof. Since we have at least three always convex hull areas. The neighboring bands of these areas build a closed 'polygon' P. Note that an edge of P is a neighboring band and it might be therefore thick. We call all point that lie either outside of P or on the convex hull of P outer-points, points that lie in a neighboring band except of the convex hull of P, points on P and the rest inner points. Since all areas in $\mathcal{F}$ are closed, any area that contains outer points must contain an always convex hull point, which contradicts our assumption that there are no partly convex hull areas. In particular no dependent convex hull area contains outer points. Since all inner points are never convex hull points, a dependent convex hull area must contain at least a point on P and therefore it must intersect with at least one neighboring band.

Lemma 13. *Let there be no partly convex hull areas in $\mathcal{F}$. Let $\mathcal{F}$ contain at least three always convex hull areas. If A is a dependent convex hull area intersecting with the neighboring band $N_{B,C}$, then $\{A, B, C\}$ is a witness set.*

Proof. Assume that we don't need to update A, B or C. Since A is a dependent convex hull area we must update other areas to change the status of A. Since $\mathcal{F}$ contains at least three always convex hull areas, without updating B or C the area A will not become an always convex hull area. Hence we must update an area D such that A becomes a never convex hull area. Therefore the updated trivial area D' must intersect with $N_{B,C}$ and it must be dependent convex hull. We are now in a similar situ ation as before. Since we don't update B or C we have to update an area E such that D' becomes a never convex hull area, but again the updated trivial area E' lies in $N_{B,C}$. Therefore we have to update at some point the area B or C.

Lemma 14. *If there are no partly convex hull areas in $\mathcal{F}$ and $\mathcal{F}$ contains at least one dependent convex hull area, there exists a witness set of size at most 3.*

Proof. Case 1) $\mathcal{F}$ contains one always convex hull area A. If the area A is not updated other updates must lead to the situation where the are only three areas in the collection of areas without counting multiple trivial area. Hence A and any two non-trivial areas in $\mathcal{F}$ build a witness set.

Case 2) $\mathcal{F}$ contains two always convex hull areas A and B. Similar to Case 1). If the areas A and B are not updated, other updates must lead to the situation in which there are only three areas, without counting multiple trivial areas. Hence A, B and any non-trivial are in $\mathcal{F}$ build a witness set.

Case 3) $\mathcal{F}$ contains three or more always convex hull areas. Let D be a dependent convex hull area in $\mathcal{F}$. By Lemma 12 the area D must intersect with at least the neighboring band $N_{B,C}$ where B and C are always convex hull areas in $\mathcal{F}$. By Lemma 13 the set $\{D, B, C\}$ is a witness set.

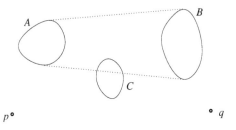

Fig. 6.

By Lemma 14, Lemma 10 and Lemma 1 we have the following Theorem.

Theorem 2. *Under the restriction to the closure of open, connected areas or trivial areas, and every non-empty intersection of two non-trivial areas contains an ϵ-ball, the witness algorithm for the convex hull problem is 3-update optimal. Furthermore, this is the best possible.*

The following example, Figure 6 shows that there is no algorithm which is c-update optimal for $c < 3$. The areas A and B are always on the convex hull. However for any strategy to determine the status of C there exists a configuration such that this strategy requires to update all three areas, but starting with updating the last one would have given the answer directly. Since we can construct a set of areas that consists entirely of triples following the same pattern, there can not exist an c-update optimal algorithm with c less than three.

A similar situation as described in Remark 1 can be used to shows that there is no update-optimal algorithm for the convex hull problem with arbitrary areas.

References

1. A. Borodin and R. El-Yaniv, *"Online Computation and Competitive Analysis,"* Cambridge University Press, 1998.
2. T. Feder, R. Motwani, R. Panigrahy, C. Olston, and J. Widom, "Computing the Median with Uncertainty," *Proc 32nd ACM STOC,* 602-607, 2000.
3. T. Feder, R. Motwani, L. O'Callaghan, C. Olston and R. Panigrahy, "Computing Shortest Paths with Uncertainty," *Proc 20th STAC,* LNCS 2607, 355-366.
4. J. Basch, L. Guibas and J. Hershberger, "Data Structures for Mobile Data," *Proc 8th ACM-SIAM SODA.*
5. S. Kahan, "A Model for Data in Motion," STOC 91, 267-277.
6. S. Khanna and W.-C. Tan, "On Computing Functions with Uncertainty," *Proc 20th ACM PODS,* 171-182.
7. C. Olston and J. Widom, "Offering a Precision-Performance Tradeoff for Aggregation Queries over Replicated Data," *Proc 26th VLDB,* Morgan Kempmann, 144-155.

Maximizing the Guarded Boundary of an Art Gallery Is APX-Complete

Euripides Markou[1], Stathis Zachos[1,2], and Christodoulos Fragoudakis[1]

[1] Computer Science, ECE, National Technical University of Athens, Greece
[2] CIS Department, Brooklyn College, CUNY, USA
{emarkou, zachos, cfrag}@cs.ntua.gr

Abstract. In the Art Gallery problem, given is a polygonal gallery and the goal is to guard the gallery's interior or walls with a number of guards that must be placed strategically in the interior, on walls or on corners of the gallery. Here we consider a more realistic version: exhibits now have *size* and may have different costs. Moreover the meaning of guarding is relaxed: we use a new concept, that of *watching* an expensive art item, i.e. overseeing a *part* of the item. The main result of the paper is that the problem of maximizing the total value of a guarded weighted boundary is APX-complete. This is shown by an appropriate gap-preserving reduction from the MAX-5-OCCURRENCE-3-SAT problem. We also show that this technique can be applied to a number of maximization variations of the art gallery problem. In particular we consider the following problems: given a polygon with or without holes and k available guards, maximize a) the *length of walls* guarded and b) the *total cost* of paintings *watched* or *overseen*. We prove that all the above problems are APX-complete.

1 Introduction

In the Art Gallery problem (as posed by Victor Klee during a conference in 1976), we are asked to place a minimum number of guards in an art gallery so that every point in the interior of the gallery can be seen by at least one guard.

Besides its application of guarding exhibits in a gallery, the Art Gallery problem has applications in wireless communication technology (mobile phones, etc): place a minimum number of stations in a polygonal area so that any point of the area can communicate with at least one station (two points can communicate if they are mutually visible).

Many variations of the Art Gallery problem have been studied during the last two decades [2,3,4]. These variations can be classified with respect to where the guards are allowed to be placed (e.g. on vertices, edges, interior of the polygon) or whether only the boundary or all of the interior of the polygon needs to be guarded, etc. Most known variations of this problem are NP-hard. Related problems that have been studied are MINIMUM VERTEX/EDGE/POINT GUARD for polygons with or without holes (APX-hard and O($\log n$)-approximable [1,5,6]) and MINIMUM FIXED HEIGHT VERTEX/POINT GUARD ON TERRAIN ($\Theta(\log n)$-approximable [5,6,8]). In [13] the case of guarding the walls (and not necessarily

R. Petreschi et al. (Eds.): CIAC 2003, LNCS 2653, pp. 24–35, 2003.
© Springer-Verlag Berlin Heidelberg 2003

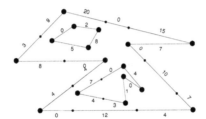

Fig. 1. A weighted polygon

every interior point) is studied. In [14] the following problem has been introduced: suppose we have a number of valuable treasures in a polygon P; what is the minimum number of mobile (edge) guards required to patrol P in such a way that each treasure is always visible from at least one guard? In [14] they show NP-hardness and give heuristics for this problem. In [15] weights are assigned to the treasures in the gallery. They study the case of placing one guard in the gallery in such a way that the sum of weights of the visible treasures is maximized. Recent (non-)approximability results for art gallery problems can be found in [1,2,3,4,5,8]. For a nice survey of approximation classes and important results the reader is referred to [11].

Here we consider the MAXIMUM VALUE VERTEX GUARD problem: A polygon without holes is given with weighted disjoint line segments on its boundary (see figure 1); an integer $k > 0$ is also given. The goal is to place at most k guards on vertices of the polygon so that the total weight of line segments visible by the guards is maximized. If we think of the weighted line segments as paintings on the walls of an art gallery then we have a realistic abstraction of the problem of guarding a maximum total value of paintings that takes into account the fact that paintings actually occupy parts of the walls, not merely points. Another possible application of this problem is the illumination of a maximum number of paintings in a gallery. Again, a painting must be totally visible from light sources in order to consider it illuminated. There are also important applications in wireless communication networks: An interpretation of weighted line segments are inhabited areas. The polygon models the geographical space. The weight interpretation is the population of an area. Imagine a number of towns lying on the boundary of a polygonal geographical area. The goal is to place at most k stations such that the total number of people that can communicate is maximized. Moreover, it could be the case that the towns are on the shore of a lake, so we can only place stations on the boundary. Similar situations may arise in various other types of landscape.

We show APX-hardness of MAXIMUM VALUE VERTEX GUARD and conclude that this problem is APX-complete since there exists a polynomial time constant-ratio approximation algorithm ([12]). Our main contribution is a gap-preserving reduction from MAX-5-OCCURRENCE-3-SAT to MAXIMUM VALUE VERTEX GUARD specially designed for weighted maximization problems. The construction part of our reduction uses some ideas from the constructions used in [1], [6] (to show NP-hardness, APX-hardness respectively of the MINIMUM

VERTEX GUARD problem). Central in our technique is a careful assignment of appropriate weights on the line segments of the constructed polygon.

Next we study a number of variations: a) the case of edge guards (guards occupying whole edges), b) the case in which our goal is to *watch* (see a part of) line segments instead of overseeing them, c) the case of maximizing the total length of the visible boundary and e) the case of polygons with holes. We prove APX-completeness for all these variations and for several of their combinations.

2   MAXIMUM VALUE VERTEX GUARD Is APX-Complete

Suppose a polygon P without holes is given with weighted disjoint line segments on its boundary. Our line segments are open intervals (a, b). The goal is to place k vertex guards maximizing the weight of the overseen boundary. The formal definition follows:

Definition 1. *Given is a polygon P without holes and an integer $k > 0$. Assume the boundary of P is subdivided into disjoint line segments with non negative weights (see figure 1). The goal of the MAXIMUM VALUE VERTEX GUARD problem is to place k vertex guards so that the total weight of the set of line segments overseen is maximum.*

We will prove that MAXIMUM VALUE VERTEX GUARD is APX-hard. We propose a reduction from MAX-5-OCCURRENCE-3-SAT problem (known to be APX-hard [10]) and we show that it is a gap preserving reduction. Let us recall the formal definition of the MAX-5-OCCURRENCE-3-SAT problem:

Definition 2. *Let Φ be a boolean formula given in conjuctive normal form, with each clause consisting of at most 3 literals and with each variable appearing in at most 5 clauses. The goal of MAX-5-OCCURRENCE-3-SAT problem is to find a truth assignment for the variables of Φ such that the number of satisfied clauses is maximum.*

2.1   Construction Part of the Reduction

For every literal, clause and variable of the given boolean expression, we construct a corresponding pattern as shown in figure 2. Figure 2a shows a clause

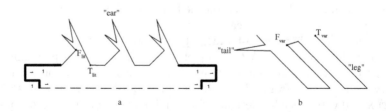

Fig. 2. a) A clause pattern with 3 literal patterns, b) a variable pattern

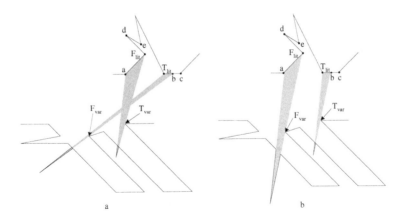

Fig. 3. a) Two spikes corresponding to an occurrence of a positive literal in a clause. Both spikes and the "ear" are overseen by two guards placed, e.g., on F_{var} (oversees left spike and "tail") and on F_{lit} (oversees right spike and "ear"). b) two spikes corresponding to an occurrence of a negative literal in a clause. Both spikes and the "ear" are overseen by two guards placed, e.g., on T_{var} (oversees right spike and "tail") and on F_{lit} (oversees left spike and "ear").

pattern with 3 literal "ear" patterns. It is possible to oversee the whole literal pattern with one vertex guard *only* if she is placed on vertex F_{lit} or T_{lit}. Figure 2b shows a variable pattern with two "legs" and a "tail". Variable patterns are augmented with additional spikes described below (see figure 3). Finally we add an "ear" pattern in the upper left corner of the polygon and the construction is complete. A guard on vertex w oversees both "legs" of every variable pattern. An example is shown in figure 4.

For every occurrence of a literal in the boolean expression, i.e. for every literal "ear" pattern, we add two "spikes" to the corresponding variable pattern: if it is a positive (negative) literal, we add the two spikes as shown in figure 3a (3b). The spike which is overseen by vertex F_{lit} (T_{lit}) is called FALSE (TRUE) spike. Notice in figure 3 that the base of the FALSE spike is the line segment (a, F_{lit}), whereas the base of the TRUE spike is (T_{lit}, b) and not (T_{lit}, c). The purpose of this is that no vertex of the clause side (see figure 4) can oversee more than one spike (in the variable side).

Three guards are necessary and sufficient in order to oversee a literal "ear", its corresponding variable pattern (two "legs" and a "tail") and its corresponding spikes. One of them is placed on vertex w and oversees the "legs" of the variable pattern. The other two are placed on vertices: i) $\{F_{var}, F_{lit}\}$, or $\{T_{var}, T_{lit}\}$, for positive literals, or ii) $\{F_{var}, T_{lit}\}$, or $\{T_{var}, F_{lit}\}$, for negative literals. We assign value 8 to every edge of the polygon, except the "cheap" edges of the clause patterns depicted in figure 2a, to which we assign value 1. We set the number of available guards $k = l + n + 1$, where l is the number of occurrences of literals and n is the number of variables of the boolean expression.

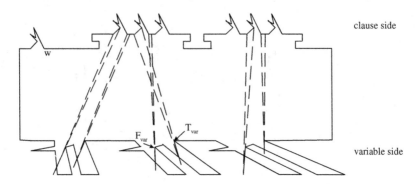

Fig. 4. Resulting polygon

2.2   Transformation of a Feasible Solution

Suppose a truth assignment for the boolean expression is given. We will construct a guard placement that corresponds to the given truth assignment. We place $k = l + n + 1$ guards on vertices of the polygon that we constructed in section 2.1, as follows: We place in each variable pattern a guard on vertex F_{var} (T_{var}), if the truth value of the corresponding variable is FALSE (TRUE). We place in each literal pattern a guard on vertex F_{lit} (T_{lit}), if the truth evaluation of the literal is FALSE (TRUE). Finally we place a guard in the additional "ear" pattern, on vertex w. Thus, every literal pattern is overseen. Furthermore, every variable pattern is overseen by guards placed as described. The "legs" of variable patterns are overseen by the guard on vertex w.

Conversely, given a placement of $l + n + 1$ guards on the resulting polygon which is an instance of MAXIMUM VALUE VERTEX GUARD we will construct a corresponding truth assignment for the original MAX-5-OCCURRENCE-3-SAT instance. First we modify the placement of guards by placing a) only one guard in every variable pattern on one of the vertices F_{var} or T_{var}, b) only one guard in every literal pattern on vertex F_{lit} (T_{lit}) if the corresponding TRUE (FALSE) spike of the variable pattern is overseen by its guard, c) one guard in the additional "ear" pattern on vertex w. In more details: given a placement of $k = l + n + 1$ guards with a total overseen boundary value B, we will modify the guard placement so that the total value overseen is $\geq B$, and so that with the exception of some "cheap" edges with weight 1, the modified guard placement achieves: a) full overseeing of all polygon edges and b) "consistent" placement on two vertices out of the four $F_{lit}, T_{lit}, F_{var}, T_{var}$ for all literals. Guard placement follows:

i) We place one guard on vertex w of the additional "ear" pattern.

ii) For every variable pattern: a) If there is only one guard in the pattern placed on a vertex which oversees a spike, we place her on F_{var} (T_{var}) if F_{var} (T_{var}) oversees the same spike. b) In all other cases (no guards, one guard overseeing no spikes, at least two guards) we place one guard on F_{var} (T_{var}) if F_{var} (T_{var}) oversees more FALSE spikes than T_{var} (F_{var}).

iii) For every literal we place one guard on F_{lit} (T_{lit}) if the corresponding FALSE (TRUE) spike of the variable pattern is not overseen by the guard placed in the variable pattern.

We will prove in section 2.3 (see Lemma 2) that the total value overseen is at least B.

Now we can construct a truth assignment as follows: assign TRUE (FALSE) to a variable if the corresponding variable pattern has a guard on vertex T_{var} (F_{var}).

2.3 Analysis of the Reduction

Let I be an instance of MAX-5-OCCURRENCE-3-SAT with n variables, l occurrences of literals and m clauses ($l \leq 3m$). Let I' be the instance of MAXIMUM VALUE VERTEX GUARD (constructed as in 2.1) with $k = l + n + 1$. Let M be the total value of the boundary.

Lemma 1. If $OPT(I) = m$ then $OPT(I') = M$.

Proof. Suppose there exists a truth assignment such that all m clauses are satisfied. If we place $l + n + 1$ guards in the polygon as in 2.2, then it is easy to see that the whole boundary of the polygon is overseen. So the total value of overseen edges is M.

Note that Lemma 1 is true no matter what the values of the cheap edges are. However we must carefully choose the values of cheap edges in order to prove Lemma 2. We want to find an optimal placement of guards in which for many clause patterns at least one of the T_{lit} vertices is occupied by a guard. Thus the values of cheap edges should not be 0. We also want to cover all non-cheap edges possibly leaving some cheap ones uncovered. For every false clause of the boolean formula cheap edges in the corresponding clause pattern will be left uncovered.

Lemma 2. If $OPT(I') \geq M - 8\epsilon m$ then $OPT(I) \geq m(1 - \epsilon)$.

Proof. Suppose there exists an $\epsilon > 0$ and a placement of the $l + n + 1$ guards in I' so that the total value of overseen boundary is at least $M - 8\epsilon m$. After the modification of guard placement described in 2.2, $k = l + n + 1$ guards oversee the whole boundary except possibly some "cheap" edges and the total value overseen is at least $M - 8\epsilon m$: Notice that if we place a guard on vertex F_{lit} or T_{lit} of an "ear" which has no guards we certainly increase the overseen value by at least 16, because edges (F_{lit}, d) and (d, e) can not be overseen by any outside guard. Similarly a guard placed on F_{var} or T_{var} of a variable pattern that has no guards, certainly increases the overseen value by at least 16 (namely weight of the two "tail" edges).

We will discuss two cases pertaining to guard placement in "ears":

a) The original placement had two guards on vertices T_{lit} and F_{lit} of a literal "ear" pattern and after the modification, one guard was placed on vertex F_{lit} of the pattern. The total value may have been decreased by at most 8 (because "cheap" edges may now be missed) but it is increased by at least 16 (because the free guard was placed in an unguarded pattern).

b) The original placement had one guard on vertex T_{lit} of a literal "ear" pattern and after the modification she was moved to vertex F_{lit} of the pattern:

i) If the corresponding FALSE spike was not overseen in the original placement (by a guard in the variable pattern), the total value may have been decreased by at most 8 (because "cheap" edges may now be missed) but it is increased by at least 16 (because the FALSE spike is now overseen by the guard on F_{lit}).

ii) If the corresponding FALSE spike was overseen in the original placement (by a guard g in the variable pattern), then it must be the case that the variable pattern had originally at least two guards and after the modification, guard g was removed and placed in another pattern because there was another guard that was overseeing the most FALSE spikes in the variable pattern. The guard g was overseeing at most 2 FALSE spikes because a variable pattern has at most 5 FALSE spikes, since a variable appears in at most 5 clauses of the boolean formula. Thus, for every variable pattern, guards have been moved from vertex T_{lit} to F_{lit} in at most two literal patterns. The total value may have been decreased by at most 16 (because "cheap" edges of two clauses may now be missed) but it is increased by at least 16 (because at least one free guard was placed in an unguarded pattern).

We can now construct a truth assignment for I as in 2.2 that leaves at most ϵm clauses unsatisfied that correspond to ϵm clause patterns not overseen by any guard in I'.

From Lemma 1 and the contraposition of Lemma 2 the following theorem holds:

Theorem 1. *Let I be an instance of* MAX-5-OCCURRENCE-3-SAT *problem with n variables, m clauses and $l \leq 3m$ occurrences of literals. Let I' be the instance of* MAXIMUM VALUE VERTEX GUARD *problem (constructed as in 2.1) with $k = l + n + 1$. Let M be the total value of the boundary of the polygon. Then:*

- $OPT(I) = m \rightarrow OPT(I') = M$
- $OPT(I) \leq m(1 - \epsilon) \rightarrow OPT(I') \leq M - 8\epsilon m$

Thus our reduction is gap-preserving [10].

In [9,10] it was proved that the MAX-5-OCCURRENCE-3-SAT problem with parameters m and $(1 - \epsilon)m$ for some $\epsilon > 0$, where m denotes the number of clauses in instance I, is NP-hard to decide.

Therefore, we obtain that unless $P = NP$, no polynomial time approximation algorithm for MAXIMUM VALUE VERTEX GUARD can achieve an approximation ratio of $\frac{M}{M-8\epsilon m}$.

Considered that $M = nV + lL + 2lS + mC + E$ where V denotes the total value of a variable pattern ("legs", "tail", "leg-edges" between spikes, plus one edge that links the variable pattern with the next one on the right : $104 \leq V \leq 168$), L denotes the total value of a literal pattern ("ear" : $L = 40$), S denotes the total value of a spike pattern ($S = 16$), C denotes the total value of a clause pattern without "ears" plus one edge that links the clause pattern with the next one on the right ($16 \leq C \leq 32$) and E denotes the total value of the additional ear pattern and the remaining edges of the polygon ($E = 80$), then:

$$M \leq 3mV + 3mL + 6mS + mC + E$$

With a few calculations it turns out:

$$\frac{M}{M - 8\epsilon m} = \frac{1}{1 - \frac{8\epsilon m}{M}} \geq \frac{1}{1 - \frac{8\epsilon}{3V + 3L + 6S + C + E}} \geq 1 + \epsilon'$$

for some ϵ' that depends on ϵ. Therefore:

Theorem 2. MAXIMUM VALUE VERTEX GUARD *is APX-hard.*

On the other hand the MAXIMUM VALUE VERTEX GUARD problem can be approximated within a constant ([12]). Therefore:

Corollary 1. MAXIMUM VALUE VERTEX GUARD *is APX-complete.*

3   A Bunch of APX-Complete Art Gallery Problems

In this section we propose appropriate modifications of the reduction of section 2 in order to show APX-hardness for a number of variations of MAXIMUM VALUE VERTEX GUARD. We also give constant ratio approximation algorithms for these problems (where not already known), thus showing them to be APX-complete.

The case in which guards are placed on edges (guards occupying whole edges), is called MAXIMUM VALUE EDGE GUARD problem. A guard which is occupying a whole edge, can be thought of as a mobile guard able to move on the edge.

Proposition 1. MAXIMUM VALUE EDGE GUARD *is APX-hard.*

Proof. We show the result by a reduction from MAX-5-OCCURRENCE-3-SAT to MAXIMUM VALUE EDGE GUARD. The reduction follows the one in section 2 using modified literal and variable patterns, as shown in figure 5. It is not hard to check that the properties mentioned in Theorem 1 hold here as well.

For the rest of our problems we consider both vertex-guard and edge-guard versions and we use the corresponding construction for our reductions, i.e. the one of section 2.1 for the vertex-guard problems and the one used in Proposition 1 (with the modified literal and variable patterns) for the edge-guard problems. All the reductions are from MAX-5-OCCURRENCE-3-SAT to the problem in hand.

Now we will relax the meaning of guarding: "watching a valuable painting", i.e. "overseeing a part of it" instead of "overseeing all of it".

Proposition 2. *The watching versions of* MAXIMUM VALUE VERTEX GUARD *and* MAXIMUM VALUE EDGE GUARD *problems are APX-hard.*

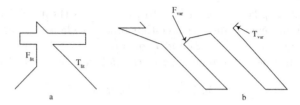

Fig. 5. a) A literal pattern and b) a variable pattern for edge guard problems

Proof. Let us describe a reduction from MAX-5-OCCURRENCE-3-SAT to the watching version of MAXIMUM VALUE VERTEX/EDGE GUARD. We first construct the polygon using the appropriate gadgets (depending on the kind of guards as explained above). We then discretize the boundary using the Finest Visibility Segmentation (FVS) described in [12]. Let us recall this technique: we use the visibility graph $V_G(P)$. By extending edges of $V_G(P)$ inside P up to the boundary of P we obtain a set of points FVS on the boundary of P (FVS includes of course all corners of P) (see Figure 6a). There are $O(n^2)$ points in FVS and these points are endpoints of line segments with the following property: for any vertex y, a segment (a, b) defined by consecutive FVS points is visible by y iff it is watched by y. Furthermore (a, b) is watched (and visible) by an edge e iff it is watched by any point in $FVS \cap e$. Thus we can find the set of line segments $E'(v)$ ($E'(e)$) which are watched by a vertex v (edge e) within polynomial time.

Every edge in a clause pattern will be subdivided into $O(n)$ FVS segments, because it can be watched only by vertices in variable patterns. Let $\delta > 0$ be an integer such that the number of FVS segments in any of the (previously) "cheap" edges of a clause pattern is at most δn. We assign value 1 to every FVS segment which belongs to a (previously) "cheap" edge of a clause pattern. We assign value $8\delta n$ to every other segment. The properties of Theorem 1 hold (details are omitted for brevity).

Consider now the following problem:

Definition 3. *Given is a polygon P without holes and an integer $k > 0$. Let $L(b)$ be the euclidean length of the line segment b. The* MAXIMUM LENGTH VERTEX/EDGE GUARD *problem asks to place k vertex (edge) guards so that the euclidean length of the* overseen *part of P's boundary is maximum.*

Proposition 3. MAXIMUM LENGTH VERTEX/EDGE GUARD *is APX-hard.*

Proof. For the construction part of the reduction, we construct the polygon using the gadgets for vertex-guard or edge-guard version with the following additional modification: we make sure that the length of every (previously) "cheap" edge in a clause pattern is designed at least 8 times shorter than any other edge of the polygon. Now the properties of Theorem 1 hold here as well (again details are omitted).

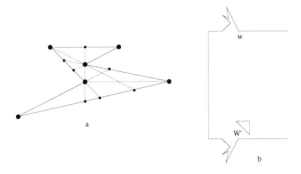

Fig. 6. a) Subdividing the boundary into line segments with endpoints in FVS, b) the left part of the polygon with the hole

All these problems may also appear in polygons with holes. Holes in polygons are useful because they give us the chance to model reality better (holes represent obstacles) and to place guards in the interior of the polygon (on predefined places), on vertices or edges of the holes. We remind the reader that for MINIMUM VERTEX/EDGE GUARD for polygons with holes, no polynomial time approximation algorithm can guarantee an approximation ratio of $\frac{1-\epsilon}{12} \ln n$ for any $\epsilon > 0$, unless $NP \subseteq TIME(n^{O(\log \log n)})$ ([7,5]).

Proposition 4. *The following problems are all APX-hard for polygons with holes.*

- *The* **overseeing** *version of* MAXIMUM VALUE VERTEX/EDGE GUARD
- *The* **watching** *version of* MAXIMUM VALUE VERTEX/EDGE GUARD
- MAXIMUM LENGTH VERTEX/EDGE GUARD

Proof. In the construction part of the corresponding reduction for every one of the above problems we add a hole and another "ear" pattern in the left lower corner of the polygon as shown in figure 6b. Theorem 1 again holds.

For the problems of Propositions 1-4, polynomial time constant ratio approximation algorithms are presented in [12]. Hence:

Theorem 3. *The following problems are all APX-complete for polygons with or without holes.*

- *the* **overseeing** *version of* MAXIMUM VALUE VERTEX/EDGE GUARD
- *The* **watching** *version of* MAXIMUM VALUE VERTEX/EDGE GUARD
- MAXIMUM LENGTH VERTEX/EDGE GUARD

Another variation of the MAXIMUM VALUE VERTEX/EDGE GUARD problem is the maximization of the total value of overseen valuable paintings where only the dimensions of the paintings are given. So the goal is to place vertex/edge guards as well as to place the given paintings on the boundary of the polygon. The problem called MAXIMUM VALUE VERTEX/EDGE GUARD PP is also APX-complete ([16]).

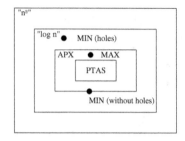

Fig. 7. Classifying Art Gallery problems in approximation classes. We use: "n^ϵ" to denote the class of problems with $O(n^\epsilon)$ approximation ratio, "$\log n$" for the class of problems with $O(\log n)$ approximation ratio, APX for the class of problems with constant approximation ratio and PTAS for the class of problems with an infinitely close to 1 constant approximation ratio.

4 Conclusions

We have proved that overseeing a maximum value part of a weighted boundary of a polygon without holes, using at most k vertex guards (MAXIMUM VALUE VERTEX GUARD) is APX-complete. We have also proved that the variations involving i) edge guards (MAXIMUM VALUE EDGE GUARD), ii) polygons with holes and iii) watching instead of overseeing the boundary, are APX-complete. In addition, we have shown that MAXIMUM LENGTH VERTEX GUARD and MAXIMUM LENGTH EDGE GUARD for polygons with or without holes are APX-complete. We end up with a hierarchy of Art Gallery problems which is shown in figure 7.

Maximization Art Gallery problems for polygons with or without holes that we studied (*MAX* in figure 7) are APX-hard while at the same time they have constant approximation ratios (thus APX-complete). Minimization Art Gallery problems for: a) polygons with holes (*MIN holes* in figure 7) are $\log n$-hard and have $O(\log n)$ approximation ratios (thus $\log n$-complete), b) polygons without holes (*MIN without holes* in figure 7) are APX-hard and have $O(\log n)$ approximation ratios but it is not known whether they have constant approximation ratios or whether they are $\log n$-hard.

We have shown that our gap-preserving reduction can be applied with minor modifications to a number of problems. New elements of problems studied here are: a) **weighted line segments** of the polygon's boundary and b) the useful and promising concept of **watching** line segments as opposed to completely **overseeing** them. Interesting open problems arise if we consider all the above problems in the case where exhibits may lie in the interior of the polygon.

References

1. Lee, D., Lin, A., Computational complexity of art gallery problems, IEEE Trans. Inform. Theory 32, 276-282, 1986.
2. O'Rourke, J., Art Gallery Theorems and Algorithms, Oxford Univ. Press, New York, 1987.

3. Shermer, T., Recent results in Art Galleries, Proc. of the IEEE, 1992.
4. Urrutia, J., Art gallery and Illumination Problems, Handbook on Comput. Geometry, 1998.
5. Eidenbenz, S., (In-)Approximability of Visibility Problems on Polygons and Terrains, PhD Thesis, ETH Zurich, 2000.
6. Eidenbenz, S., Inapproximability Results for Guarding Polygons without Holes, Lecture notes in Computer Science, Vol. 1533 (ISAAC'98), p. 427-436, 1998.
7. Eidenbenz, S., Stamm, C., Widmayer, P., Inapproximability of some Art Gallery Problems, Proc. 10th Canadian Conf. Computational Geometry (CCCG'98), pp. 64-65, 1998.
8. Ghosh, S., Approximation algorithms for Art Gallery Problems, Proc. of the Canadian Information Processing Society Congress, pp. 429-434, 1987.
9. Arora, S., Probabilistic Checking of Proofs and the Hardness of Approximation Problems, PhD thesis, Berkeley, 1994.
10. Arora, S., Lund, C., Hardness of Approximations; in: Approximation Algorithms for NP-Hard Problems (Dorit Hochbaum ed.), pp. 399-446, PWS Publishing Company, 1996.
11. Hochbaum, D., Approximation Algorithms for NP-Hard Problems, PWS, 1996.
12. Markou, E., Fragoudakis, C., Zachos, S., Approximating Visibility Problems within a constant, 3rd Workshop on Approximation and Randomization Algorithms in Communication Networks, Rome pp. 91-103, 2002.
13. Laurentini A., Guarding the walls of an art gallery, The Visual Computer Journal, (1999) 15:265-278.
14. Deneen L. L., Joshi S., Treasures in an art gallery, Proc. 4th Canadian Conf. Computational Geometry, pp. 17-22, 1992.
15. Carlsson S., Jonsson H., Guarding a Treasury, Proc. 5th Canadian Conf. Computational Geometry, pp. 85-90, 1993.
16. Markou, E., Zachos, S., Fragoudakis, C., Optimizing guarding costs for Art Galleries by choosing placement of art pieces and guards (submitted)

An Improved Algorithm for Point Set Pattern Matching under Rigid Motion

Arijit Bishnu, Sandip Das, Subhas C. Nandy, and Bhargab B. Bhattacharya

Indian Statistical Institute, Calcutta 700 108, India

Abstract. This paper presents a simple algorithm for the *partial point set pattern matching* in 2D. Given a set P of n points, called sample set, and a query set Q of k points ($n \geq k$), the problem is to find a matching of Q with a subset of P under rigid motion. In other words, whether each point in Q is matched with corresponding point in P under translation and/or rotation. The proposed algorithm requires $O(n^2)$ space and $O(n^2 \log n)$ preprocessing time, and the worst case query time complexity is $O(k\alpha \log n)$, where α is the maximum number of equidistant pairs of points. For a set of n points, α may be $O(n^{4/3})$ in the worst case. Experimental results on random point sets and fingerprint databases show that it needs much less time in actual practice. The algorithm is then extended for checking the existence of a matching among two sets of line segments under rigid motion in $O(kn \log n)$ time, and locating a query polygon among a set of sample polygons in $O(kn)$ time under rigid motion.

1 Introduction

In computer vision and related applications, point sets represent some spatial features like spots, corners, lines, curves in various images pertaining to fingerprints, natural scenery, air traffic, astronomical maps, etc. Pose estimation involves assessment of object position and orientation relative to a model reference frame. In many problems of pattern recognition, such as registration and identification of an object, a suitably chosen set of points may efficiently preserve desired attributes of the object. In all such cases, the problem can be transformed to matching point sets with templates. For practical applications of geometric pattern matching, see [12,16].

The objective of the *point set pattern matching* problem is to determine the resemblance of two point sets P and Q (representing two different objects), under different transformations. The most simple kind of transformation is translation. If rotation is allowed along with translation, it is called rigid motion or congruence. Other transformations include reflection and scaling. The latter refers to magnifying (or reducing) the object by a certain factor π. Combination of rotation, translation and scaling is called similarity. Under these transformations, the problem can be classified into three groups [4]:

R. Petreschi et al. (Eds.): CIAC 2003, LNCS 2653, pp. 36–45, 2003.

Exact point set pattern matching: A method for finding congruences between two sets P and $Q \subset \mathbb{R}^d$, $|P| = |Q| = n$ is reported in [6]. In [7], it is shown that exact point pattern matching can be easily be reduced to string matching [19].

Approximate point set pattern matching: The approximate point set pattern matching is more realistic in many actual applications [4,5,15]. Given two finite sets of points P and Q, the problem is to find an approximate matching under some transformation, i.e., for each point $q_i \in Q$, find its match, $p_i \in P$ such that q_i lies in the ϵ-neighbourhood of p_i (a circle of radius ϵ centered at p_i).

Partial point set pattern matching: Let $|P| = n$, and $|Q| = k$, $n \geq k$, the problem is to ascertain whether a subset of P matches with Q under some transformation. Here P is referred to as the *sample set*, and Q is a *query set*. In one dimension, the problem can be solved by sorting the points in both P and Q with respect to their coordinates on the respective real lines, and then performing a merge like pass among the two sets by considering the distances of consecutive members in the respective sets. Using the same idea, Rezende and Lee [23] showed that given a query set Q of k points, the existence of its match in a sample set P of n ($\geq k$) points in $\mathbb{R}^d$ can be reported in $O(kn^d)$ time under translation, rotation and scaling. The implementation of their algorithm needs the circular ordering of points in $P \setminus \{p_i\}$ around each point $p_i \in P$; this can be done in $O(n^d)$ time for all the points in P [20]. For testing the congruence in 2D (i.e., if only translation and rotation are considered), the time complexity of the best known algorithm is $O(kn^{4/3}\log n + \mathcal{A})$ [3], where $\mathcal{A}$ is the time complexity of locating r-th smallest distance among a set of n points in the plane. In [2], it is shown that $\mathcal{A}$ is $O(n^{4/3}\log^{8/3} n)$ in the average case, and $O(n^{4/3+\epsilon})$ in the worst case. However, determination of $\mathcal{A}$ needs parametric searching technique [22], which looks difficult to implement and not very efficient in practice [1,21].

A detailed survey on point set pattern matching appears in [4,13].

Many applications in pattern recognition and computer vision (e.g., fingerprint minutiae matching for access control mechanisms) require fast congruence checking. There may be more than one sample set. Given a query point set, we need to report the sample sets whose at least one k-subset matches with Q in an efficient manner.

We present a simple and easy to implement algorithm for the partial point set pattern matching under translation and rotation. Our proposed algorithm creates efficient data structure for each sample point set in the preprocessing phase, and it leads to an efficient query algorithm which does not need the complicated parametric searching technique. The time and space complexity of the preprocessing are $O(n^2 \log n)$ and $O(n^2)$ respectively, and the query time complexity is $O(\alpha k \log n)$ in the worst case, where α is the number of equidistant pairs of points in the set P. In [24], it is shown that α, the maximum number of equidistant pairs in a point set of size n, may be $O(n^{4/3})$ in the worst case. Thus for the applications where repeated query needs to be answered, our algorithm is an improvement over the best known existing algorithm where the query time complexity is $O(n^{4/3}\log^{8/3}n + k\alpha\log n)$ [3], and which uses parametric searching technique. Our empirical evidence shows that α is much less than $O(n^{4/3})$. We have implemented both our algorithm, and the algorithm proposed in [23] and run on random point sets and also in some actual applications, like fingerprint matching. In all cases, our algorithm performs much better than that of [23]. We also show that our technique can be extended for checking the congruence of line segments and polygons.

2 Preliminaries

Let $P = \{p_1, p_2, \ldots, p_n\}$ be the sample set and $Q = \{q_1, q_2, \ldots, q_k\}$ be a query set, $k \leq n$. The objective is to find a subset of size k in P that matches Q under translation and/or rotation. The most trivial method is inspecting all possible $\binom{n}{k}$ subsets of P for a match with Q. The time complexity is improved in [23], by extending the concept of *circular sorting* [20] to higher dimensions, which facilitates an orderly traversal of the point sets. It may be noted that the points in $\mathbb{R}^d$, $d \geq 2$, lack a total order. A characterization of canonical ordering of a point set in $\mathbb{R}^d$ is first reported in [14]. In [23], the authors pointed out that the method given in [14], though elegant, may not be of much use in subset matching. But, the idea of canonical ordering of [14] can be extended to the concept of circular sorting, that basically imposes a partial order on the points. The algorithm in [23] stores the circular order of the points in $P \setminus \{p_i\}$ with each point $p_i \in P$. Then it selects a query point, say $q_1 \in Q$, arbitrarily, and computes the circular order of the points in $Q \setminus \{q_1\}$ around q_1. It anchors q_1 with each point in P, and rotates Q to identify a k-subset match in P. This scheme can detect a match (if any) under translation, rotation and scaling; the space and query time complexities are $O(n^2)$ and $O(n^2 + k\log k + kn(n-1)) = O(kn^2)$ respectively.

3 Proposed Algorithm

We now propose a new scheme of detecting a match under congruence (translation and rotation only) by selectively choosing a few points in P for anchoring with q_1. We adopt a preprocessing method on the sample set and use the following facts to design an efficient query algorithm with reduced number of anchorings.

Fact 1. *As the distances amongst the points in a set are preserved under translational or rotational transformations, a sufficient condition for a match of Q with a k-subset of P under translation and/or rotation is that all the $\binom{k}{2}$ distances in the point set Q should occur among the $\binom{n}{2}$ distances in the point set P.*

Fact 2. *[24] For a sample set P of n points, the number of equidistant pairs of points is $O(n^{4/3})$ in the worst case.*

3.1   Preprocessing

Let $P = \{p_1, p_2, \ldots, p_n\}$ be a sample set, where each point in P is assigned a unique label. In the entire text, we shall denote a line joining a pair of points a and b by (a, b), the corresponding line segment by $\overline{ab}$, and the length of the line segment $\overline{ab}$ by $\delta(\overline{ab})$.

The distances of all $\binom{n}{2}$ pairs of points are computed, and an AVL-tree $\mathcal{T}$ [18] is created with all the distinct distances. Each node of $\mathcal{T}$ is attached with a pointer to an array χ_δ, whose each element is the pair of points separated by distance δ. Each element of this array contains a triple (p_i, p_j, θ_{ij}), where p_i and p_j denote the labels of the pair of points contributing distance δ, and θ_{ij} denotes the angle of the line (p_i, p_j) with the X-axis.

We need another data structure $\mathcal{S}$, which is an array containing n elements; its i-th element (corresponding to the point $p_i \in P$) holds the address of the root of an AVL-tree $\mathcal{S}_i$, containing exactly $n - 1$ nodes corresponding to the points in $P \setminus \{p_i\}$. The nodes in $\mathcal{S}_i$ are ordered with respect to the tuples (θ_{ij}, r_{ij}), where $r_{ij} = \delta(\overline{p_i p_j})$ and $\theta_{ij} =$ the angle made by (p_i, p_j) with the X-axis.

The total space used by $\mathcal{T}$ is $O(n^2)$. The structure $\mathcal{S}$ also needs $O(n^2)$ space since it has n elements, and each $\mathcal{S}_i$ takes $O(n)$ space. The time complexity for the preprocessing is dominated by sorting the $\binom{n}{2}$ distances, which is $O(n^2 \log n)$ in the worst case.

3.2   Query Processing

Given a set of points $Q = \{q_1, q_2, \ldots, q_k\}$, the query is to ascertain whether a k-subset of P matches with Q under translation and/or rotation. We select any two points, namely q_1 and q_2, from the set Q, and search whether the distance $\delta(\overline{q_1 q_2})$ is in $\mathcal{T}$. If not, then no k-subset of P matches with Q (by Fact 1). If such an entry is found, the following steps are needed:

Let $\delta(\overline{q_1 q_2}) = d$. We consider each member in χ_d separately. Assume that $\delta(\overline{p_i p_j}) \in \chi_d$ is under processing. We anchor the line segment $\overline{q_1 q_2}$ with the line segment $\overline{p_i p_j}$ with q_1 anchored at p_i, (if it fails we would anchor q_2 with p_i) and search in $\mathcal{S}_i$ (data structure corresponding to p_i in $\mathcal{S}$) to identify the presence of a match.

For each point $q_\alpha \in Q$, the tuple $(\angle q_\alpha q_1 q_2, \delta(\overline{q_1 q_\alpha}))$ is searched in $\mathcal{S}_i$. If such an element $(\angle p_\ell p_i p_j, \delta(\overline{p_i p_\ell}))$ is found (i.e., the triangle $\triangle p_\ell p_i p_j$ in P is congruent with the triangle $\triangle q_\alpha q_1 q_2$ in Q), p_ℓ matches with q_α. If all the points in Q are matched with k points of P, then Q matches with P (see Figure 1 for an illustration).

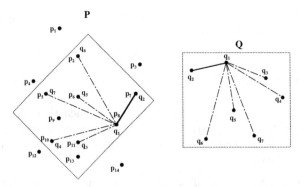

Fig. 1. Illustration of the match with the bold lines showing anchors

Query time complexity

Searching for the distance $\delta(\overline{q_1q_2})$ in $\mathcal{T}$ needs $O(\log n)$ time. If the search is successful, i.e., $\delta(\overline{q_1q_2}) = d$ (corresponding to a node of $\mathcal{T}$), then one needs to consider each member $\overline{p_ip_j} \in \chi_d$. The line segment $\overline{q_1q_2}$ is anchored with the line segment $\overline{p_ip_j}$, and for each point $q_\alpha \in Q$, the tuple $(\angle q_\alpha q_1 q_2, \delta(\overline{q_1q_\alpha}))$ is searched in $\mathcal{S}_i$, which needs $O(\log n)$ time. For $k - 2$ points in Q, the total time required is $O(k\log n)$. If the match fails, the same steps are repeated by anchoring $\overline{q_1q_2}$ with $\overline{p_jp_i}$, i.e., searching in S_j.

In the worst case, a match has to be checked for all the elements of χ_δ, i.e., for all pairs of points in P having their distances equal to δ. As the number of elements in χ_δ can be at most $O(n^{4/3})$ (see Fact 2), the worst case query time complexity of this algorithm is $O(kn^{4/3}\log n)$.

4   Experimental Results and Applications

4.1   Experiments on Randomly Generated Point Sets

We have implemented the proposed algorithm as well as the one reported in [23] on a Sun Ultra-5_10, Sparc, 233 MHz; the OS is SunOS Release 5.7 Generic. Table 1 illustrates the experimental performance of the two algorithms. We considered K sample sets $P_i, i = 1, 2, \ldots, K$, each of size ranging from 50 to 80. Thereafter, a particular sample set P_i is selected at random; a sub-set from that set is then chosen which under random translation and rotation forms the query set Q. Thus, a match of Q with at least one P_i is ensured. Next, both the algorithms for partial point set pattern matching are executed with each of the K sample sets. The time reported is averaged over a number of experiments for different values of K. The drastic improvement in the execution time, as reported in the last column of Table 1, is due to the fact that, in general, the maximum

Table 1. Comparative results with respect to CPU Time

Number of sample sets, K	CPU Time in μ sec.		
	Rezende and Lee's method [23]	Proposed method	% saving
10	39,696,066	1,280,111	96.70
20	85,016,140	3,457,594	95.90
50	228,242,586	11,784,255	94.83
100	548,033,187	19,754,415	96.39
200	1075,977,761	39,225,959	96.35

number of equidistant pair of points is very small compared to its theoretical upper bound (see Table 2).

Table 2. Maximum number of equidistant pairs in a point set

(Experiment is performed for randomly generated point sets for different values of n)

number of points (n)	100	200	500	1000
maximum number of equidistant pairs observed	4	6	20	53

4.2   Experiment with Real Fingerprint Minutiae

The proposed algorithm is also tested using real fingerprint minutiae, extracted from fingerprints in the NIST 14 sdb [10,26]. The one pixel thick ridge lines are extracted from a grayscale fingerprint image using the algorithm reported in [8]. Thereafter, minutiae that are terminations and bifurcations on the ridge lines, are extracted using the method in [11] (see Figure 2). We analyzed 100 images of the NIST14 sdb for extracting minutiae. A fingerprint image is then chosen at random; its minutiae are extracted and searched in the database using our proposed algorithm, and always a match is found. The preprocessing stage for the database containing 100 images requires 2.3 seconds on an average, and the search for a particular minutiae point set in the database takes on an average 0.7 seconds.

5   Line Segment Matching

The algorithm can be tailored for checking the congruence of a query set Q among the members in a sample set P where the members in both P and Q are line segments of non-zero length.

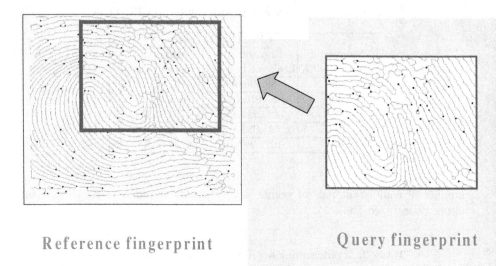

Reference fingerprint   Query fingerprint

Fig. 2. An example of a subset matching in a fingerprint

5.1   Preprocessing

The $\mathcal{T}$ data structure for this problem is an AVL-tree containing the length of the line segments in the sample set P. The members in $\mathcal{T}$ are all distinct. If more than one line segment are of the same length δ, then they are stored in the structure χ_δ attached to the corresponding node of $\mathcal{T}$. Each entry of χ_δ corresponds to a line segment $\ell = \overline{p_i p_m} \in P$; it contains a 4-tuple $\{p_i, p_m, ptr_1, ptr_2\}$, where ptr_1 and ptr_2 point to two AVL-trees $\mathcal{S}_i$ and $\mathcal{S}_m$ corresponding to the points p_i and p_m respectively.

Unlike the earlier problem, here we don't need to maintain the array $\mathcal{S}$, but we need to maintain $\mathcal{S}_i$ for each point p_i, which is an end-point of a line segment in P. Let p_i be an end point of a line segment $\ell \in P$ (the other end point of ℓ is say p_m). $\mathcal{S}_i$ is an AVL-tree containing exactly $n-1$ nodes, corresponding to the line segments in $P \backslash \{\ell\}$. Each node in $\mathcal{S}_i$ corresponds to a line segment, say $\ell' = \overline{p_j p_k}$ (i.e. with end points p_j and p_k), and it contains a 4-tuple $\{(\theta, d), (\theta^*, d^*)\}$, where $\theta = \angle p_j p_i p_m$ and $\theta^* = \angle p_k p_i p_m$, and d and d^* are the distances of p_j and p_k from the point p_i.

The preprocessing time and space complexity is dominated by the time and space required for constructing $\mathcal{S}_i$ for the end points of all line segments in P, which are $O(n^2 \log n)$ and $O(n^2)$ respectively.

5.2   Query

During a query operation, a line segment, say $\ell_1 = \overline{q_1q_2} \in Q$, is chosen from the query set. Let $\delta(\overline{q_1q_2}) = d$ and if the search for d (the length of l_1) is successful in $\mathcal{T}$, then for each line segment in χ_d, we need to check for a match.

We anchor $\overline{q_1q_2}$ with each member $\overline{p_ip_j} \in \chi_d$ (if the match fails, we need to anchor $\overline{q_1q_2}$ with $\overline{p_jp_i}$), and check the presence of the line segments in $Q \setminus \{\ell_1\}$ with those of P. In other words, for each line segment $\ell = \overline{qq'} \in Q \setminus \{\ell_1\}$, we compute a 4-tuple $\{(\phi, \delta), (\phi^*, \delta^*)\}$, where $\phi = \angle q_2q_1q$, $\phi^* = \angle q_2q_1q'$, $\delta = \delta(\overline{q_1q})$ and $\delta^* = \delta(\overline{q_1q'})$, and search in $\mathcal{S}_i$ to identify the presence of that 4-tuple. If it fails, the same search is performed in $\mathcal{S}_j$. This checking needs $O(k\log n)$ time. The number of elements in χ_δ may be at most n. Thus, the worst case query time is $O(kn\log n)$.

6   Polygon Matching

Here the sample set consists of a set of simple polygons with a total of n vertices, and the query set is a simple polygon of k vertices. The polygons in the sample set may overlap. The objective is to report the congruence, if any, of the query polygon to a polygon in the sample set. The data structure $\mathcal{T}$ for this problem is very similar to the line segment matching problem; it contains distinct edge lengths considering all the polygons. If many edges are of same length (say δ), they are stored in χ_δ data structure attached to δ in $\mathcal{T}$. As in the earlier problem, each element of χ_δ corresponds to an edge of a polygon, and it contains a pair of pointers indicating its presence in another data structure $\mathcal{S}$. In the $\mathcal{S}$ data structure, we store the polygons using doubly connected circular link list. The space complexity of this data structure is $O(n)$. The preprocessing time complexity is dominated by the time of creation of $\mathcal{T}$, which needs a sorting of the edges of all the polygons with respect to their lengths, which is $O(n\log n)$.

During the query, we choose one edge from the query polygon. Let its length be δ. We consider elements in χ_δ one by one; for each element e, we reach it in $\mathcal{S}$ using the pointer field attached to e in χ_δ. Next, we traverse the circular link list for Q and the polygon in $\mathcal{S}$ containing the edge e for a detailed match. Thus, if $|Q| = k$, then the detailed match takes $O(k)$ time. As the number of edges of length δ may be $O(n)$ in the worst case, the query time complexity is $O(\log n + kn) = O(kn)$.

7   Conclusion

A simple algorithm for point set pattern matching under congruence is proposed. Given a sample set of n points, our algorithm preprocesses it in $O(n^2\log n)$ time to create a data structure of size $O(n^2)$. Next, given a query set of arbitrary size, say k, it can locate a k-subset of the sample set which matches Q under congruence in $O(k\alpha \log n)$ worst case time, where α is the maximum number of equidistant pair in P. Though the worst case value of α is $O(n^{4/3})$, the empirical results show that it is much less in general. Experimental results show that our algorithm outperforms the algorithm proposed in [23]. It needs to mention that, another algorithm for testing the congruence between two sets is proposed in [3] whose worst case time complexity is same. But that algorithm is not efficient for multiple query on the same sample set, and it uses parametric searching which is not practical for implementation.

The algorithm is then extended for checking congruence among two sets of line segments in $O(kn \log n)$ time, and locating a query polygon among a set of sample polygons in $O(kn)$ time under congruence. We also like to mention the followings:

If the query set consists of more than one polygon with k edges in total, then also the congruence can be checked in $O(kn)$ time.

If we have a planar map consisting of straight line segments, where a point can appear as end point of many lines, then a minor modification of the aforesaid data structure can identify a piecewise linear curve (may be closed or open), in $O(kn \log m)$ time, where m denotes the maximum number of lines incident on a point in the planar map. The preprocessing time and space complexity remains same.

Acknowledgment

We thank Mr. Arindam Biswas for his help in the experiments.

References

1. P. K. Agarwal and M. Sharir, *Efficient Algorithm for Geometric Optimization*, ACM Computing Surveys, vol. 30, No. 4, pp. 412-458, 1998.
2. P. K. Agarwal, B. Aronov, M. Sharir and S. Suri, *Selecting Distances in the Plane*, Algorithmica, vol. 9, pp. 495-514, 1993.
3. T. Akutsu, H. Tamaki, and T. Tokuyama, *Distribution of Distances and Triangles in a Plane Set and Algorithms for Computing the Largest Common Point Sets*, Discrete Computational Geometry, vol. 20, pp. 307-331, 1998.
4. H. Alt and L. J. Guibas, *Discrete Geometric Shapes: Matching, Interpolation, and Approximation - A Survey*, Technical Report B 96-11, Freie Universität Berlin, 1996.
5. H. Alt, B. Behrends, and J. Blömer, *Approximate Matching of Polygonal Shapes*, in Proc. ACM Symposium on Computational Geometry, pp. 186-193, 1991.
6. H. Alt, K. Mehlhorn, H. Wagener, and E. Welzl, *Congruence, Similarity and Symmetries of Geometric Objects*, Discrete Computational Geometry, vol. 3, pp. 237-256, 1988.
7. M. D. Atkinson, *An Optimal Algorithm for Geometrical Congruence*, J. Algorithms, vol. 8, pp. 159-172, 1987.

8. A. Bishnu, P. Bhowmick, J. Dey, B. B. Bhattacharya, M. K. Kundu, C. A. Murthy, and T. Acharya, *Combinatorial Classification of Pixels for Ridge Extraction in a Gray-scale Fingerprint Image*, accepted in: The 3rd Indian Conference on Computer Vision, Graphics and Image Processing, Ahmedabad, India, Dec. 2002.

9. J. Beck and J. Spencer, *Unit Distances*, J. Combinatorial Theory A, vol. 37, pp. 231-238, 1984.

10. G. T. Candela, P. J. Grother, C. I. Watson, R. A. Wilkinson, and C. L. Wilson, *PCASYS - A Pattern-Level Classification Automation System for Fingerprints*, NISTIR 5647, National Institute of Standards and Technology, August 1995.

11. A. Farina, Zs. M. Kovacs-Vajna, and A. Leone, *Fingerprint Minutiae Extraction from Skeletonized Binary Images*, Pattern Recognition, vol. 32, pp. 877-889, 1999.

12. D. Forsyth, J. L. Mundy, A. Zisserman, C. Coelho, A. Heller, and C. Rothwell, *Invariant Descriptors for 3-D Object Recognition and Pose*, IEEE Trans. PAMI, vol. 13, pp. 971-991, 1991.

13. M. Gavrilov, P. Indyk, R. Motwani, and S. Venkatasubramanian, *Geometric Pattern Matching: A Performance Study*, in Proc. ACM Symposium on Computational Geometry, pp. 79-85, 1999.

14. J. E. Goodman and R. Pollack, *Multidimensional Sorting*, SIAM Journal on Computing, vol. 12, pp. 484-507, 1983.

15. M. T. Goodrich, J. S. B. Mitchell, and M. W. Orletsky, *Approximate Geometric Pattern Matching under Rigid Motions*, IEEE Trans. PAMI, vol. 21, no. 4, pp. 371-379, April, 1999.

16. R. M. Haralick, C. N. Lee, X. Zhuang, V. G. Vaidya, and M. B. Kim, *Pose Estimation from Corresponding Point Data*, IEEE Trans. SMC, vol. 19, pp. 1426-1446, 1989.

17. S. Jozsa and E. Szemeredi, *The Number of Unit Distances on the Plane*, in: *Infinite and Finite Sets* Coll. Math. Soc. J. Bolyai, vol. 10, pp. 939-950, 1973.

18. D. E. Knuth, The Art of Computer Programming, Vol. 3: Sorting and Searching, Addison-Wesley Publishing Company, Inc., 1973.

19. D. E. Knuth, Jr. J. H. Morris, and V. R. Pratt, *Fast Pattern Matching in Strings*, SIAM J. on Computing, vol. 6, no. 2, pp. 240-267, 1977.

20. D. T. Lee and Y. T. Ching, *The Power of Geometric Duality Revisited*, Inform. Process. Lett., vol 21, pp. 117-122, 1985.

21. J. Matousek, *On Enclosing k Points by a Circle*, Information Processing Letters, vol. 53, pp. 217-221, 1995.

22. N. Megiddo, *Applying Parallel Computation Algorithms in the Design of Serial Algorithms*, J. Assoc. Comput. Mach, vol 30, pp. 852-865, 1983.

23. P. J. Rezende and D. T. Lee, *Point Set Pattern Matching in d-dimensions*, Algorithmica, vol. 13, pp. 387-404, 1995.

24. L. Szekely, *Crossing numbers and Hard Erdős Problems in Discrete Geometry*, Combinatorics, Probability and Computing, vol. 6, pp. 353-358, 1997.

25. J. Spencer, E. Szemeredi, and W. T. Trotter, *Unit Distances in the Euclidean Plane*, in: *Graph Theory and Combinatorics* (B. Bollobas ed.) Academic Press, London, pp. 293-308, 1984.

26. C. I. Watson, *Mated Fingerprint Card Pairs 2*, Technical Report Special Database 14, MFCP2, National Institute of Standards and Technology, September 1993.

Unlocking the Advantages
of Dynamic Service Selection and Pricing

Bala Kalyanasundaram[1]*, Mahe Velauthapillai[1] **, and John Waclawsky[2] ***

[1] Georgetown University, 37th and O Streets, NW, Washington, DC 20057, USA
kalyan@cs.georgetown.edu
[2] Cisco Systems, 7025 Kit Creek Rd, Research Triangle Park, NC 27709, USA
mahe@cs.georgetown.edu

Abstract. Bandwidth limitations, resource greedy applications and an increasing number of voice and data users are straining the air interface of wireless networks. Hence, novel approaches and new algorithms to manage wireless bandwidth are needed. This paper unlocks the potential to improve the performance of overall system behavior by allowing users to change service level and/or service provider for a (small) price. The ability to dynamically re-negotiate service gives the user the power to control QoS while minimizing usage cost. On the other hand, the ability to dynamically change service level pricing allows the service providers to better manage traffic, improve resource usage and most importantly maximize their profit. This situation provides a surprising win-win situation for BOTH the service providers AND the users. In this paper, we present easy to implement online algorithms to minimize the overall usage cost to individual mobile users.

1 Introduction

Commercial wireless data services are beginning to proliferate. Profit is the typical motive for their creation and deployment. In the pursuit of profit, future wireless networks will attempt to support a wide range of applications. It is clear that controlling and allocating bandwidth for individual users with potentially bursty data flows over constrained, error prone and costly communication channels (such as mobile wireless) is increasingly important.

One way to better manage traffic is to allow bandwidth requests to closely track application demands, such as browser needs and explicit user requests, along with the ability to charge for more capacity. From the service provider's perspective, it is advantageous to increase the price when the demand is high and decrease the price when the demand is low. However, due to the bursty nature of the traffic, it is unrealistic to use statically fixed peak and non-peak hours. Instead, it is better to allow the network to dynamically change the pricing

---

* Supported in part by NSF under grant CCR-0098271, Airforce Grant, AFOSR F49620-02-1-0100 and Craves Family Professorship funds.
** Supported in part by a gift from AT&T and McBride Endowed Chair funds.
*** *jgw@cisco.com*

R. Petreschi et al. (Eds.): CIAC 2003, LNCS 2653, pp. 46–57, 2003.
© Springer-Verlag Berlin Heidelberg 2003

scheme. This is called dynamic pricing [7]. Recall that pricing has drastically influenced the growth of many industries, including network and communication industries [3,2,9,10,11,1]. Examples of dynamic pricing include peak/off-peak pricing, usage based pricing, priority based pricing, congestion based pricing [5,8,10], popularity based pricing (see [6,7,9]).

The dynamic pricing model allows for users to re-negotiate the service level they are accessing and change service providers in response to advertised price fluctuations. The ability to re-negotiate between/within service providers gives the user the ability to abandon or complete a task with low cost. Another opportunity for re-negotiation arises when a user has low bandwidth requirements but occasionally requires something more.

The problem from a user's perspective is that mobility exposes them to, not only changing network conditions but, changing service providers with differing service levels and pricing structures. In the future, mobile users are likely to be besieged by choice.

In this paper we propose an online algorithm that exploits choice and leverages dynamic pricing to improve service for these users. In addition, these algorithms can be applied to determine good transmission rate with the minimum cost for mobile users and/or work cooperatively with other network algorithms that set costs and broadcast variable pricing information.

Why wouldn't the user always choose the exact bandwidth/service-level required for QoS from the cheapest service provider? (Un)Fortunately, there is a small cost to change service levels and a larger cost to change service providers. Moreover, the cheapest service provider may increase the price within a short time (i.e., dynamic pricing). Since it costs to change providers, tracking the cheapest service provider may not be a good choice for the user.

The objective function of our algorithm is to minimize the overall cost for the user. Observe that the overall cost includes both the cost of requested bandwidth/level as well as the costs for re-negotiation.

1.1 Definitions, Notations and Our Model

We now define our model that is an abstraction of a network (wired/wireless) that is capable of bandwidth re-negotiation.

Service Providers: We assume that there are K service providers (say P_1, P_2, ..., P_K) and each employ dynamic pricing. But, the set of available (or suitable) providers for the user is constantly changing since the user can be mobile. So, at any moment in time, a user can choose a service provider from the set of available providers which is a subset of these K providers.

A service provider may offer different levels of service. For instance, the service provider may divide the bandwidth into channels and tome slots. For better quality, more time slots (or codes for CDMA) are needed, which will result in higher cost. We have made the following assumptions to reflect these facts:

Discretization: Within a single service provider P_i, traffic may be carried at B_i different levels. The number B_i of levels is a measure of the granularity of the service offering and may differ from provider to provider.

Let $\ell_1^i, \ell_2^i, \ldots, \ell_{B_i}^i$ be the levels of quality of service of the service provider P_i. We also assume that there is a linear order $>$ among the levels and we number the levels such that $\ell_{j+1}^i > \ell_j^i$ for all j.

In order to generalize our results, we assume that the quality of the service at level ℓ_i^* need NOT be the same for two different providers. However, within a service provider, we impose a natural restriction that the cost function is monotonically increasing with increasing quality of service. We define the capacity of a level for a service provider to be the maximum number of customers that the provider can support for that level.

Cost Function: Let $L_k = \{\ell_1^k, \ell_2^k, \ldots, \ell_{B_k}^k\}$, and T be the time domain. Also, let N^+ be the set of positive integers including zero. Then the cost function C_k for the service provider P_k is defined to be:

$C_k : L_k \times T \longrightarrow N^+$ with the condition that for all time t and for all $\ell_i^k < \ell_j^k$ $[C_k(\ell_i^k, t) \leq C_k(\ell_j^k, t)]$.

Minimum QOS: Assume that the user demands a minimum quality of service (or level) $qos_k(t) \in L_k$ for service at time t if the user chooses P_k as the service provider.

We assume that a user (or the mobile device) has access to current cost of any level of the potential service provider (e.g. through wireless control channel). As a consequence, the user, given his/her current selection of service provider(s), can calculate what (s)he has to pay thus far. However, due to dynamic pricing, the user has no knowledge of the future costs.

A mobile user may re-negotiate (1) to increase or decrease the level of service within a provider or (2) jump to a new service provider due to increasing price or dropped call.

In an ideal world, the user re-negotiates with the network administrator as often as necessary so that the negotiated bandwidth exactly matches with the minimum requirements, thus incurring absolute minimum cost (we call it minimal or required bandwidth cost). But, whenever the user tries to re-negotiate, the request goes through the admission control mechanism of the network which decides whether to support it or not. As a result, re-negotiation takes time and communication for both user and the network. This re-negotiation process becomes a severe bottleneck if the network is servicing many users who re-negotiate often. One way to solve this problem is to associate a fixed cost r for each re-negotiation within a provider and a fixed cost R for changing providers. Moreover, one can also assume that the re-negotiation cost as well as the bandwidth cost are expressed in the same unit (say US dollars).

Formal Problem Definition: At time t, given $C_x(\ell_i^x, t)$ and $qos_x(t)$ (for all x) online (we denote it as input I), let $A(I, t) = (P_x, \ell_y^x)$ be the selection of online algorithm A where P_x is the selected provider with $\ell_y^x \geq qos_x(t)$ be the selected level of service.

At time t, we associate two types of cost, re-negotiation cost and total bandwidth cost to online algorithm A. Let $Rcost(A, t)$ be the cost of re-negotiation, if it happens, for the algorithm A at time t. If no re-negotiation takes place at time t then $Rcost(A, t) = 0$. Otherwise, $Rcost(A, t)$ will depend upon the type of re-negotiation that takes place at time t. Typically, we associate a fixed cost for each type of re-negotiation. Let $Bcost(A, t)$ be the total bandwidth cost for using the requested level of service by A at time t. That is, $Bcost(A, t)$ is defined to be $C_x(\ell_y^x, t)$ where $A(I, t) = (P_x, \ell_y^x)$. Finally, let $Mcost(t) = min_x \ C_x(qos_x(t), t)$ be the minimum cost required to maintain the minimum quality of service.

Given the input I, the goal of the algorithm A is to minimize

$$Cost(A, I) = \sum_t [Rcost(A, t) + Bcost(A, t) - Mcost(t)].$$

In this paper, we prove optimality of the following competitive ratio, where A is our algorithm and OPT is the offline optimal algorithm.

$$\max_I \ \frac{Cost(A, I)}{Cost(OPT, I)}.$$

For algorithm A, we define excess bandwidth cost at time t to be $Bcost(A, t) - Mcost(t)$. We also define "total cost" to be the sum of "total bandwidth cost" and "re-negotiation cost". Analogously, we define "excess cost" to be the sum of "excess bandwidth cost" and "re-negotiation cost". It is not hard to see that the excess cost measure is better than total cost measure.

1.2   Summary of Results

Our major result is an easy to implement online randomized algorithm, we call it KVW2-algorithm, that a user or a mobile device can use to negotiate with available service providers for appropriate level of service to maintain a minimum QoS and minimize overall cost. Pricing and availability information as well as the required (dynamically changing) QoS is also fed to the algorithm online.

In section 2 of the paper we consider a simpler case where there is only one service provider with B levels of service. We present a **deterministic** online algorithm called KVW algorithm. This algorithm is not practical since it does not take into account the following implementation details: (1) capacity limitation of a service provider, (2) price of a level may vary arbitrarily, (3) service levels may not match across different service providers, and (4) available service providers may change due to the mobility of the user or dropped calls.
We show that KVW-algorithm is the best online algorithm by proving the following theorem.

Theorem 1.
(a) The competitive ratio of KVW-algorithm is $O(\log B)$.
(b) The competitive ratio of every deterministic online algorithm is $\Omega(\log B)$.

We then show that randomization does **not** improve the performance.

Theorem 2. *The competitive ratio of every randomized algorithm is $\Omega(\log B)$.*

We then consider the case where either cost or requirement but not both are fully known to the algorithm. Surprisingly, we show that the additional information does not significantly improve the performance of online algorithms.

Theorem 3. *Assume that the lowest service level gets a cost of 1 and the highest level gets a cost of M. Also assume that the cost does not change over time. But the minimum requirements are given online. The competitive ratio of every online algorithm is $\Omega(\log M / \log \log M)$.*

Theorem 4. *Assume that the bandwidth requirements are given in advance while the cost is given online. The competitive ratio of every online algorithm is $\Omega(\log B)$.*

In section 3, we show how to handle the first two restrictions, namely limited capacity and price fluctuations, by presenting a practical version of KVW algorithm for the single service provider. We prove that the practical online algorithm is also optimal under *r-sensible* pricing scheme which we will introduce later in this paper. Intuitively, under such *r-sensible* pricing scheme the current cost of any unavailable level is at least r more than any available level.

Theorem 5. *Under r-sensible pricing scheme,*
(a) The competitive ratio of practical KVW-algorithm is $O(\log B)$.
(b) The competitive ratio of every deterministic online algorithm is $\Omega(\log B)$.

In section 4, we consider the case of multiple service providers where the pool of available service providers changes dynamically. We consider the possibility that service levels may not match across different service providers. Using practical KVW-algorithm as a sub-routine, we present a randomized online algorithm, we call it KVW2-algorithm, to remove all of the four restrictions. Here $B = max_i \ B_i$.

Theorem 6. *Assume oblivious adversary that does not know the random bits used by the online algorithm. Also, assume either unbounded capacity or R-sensible pricing scheme, where R is the fixed cost of moving from one provider to another.*
(a) The competitive ratio of modified KVW2-algorithm is $O((\log_2 B) (\log K))$.
(b) The competitive ratio of any randomized online algorithm is $\Omega(\log_2 B + \log K)$.

Finally, we point out that there are no good deterministic algorithms. This shows that randomization **does help** significantly when we have multiple service providers.

Theorem 7. *The competitive ratio of any deterministic online algorithm is $\Omega(K + \log B)$.*

2   Single Service Provider

In this section, we assume that there is only one service provider. Without loss of generality, let us assume that the service levels are $1, 2, \ldots, B$. Let c be the smallest integer such that $2^c \geq B$.

2.1   KVW-Algorithm

We now present a deterministic algorithm to perform dynamic re-negotiation when price and requirements are given online.

However, this algorithm is restricted in the following way: (1) service provider has **unlimited** capacity, and (2) price of a level does **not** exceed r, the **fixed** cost for re-negotiation. For the ease of understanding the algorithm, we first provide some intuition. Imagine that service levels are integers from 1 through B. Our algorithm maintains a window [minlevel, maxlevel] which is initially [qos[0],B+1]. Typically, our algorithm negotiates for a level (called *currentlevel*) in the middle of the window and continues with level of service until either the accumulated excess cost during this window period exceeds r, or minimum required level of service exceeds *currentlevel*. At this time, our algorithm shrinks the window by at least half and chooses either the top or bottom half depending on the event.

Observe that the excess cost incurred by our algorithm during any fixed window period is at most r. After $O(\log B)$ resizing of the window, it collapses and gets reset to [qos(t),B+1] again. During this $O(\log B)$ iterations of resizing, either offline optimal re-negotiated or incurred a cost of r.

```
** t is current time; qos(t) is continuously fed **
    maxlevel= B+1
    minlevel= qos(t)
    currentlevel = qos(t)
    excesscost = 0
* excesscost = incurred excess cost during the current window **
while not(done) {
    while (maxlevel > minlevel){ ** one move **
        Wait Until
            Case (qos(t) > currentlevel) ** up move **
                minlevel = qos(t)
                maxlevel = max(maxlevel,qos(t))
            Case (excesscost is at least r) ** down move **
                maxreqlevel = max. requested level during this move
                minlevel = max(maxreqlevel,minlevel)
                maxlevel = currentlevel
            End of Case
        End of Wait
        excesscost = 0
        currentlevel = FLOOR[(maxlevel+minlevel)/2]
        Re-negotiate for currentlevel
```

```
    } ** end of inner while **
    maxlevel= B+1; minlevel= qos(t)
} ** end of outer while **
```

Lemma 1. *The body of the inner while loop of* KVW-ALGORITHM *is executed consecutively* $O(c)$ *times.*

Lemma 2. *After* $2i$ *iterations of the outer while loop, off-line optimal incurs a cost of at least* ir.

We are now ready to prove Theorem 1.

Proof. (sketch of Theorem 1)
The upper bound, that is part (a), follows from Lemma 1 and Lemma 2.

In order to prove part (b), we prove that the competitive ratio of every online algorithm is $\Omega(\log B)$ even in the *total cost* model. It is not hard to observe that this lower bound is strictly stronger than the lower bound for *excess cost* model.

The adversary works in stages. Let A be the given online deterministic algorithm. For a stage, based on the moves of A, the adversary selects $1 \le i \le B$ and sets the cost of the levels $\ell_1, \ell_2, \ldots, \ell_i$ to be zero and the cost of the remaining levels to be 1. In addition, the level of no request during this stage exceeds ℓ_i. Thus, by selecting ℓ_i, the off-line optimal does not incur a cost. We will now describe the moves of the adversary that forces A to make $\Omega(\log B)$ moves during the stage where each move costs r.

The adversary maintains two levels *maxlevel* (initially B) and *minlevel* (initially 1) and the current request level is *minlevel*. The adversary also maintains the invariant *maxlevel* $\ge \ell_i \ge$ *minlevel*. Suppose the algorithm A requests for level *currentlevel* at the time of re-negotiation. If *currentlevel* $>$ (*maxlevel*+*minlevel*)/2, then the adversary resets *maxlevel* = *currentlevel* − 1 and waits until A re-negotiates. If A never re-negotiates, then ℓ_i is any level between *maxlevel* and *minlevel*. On the other hand, if *currentlevel* $\le$ (*maxlevel*+*minlevel*)/2, then the adversary resets *minlevel* = *currentlevel* + 1 and brings a new input request with level *minlevel*. Now, the online algorithm must re-negotiate and adversary continues until either *maxlevel* = *minlevel* or A continues to incur cost at a constant rate since it does not re-negotiate. It is easy to see that A incurs a cost of $\Omega(\log B)r$.

Next we will show that randomization does not help. In contrast, we will show later that randomization does help when there are more than one service providers. We employ oblivious adversary.

Proof. (sketch of Theorem 2)
Applying Yao's technique, it suffices to fix an input distribution and show that the expected cost of every deterministic algorithm is $\Omega(\log B)$ times the excess cost of off-line optimal.

For the sake of simplicity, we drop floors and ceiling (i.e., division by 2 results in an integer). We maintain a window $(low, high)$ which is $(1, B)$ initially. We uniformly at random set the price and minimum requirement for the next time unit according to one of the following two choices:

1. Set the cost of levels 1 through $high$ to be 0 and all other levels to be r for the next interval. Minimum requirement is raised to $(low + high)/2$. The window is reset to be $((low + high)/2, high)$.

2. Set the cost of levels 1 through $-1 + (low + high)/2$ to be 0 and all other levels to be r for the next interval. Minimum requirement does not change. The window is reset to be $(low, -1 + (low + high)/2)$.

We repeat this process (for $\Omega(\log B)$ times) until the window collapses. Let A be the given deterministic algorithm. Depending on what A does, it is not hard to show that A incurs a cost of r with probability $1/2$. Therefore, the expected cost of A is $\Omega(r \log B)$. For any random sequence, it is not hard to show that off-line optimal incurs a cost of r, by setting the initial level to be a level in the window right before it collapses.

3   Single Service Provider with Limited Capacity

The base KVW-ALGORITHM assumes that the service provider has adequate (infinite) capacity to handle all of the requests. In addition, it also makes another assumption that the cost of any level is at most r.

We now show that under some reasonable conditions, we can extend the KVW-algorithm to deal with bounded capacity and arbitrary price fluctuations. Suppose the algorithm asks for level k that is not available. Since the provider denies a request, the algorithm can **not** strictly follow the strategy of KVW-algorithm. In contrast, offline optimal, knowing the future, could have asked for level k, sometime ago when it was available. There is no way any online algorithm can predict such lack of availability. Thus it is trivial to establish that the competitive ratio of any deterministic and randomized algorithm is unbounded.

When capacity limit is reached for a level, say k, an obvious solution to reduce the congestion is to reduce the level of service for those who do not need level k. This can be accomplished by raising the price of the level to be at least r more than the price of the highest available lower level. By setting the price this way, even for a short time, any customer who is on level k or higher will either pay the increased cost or cut down their level of service. We call it a *r-sensible* pricing scheme.

We have yet another problem to consider. If the minimum requested level for a user is not available, the user can dropout or continue with degraded service. Even though the offline optimal can satisfy this requests, it must then reject some other requests due to similar capacity restrictions (or allow the user to continue with degraded service). As a consequence it is reasonable to compare the cost of those inputs that are satisfied by both online and offline optimal algorithms.

Yet another attraction in dynamic pricing is to allow arbitrary price fluctuations set by the providers as a result of (free) market demand. Unfortunately,

the analysis of the KVW-algorithm assumes that the price of any level does
not exceed r, the re-negotiation cost. We fix this problem in our extension by
allowing the user to set a tolerance for the price fluctuations. Given a tolerance
for the price fluctuation, the algorithm re-negotiates so that the total cost is
small and the cost at any moment is within the tolerance level of the cost of
minimum required level. Assuming the tolerance to be r, we now present our
algorithm. This tolerance r is used in the *minfind* function. This algorithm is
still surprisingly straightforward and easy to implement.

There is a technically subtle point to be observed here. Suppose a user is
using a level k and its price is hiked to (say) ∞. This can happen to any online
algorithm. As a consequence, any online algorithm would incur ∞ cost. So it is
reasonable to assume that the price information for next time unit is available
so that the online algorithm can calculate the costs (excess or total) including
the next time unit and make decision to continue or not.

```
** t is current time; qos(t) is continuously fed **
    t=0; ** this is time and gets updated automatically
    maxlevel= minfind(qos(t),B)+1
    minlevel= qos(t)
    currentlevel = FLOOR[(minlevel+maxlevel)/2]
    Re-negotiate for currentlevel
    K = highest available level not greater than currentlevel
    currentlevel = K
    excesscost = 0
* excesscost = incurred excess cost during the current window **
while (currentlevel ≤ qos(t)) {
    while ((maxlevel > minlevel)&(currentlevel ≤ qos(t)){
        Wait Until
            Case (qos(t) > currentlevel) ** up move
                minlevel = qos(t)
                x = max(maxlevel,qos(t))
                maxlevel = minfind(qos(t),x)
            Case (excesscost is at least r) ** down move
                maxreqlevel = max. requested level during this move
                    which is at least current minlevel
                minlevel = minfind(qos(t), maxreqlevel)
                maxlevel = minfind(qos(t),currentlevel)
            End of Case
        End of Wait
        currentlevel = FLOOR[(maxlevel+minlevel)/2]
        Re-negotiate for currentlevel
        K = highest available level not greater than currentlevel
        currentlevel = K
        excesscost = 0
    } ** end of inner while **
    maxlevel= minfind(qos(t),B)+1
```

```
    minlevel= qos(t)
    excesscost = 0
} ** end of outer while **
```

We are now ready to prove Theorem 5.

Proof. (sketch of Theorem 5)
Observe that whenever our algorithm re-negotiates, there is a possibility that the available level is smaller than the one we asked for. During an iteration of the inner loop, observe that the window $[minlevel, maxlevel]$ reduces by at least 50% if the *currentlevel* is at the middle of the interval. However, *currentlevel* is either at the middle or at the lower half of the interval. Therefore, if the current move is a *down move* then the window reduces by at least 50%. On the other hand, for a *up move* the window may not reduce by much but will not increase. We call such a move a *bad up move*. Observe that a *bad up move* is a consequence of the fact that the *currentlevel* was not at the middle due to the fact that the available level was smaller than the requested level when this *currentlevel* was set. Observe that the inner while loop is executed $X + \log B$ times consecutively where X is the number of *bad up moves*. Also observe that the cost incurred by our algorithm for each iteration of the inner loop is at most $2r$.

Consider the setting of *currentlevel* $= K$ where *currentlevel* is not at the middle of the window at time t_1 and the following iteration of the loop results in a *bad up move* at time t_2. We now argue that offline optimal incurs a cost of at least r during the interval $[t_1, t_2]$. This is true if offline optimal renegotiates during this interval. Otherwise, let *opt* be the level chosen by offline optimal. Suppose *opt* $>$ *currentlevel*. Since at time t_1, the highest available level was *currentlevel*, offline optimal incurs an excess cost of r due to *sensible pricing* scheme. On the other hand, suppose *opt* $\leq$ *currentlevel*. Then at time t_2, offline optimal must re-negotiate since $qos(t_2) >$ *currentlevel* $\geq$ *opt*.

Our upperbound follows if for every consecutive two iterations of the outer while loop, there is at least one *bad up move*. We now consider the case where consecutive two iterations of the outer while loop there are no *bad up moves*. Observe that there can be at most $2 \log_2 B$ such iterations of inner while loop. So it suffices to show that offline optimal incurs a cost of r during such interval $[t_1, t_2]$. This proof is analogous to the proof of lemma 2 where we consider two consecutive iterations of the outer loop.

4   Mobile User
and Dynamically Changing Service Providers

In this section, we consider the case where there are more than one, possibly competing, service providers. In addition, when a user is mobile, it is inevitable that the pool of available service providers change dynamically. Recall that R is the **fixed** cost of changing service providers.

We now present a randomized KVW2-ALGORITHM that dynamically decides when to change a service provider. Due to technological differences in service providers, it is possible that service level i of two different providers may not be identical. In spite of all these differences, we will show that we can modify the KVW-algorithm to perform well. Interestingly, our algorithm does **not** assume that the service providers vary cost independently.

Let $\{P_1, P_2, \ldots, P_K\}$ be the set of K service providers. Also, let B_i be the number of levels of service offered by the provider P_i. For each service provider P_j, our algorithm will maintain E_j, the excess cost, since the last time it was reset to zero by KVW2-Algorithm. It is also important to observe how we measure excess cost. Given the user's minimum QoS requirement, one can calculate optimal bandwidth cost assuming re-negotiation costs (within and among service providers) are zero. The cost incurred in excess of this minimal cost is defined to be the excess cost. We will also run (or simulate) KVW-Algorithm on a selected subset of available providers. Recall that the KVW-Algorithm also maintains excess cost which is different from E_j's. Each run or simulation of KVW-Algorithm imagines that there is only one service provider and focuses on when and where to change the service levels. Thus, the excess cost calculation for KVW-Algorithm is based on the optimal solution that has only one available provider. Since the service levels of different service providers may not match, it is important to observe that minimum requirements for a user differs for different service providers.

```
    A = {P1, P2, ..., PK } ** the set of available providers
    C = { } ** the set of eliminated providers for a round
    For all Pj in A, set Ej = 0
while (not done) {
    while (A is not empty) {
        Choose uniformly at random choose Pi from A
        Run KVW-Algorithm on Pi
        For all Pj in A - { Pi }
        Simulate KVW-Algorithm on Pj
        Wait Until Ei >= 4R log Bi or
            Pi could not allocated level at least equal to qos(t)
        Z = {Pj in A: Ej >=4R log Bj or Pj couldn't allocate level
            equal or greater that minimimum QoS of the user }
        A = A - Z
        C = C UNION Z
    } End of While
    A = {P1, P2, ..., PK } ** the set of available providers
    C = { } ** the set of eliminated providers for a round
    For all Pj in A, set Ej = 0
} ** end of outer while **
```

It is important to observe that actual negotiations takes place between the user and only one service provider at any time. On the other hand, using available pricing information, the algorithm can simulate other service providers and calculate excess cost as if it were the (randomly) chosen service provider. When we analyze the algorithm, we assume oblivious adversary which does not see the random actions of the algorithm. However, it can be argued that service providers may set the price based on the traffic. Thus, the adversary may be adaptive by setting the price based on the behavior of individual users. We agree that this is true and that the service providers may target a single user. But, in practice, service providers change the price according to current demand and prior usage statistics but not based on a particular individual. Also, observe that our algorithm randomly selects a service provider from a pool. So, given a class of service providers, the choice of a user has nothing to do with how expensive or inexpensive the service provider is. Therefore any dynamic price scheme that does target a particular individual is an oblivious adversary as far as a random individual user is concerned. In other words, it is reasonable to assume that a random user faces an oblivious adversary.

Due to space limitations many of the proofs are omitted.

References

1. M. Barry, A.T. Campbell, and Veres. A. Distributed control algorithms for service differentiation in wireless packet networks. In *Proceedings of Twentieth Annual Joint Conference on IEEE Computer and Communications (IEEE INFOCOM)*, April 2001.
2. D. Gupta, D. Stahl, and A. Whinston. A Priority Pricing Approach to Manage Multi-Service Class Networks in Real Time. *MIT Workshop on Internet Economics*, 1995.
3. M.T. Hills. New Choices in Data Network Pricing. *Business Communication Review*, 25, 1995.
4. Bala Kalyanasundaram and Mahe Velauthapillai. Dynamic pricing schemes for multilevel service providers. In *Proceedings of Second International Conference on Advances in Infrastructure for E-Business, E-Science, and E-Education on the Internet*, August 2001.
5. K. Lavens, P. Key, and D. McAuley. An ECN-based end-to-end congestion-control framework: experiments and evaluation. Technical Report MSR-TR-2000-104, Microsoft Research Technical Report, October 2000.
6. J. Murphy and L. Murphy. Bandwidth Allocation By Pricing in ATM Networks. *IFIP Transactions C-24:Broadband Communications II, North-Holland*, 1994.
7. Nortel and Philips. Dynamic Charging Schemes, Network Implications. In *Project AC014: CANCAN: Contract Negotiations and Charging in ATM Networks*, July 23 1998.
8. K. Peter, D. McAuley, P Barham, and K. Lavena. Dynamics of Congestion Pricing. Technical Report MSR-TR-99-15, Microsoft Research Technical Report, February 1999.
9. D.J. Reinenger, D. Raychaudhuri, and J.Y. Hui. Bandwidth Re-negotiation for VBR Video Over ATM Networks. *IEEE Journal on Selected Areas in Communications*, 14(6), August 1996.
10. S. Shenker, D. D. Clark, D. Estrin, and S. Herzog. Pricing in Computer Networks: Reshaping the Research Agenda. *ACM Computer Communication Review*, 26:19–43, April 1996.
11. A. Whit. Dynamic Pricing 101: Business Models. *Internet World*, 6(10), May 2000.

The Relative Worst Order Ratio
for On-Line Algorithms*

Joan Boyar[1,**] and Lene M. Favrholdt[2,***]

[1] Department of Mathematics and Computer Science,
University of Southern Denmark, Odense, Denmark
joan@imada.sdu.dk
[2] Department of Computer Science, University of Copenhagen, Denmark
lenem@diku.dk

Abstract. We consider a new measure for the quality of on-line algorithms, the *relative worst order ratio*, using ideas from the Max/Max ratio [2] and from the random order ratio [8]. The new ratio is used to compare on-line algorithms directly by taking the ratio of their performances on their respective worst orderings of a worst-case sequence. Two variants of the bin packing problem are considered: the Classical Bin Packing Problem and the Dual Bin Packing Problem. Standard algorithms are compared using this new measure. Many of the results obtained here are consistent with those previously obtained with the competitive ratio or the competitive ratio on accommodating sequences, but new separations and easier results are also shown to be possible with the relative worst order ratio.

1 Introduction

The standard measure for the quality of on-line algorithms is the competitive ratio [5,10,7], which is, roughly speaking, the worst-case ratio, over all possible input sequences, of the on-line performance to the optimal off-line performance. There have been many attempts to provide alternative measures which either give more realistic results than the competitive ratio or do better at distinguishing between algorithms which have very different behaviors in practice. Two very interesting attempts at this are the Max/Max ratio [2] and the random order ratio [8].

The Max/Max Ratio. The Max/Max ratio allows direct comparison of two on-line algorithms for an optimization problem, without the intermediate comparison to OPT, as is necessary with the competitive ratio. Rather than comparing

---

* Supported in part by the Danish Natural Science Research Council (SNF) and in part by the Future and Emerging Technologies program of the EU under contract number IST-1999-14186 (ALCOM-FT).
** Part of this work was done while visiting the Computer Science Department, University of Toronto.
*** Part of this work was done while working at the Department of Mathematics and Computer Science, University of Southern Denmark.

R. Petreschi et al. (Eds.): CIAC 2003, LNCS 2653, pp. 58–69, 2003.
© Springer-Verlag Berlin Heidelberg 2003

two algorithms on the same sequence, they are compared on their respective worst case sequences of the same length. Ben-David and Borodin [2] demonstrate that for the k-server problem the Max/Max ratio can provide more optimistic and detailed results than the competitive ratio.

The Random Order Ratio. The random order ratio gives the possibility of considering some randomness of the request sequences without specifying a complete probability distribution. For an on-line algorithm $\mathbb{A}$ for a minimization problem, the random order ratio is the maximum ratio over all multi-sets of requests of the expected performance of $\mathbb{A}$ compared with OPT on a random permutation of the multi-set. If, for all possible multi-sets of requests, any ordering of these requests is equally likely, this ratio gives a meaningful worst-case measure of how well an algorithm can do. Kenyon [8] has shown that for the Classical Bin Packing Problem, the random order ratio of Best-Fit lies between 1.08 and 1.5. In contrast, the competitive ratio of Best-Fit is 1.7.

The Relative Worst Order Ratio. Attempting to combine the desirable properties of both the Max/Max ratio and the random order ratio, we define the *relative worst order ratio*, where when comparing two on-line algorithms, we consider a worst-case sequence and take the ratio of how the two algorithms do on their worst orderings of that sequence. Note that the two algorithms may have different "worst orderings" for the same sequence.

The Worst Order Ratio. Although one of the goals in defining the relative worst order ratio was to avoid the intermediate comparison of any on-line algorithm to the optimal off-line algorithm OPT, it is still possible to compare on-line algorithms to OPT. In this case, the measure is called the *worst order ratio*. Note that for many problems, the worst order ratio is the same as the competitive ratio, since the order in which requests arrive does not matter for an optimal off-line algorithm. However, for the Fair Bin Packing Problem mentioned below, the order does matter, even for OPT. The same is true for bounded space bin packing [6] where only a limited number of bins are allowed open at one time.

The Bin Packing Problems. In the *Classical Bin Packing Problem* we are given an unlimited number of unit sized bins and a sequence of items each with a non-negative size, and the goal is to minimize the number of bins used to pack all the items. In contrast, in the *Dual Bin Packing Problem*, we are given a fixed number n of unit sized bins, and the goal is to maximize the number of items packed in the n bins. A variant of Dual Bin Packing is the *Fair Bin Packing Problem*, where the algorithms have to be *fair*, i.e., to reject items only when they do not fit in any bin.

Our Results. Many results obtained are consistent with those previously obtained with the competitive ratio or the competitive ratio on accommodating sequences [4], but new separations and easier proofs are also shown to be possible with the relative worst order ratio.

   For the Classical Bin Packing Problem, First-Fit and Best-Fit are better than Worst-Fit, which is better than Next-Fit. This latter result is in contrast to the competitive ratio, where there appears to be no advantage to Worst-Fit being able to use empty space in earlier bins, since both have a competitive ratio of 2 [6]. First-Fit is still the best Any-Fit algorithm and Next-Fit is no better than any Any-Fit algorithm.

The worst order ratio for any fair deterministic algorithm for the Dual Bin Packing Problem is not bounded below by any constant. In contrast, with the relative worst order ratio, one gets constant ratios. We find that First-Fit does at least as well as any Any-Fit algorithm and is better than Best-Fit, which is better than Worst-Fit. Worst-Fit is at least as bad as any fair on-line algorithm and strictly worse than First-Fit. This contrasts favorably with the competitive ratio, where Worst-Fit is better than First-Fit [4].

Unfair-First-Fit is an algorithm for Dual Bin Packing which is not fair and does better than First-Fit when using the competitive ratio on accommodating sequences. Under the relative worst order ratio, Unfair-First-Fit is incomparable to all Any-Fit algorithms, i.e., for any Any-Fit algorithm $\mathbb{A}$ there are sequences where $\mathbb{A}$ does better and sequences where Unfair-First-Fit does better.

2   The (Relative) Worst Order Ratio

This paper considers the *worst order ratio* as well as the *relative worst order ratio*. The relative worst order ratio appears to be the more interesting measure for the two variants of bin packing studied here.

The definition of the relative worst order ratio uses $\mathbb{A}_W(I)$, the performance of an on-line algorithm $\mathbb{A}$ on the "worst ordering" of the multi-set I of requests, formally defined in the following way.

Definition 1. *Consider an on-line problem P and let I be any request sequence of length n. If σ is a permutation on n elements, then $\sigma(I)$ denotes I permuted by σ.*

For a maximization problem, $\mathbb{A}(I)$ is the value of running the on-line algorithm $\mathbb{A}$ on I, and $\mathbb{A}_W(I) = \min_\sigma \mathbb{A}(\sigma(I))$.

For a minimization problem, $\mathbb{A}(I)$ is a cost, and $\mathbb{A}_W(I) = \max_\sigma \mathbb{A}(\sigma(I))$.

Definition 2. *Let $S_1(c)$ be the statement*
    There exists a constant b such that $\mathbb{A}_W(I) \leq c \cdot \mathbb{B}_W(I) + b$ for all I
and let $S_2(c)$ be the statement
    There exists a constant b such that $\mathbb{A}_W(I) \geq c \cdot \mathbb{B}_W(I) - b$ for all I.
The relative worst order ratio $WR_{\mathbb{A},\mathbb{B}}$ of on-line algorithm $\mathbb{A}$ to algorithm $\mathbb{B}$ is defined if $S_1(1)$ or $S_2(1)$ holds. Otherwise the ratio is undefined and the algorithms are said to be incomparable.

$$\text{If } S_1(1) \text{ holds, then } WR_{\mathbb{A},\mathbb{B}} = \sup \{r \mid S_2(r)\}.$$
$$\text{If } S_2(1) \text{ holds, then } WR_{\mathbb{A},\mathbb{B}} = \inf \{r \mid S_1(r)\}.$$

Note that if $S_1(1)$ holds, the supremum involves S_2 rather than S_1, and vice versa. A ratio of 1 means that the two algorithms perform identically with

Table 1. Ratio values for minimization and maximization problems

	minimization	maximization
$\mathbb{A}$ better than $\mathbb{B}$	< 1	> 1
$\mathbb{B}$ better than $\mathbb{A}$	> 1	< 1

respect to this quality measure; the further away from 1 the greater the difference in performance. The ratio may be greater than or less than one, depending on whether the problem is a minimization or a maximization problem and on which of the two algorithms is better. These possibilities are illustrated in Table 1.

Although not all pairs of algorithms are comparable with the relative worst order ratio, for algorithms which are comparable, the measure is transitive.

Theorem 1. *The ordering of algorithms for a specific problem is transitive.*

Proof. Suppose that three algorithms $\mathbb{A}$, $\mathbb{B}$, and $\mathbb{C}$ for some on-line problem, P, are such that $\mathbb{A}$ is at least as good as $\mathbb{B}$ and $\mathbb{B}$ is at least as good as $\mathbb{C}$, as measured by the relative worst order ratio. If P is a minimization problem there exists a constant b such that for all I, $\mathbb{A}_W(I) \leq \mathbb{B}_W(I) + b$, and there exists a constant d such that for all I, $\mathbb{B}_W(I) \leq \mathbb{C}_W(I) + d$, so there exist constants b and d such that for all I, $\mathbb{A}_W(I) \leq \mathbb{C}_W(I) + b + d$, Thus $\mathbb{A}$ also performs at least as well as $\mathbb{C}$, according to the relative worst order ratio. The argument for a maximization problem is essentially the same.   □

Finally we define the worst order ratio formally:

Definition 3. *The worst order ratio $WR_{\mathbb{A}}$ of an on-line algorithm $\mathbb{A}$ is the relative worst order ratio of $\mathbb{A}$ to an optimal off-line algorithm OPT, i.e., $WR_{\mathbb{A}} = WR_{\mathbb{A},OPT}$.*

As mentioned in the introduction, if there is no restriction on the behavior of OPT, the worst order ratio is the same as the competitive ratio. This does not necessarily hold for the Fair Bin Packing Problem or bounded space bin packing, because of the restrictions on OPT's behavior. Clearly, the worst order ratio is never worse than the competitive ratio.

3   Classical Bin Packing

The Classical Bin Packing Problem is a minimization problem, so algorithms which do well compared to others have relative worst order ratios of less than 1 to the poorer algorithms.

We consider *Any-Fit algorithms*, a class of fair algorithms defined by Johnson [6], which use an empty bin, only if there is not enough space in any partially full bins. Three examples of Any-Fit algorithms are First-Fit (FF), which places an item in the first bin in which it fits, Best-Fit (BF), which places an item in one of the fullest bins in which it fits, and Worst-Fit (WF), which will place an item in a least full open bin. We also consider an algorithm, Next-Fit (NF), which is not an Any-Fit algorithm. Next-Fit is the algorithm which first attempts to fit an item in the current bin, places it there if it fits, or opens a new bin if it does not.

Most of the results we obtain with the relative worst order ratio for the Classical Bin Packing Problem are very similar to those obtained for the competitive ratio and use similar techniques. However, the relative worst order ratio separates Worst-Fit and Next-Fit, which the competitive ratio cannot. We first present those results consistent with the competitive ratio.

According to the competitive ratio for the Classical Bin Packing Problem, First-Fit is the best Any-Fit algorithm [6]. Using the same proof, one can show that this also holds for the relative worst order ratio. The idea is to consider the First-Fit packing of an arbitrary sequence. If the items are given bin by bin, any Any-Fit algorithm will produce exactly the same packing.

Theorem 2. *For any Any-Fit algorithm $\mathbb{A}$, $WR_{FF,\mathbb{A}} \leq 1$.*

Not all Any-Fit algorithms perform as well. Worst-Fit is the worst possible among the Any-Fit algorithms, and it is significantly worse than First-Fit and Best-Fit.

Theorem 3. *For any Any-Fit algorithm $\mathbb{A}$, $WR_{WF,\mathbb{A}} \geq 1$.*

Proof. Consider a request sequence I and its packing by $\mathbb{A}$. Call the bins used by $\mathbb{A}$, $b_1, b_2, \ldots, b_n$, numbered according to the order in which they were opened by $\mathbb{A}$. Let $\ell_{\mathbb{A}}(b_j)$ be the level of b_j, i.e., the sum of the sizes of the items packed in b_j. Let $\ell_{\mathbb{A}}^{\min}(j) = \min_{1 \leq i \leq j} \{\ell(b_i)\}$. Furthermore, let E_j be the set of items packed in b_j, $1 \leq j \leq n$.

Let I' be a permutation of I, where each item $e \in E_i$ appears before each item $e' \in E_j$, for $1 \leq i < j \leq n$. Consider the packing of I' produced by Worst-Fit. Let $\ell_{\mathrm{WF}}(b_j)$ be the level of b_j in this packing. We prove by induction on j that, for $1 \leq j \leq n$, $\ell_{\mathrm{WF}}(b_j) \geq \ell_{\mathbb{A}}^{\min}(j)$ and all items in $\cup_{i=1}^{j} E_i$ are packed in $b_1, \ldots, b_j$ by Worst-Fit.

The base case $j = 1$ is trivial: all items in E_1 are clearly packed in b_1, since Worst-Fit is an Any-Fit algorithm.

For $j > 1$, the induction hypothesis says that $\ell_{\mathrm{WF}}(b_i) \geq \ell_{\mathbb{A}}^{\min}(i)$, $1 \leq i \leq j-1$, and that before giving the items of E_j, b_j is empty. If Worst-Fit packs all items of E_j in b_j, the result trivially follows. Assume now, that some item in E_j is packed in some bin $b_i \neq b_j$. Since b_j is empty before giving the items of E_j and Worst-Fit is an Any-Fit algorithm, we conclude that $i < j$. By the Worst-Fit packing rule, $\ell_{\mathrm{WF}}(b_j) \geq \ell_{\mathrm{WF}}(b_i) \geq \ell_{\mathbb{A}}^{\min}(i) \geq \ell_{\mathbb{A}}^{\min}(j)$.

Now, let e_j be the first item packed by $\mathbb{A}$ in b_j, $1 \leq j \leq n$. Since $\mathbb{A}$ is an Any-Fit algorithm, e_j does not fit in b_i, $1 \leq i < j$. This means that e_j is larger than $1 - \ell_{\mathbb{A}}^{\min}(j-1)$. Since, in the Worst-Fit packing, b_i has a level of at least $\ell_{\mathbb{A}}^{\min}(i)$, this means that for each b_i and each e_j, $1 \leq i < j \leq n$, e_j does not fit in b_i. In words, for each bin b_i, the bottommost item in each bin b_j, $1 \leq i < j \leq n$, in the packing of $\mathbb{A}$ does not fit in b_i in the Worst-Fit packing. Hence, for each e_j, $1 \leq j \leq n$, Worst-Fit must open a new bin, i.e., Worst-Fit uses the same number of bins as $\mathbb{A}$. $\square$

Using Johnson's results and techniques [6], one can show that the relative worst order ratio of Worst-Fit to either First-Fit or Best-Fit is 2.

Theorem 4. $WR_{WF,FF} = WR_{WF,BF} = 2.$

Proof. Since $FF_W(I) \leq BF_W(I)$ for all I, by the proof of Theorem 2, it is only necessary to compare Worst-Fit and Best-Fit. The above theorem shows that $WR_{WF,BF} \geq 1$, so in order to prove a lower bound of 2, it is sufficient to find a family of sequences I_n, with $\lim_{n \to \infty} WF_W(I_n) = \infty$, where there exists a constant b such that for all I_n, $WF_W(I_n) \geq 2BF_W(I_n) - b$. The family of sequences used in [6] to bound Worst-Fit's competitive ratio works here. Let $0 < \varepsilon \leq \frac{1}{2n}$. Consider the sequence I_n with pairs $(\frac{1}{2}, \varepsilon)$, for $i = 1...n$. In this order, Worst-Fit will pack all of the pairs, one per bin, using n bins. Best-Fit will pack the small items all in one bin, regardless of the ordering, using only $\lceil \frac{n+1}{2} \rceil$ bins. Thus, $WF_W(I_n) = n \geq 2\lceil \frac{n+1}{2} \rceil - 2 = 2BF_W(I_n) - 2$, so the relative worst order ratio is at least 2.

The relative worst order ratio of Worst-Fit to either First-Fit or Best-Fit is at most 2, since Worst-Fit's competitive ratio is 2 [6]. □

Now we consider Next-Fit, which is not an Any-Fit algorithm. Next-Fit is strictly worse than Worst-Fit and all other Any-Fit algorithms. This result is in contrast to the competitive ratio where Next-Fit and Worst-Fit both have ratios of 2 [6].

Theorem 5. *For any Any-Fit algorithm* $\mathbb{A}$, $WR_{NF,\mathbb{A}} = 2$.

Proof. To see that $WR_{NF,\mathbb{A}} \geq 1$, consider any request sequence I and its packing by $\mathbb{A}$. Create a new sequence I' from I by taking the items bin by bin from $\mathbb{A}$'s packing, starting with the first bin, in the order they were placed in the bins, and concatenate the contents together to form the sequence I'. Next-Fit also has to open a new bin for the first item put in each bin, so it ends up with the same configuration. Hence, for all I, $NF_W(I) \geq \mathbb{A}_W(I)$, giving a ratio of at least one.

Since $WR_{NF,\mathbb{A}} \geq 1$, to prove the lower bound of 2 it is sufficient to find a family of sequences I_n, with $\lim_{n \to \infty} NF_W(I_n) = \infty$, where there exists a constant b such that for all I_n, $NF_W(I_n) \geq 2\mathbb{A}_W(I_n) - b$. Let $0 < \varepsilon \leq \frac{1}{n-1}$. Consider the sequence I_n with $n-1$ pairs $(\varepsilon, 1)$. In this order, Next-Fit will pack each item in a new bin, whereas $\mathbb{A}$ will pack all of the small items in the same bin. Thus, $NF_W(I_n) = 2(n-1) = 2\mathbb{A}_W(I_n) - 2$, so $WR_{NF,\mathbb{A}} \geq 2$.

For the upper bound, note that the relative worst order ratio of Next-Fit to any Any-Fit algorithm is at most 2, since Next-Fit's competitive ratio is 2 [6]. □

4 Dual Bin Packing

When switching to the Dual Bin Packing Problem, which is a maximization problem, algorithms which do well compared to others have relative worst order ratios greater than 1, instead of less than 1, to the poorer algorithms.

First-Fit and Best-Fit are defined in the same manner for the Dual Bin Packing Problem as for the Classical Bin Packing Problem, though clearly they reject items which do not fit in any of the n bins. If one uses the same definition for Worst-Fit for Dual Bin Packing as for Classical Bin Packing, one can again show that it is the worst Any-Fit algorithm. However, we use a more natural definition for the fixed number of bins, where Worst-Fit always places an item in a least full bin. For the first n items, the least full bin will be empty, so the Worst-Fit we consider here is not an Any-Fit algorithm. (Note that this definition is not at all natural for the Classical Bin Packing Problem, since a new bin would be opened for every item. This variant of Worst-Fit is clearly the worst possible algorithm for the classical problem.)

The result we obtain for the worst order ratio for the Dual Bin Packing Problem is similar to that obtained previously with the competitive ratio, while most of the results we obtain for the relative worst order ratio are similar to those for the competitive ratio on accommodating sequences. The proof that First-Fit and Best-Fit are strictly better than Worst-Fit, which is true with the relative worst order ratio and the competitive ratio on accommodating sequences, but not with the competitive ratio [4], is much easier using the relative worst order ratio.

Computing the worst order ratio for deterministic algorithms for Fair Bin Packing, not surprisingly, gives similar results to the competitive ratio [4] — very pessimistic results.

Theorem 6. *The worst order ratio for any deterministic algorithm for the Fair Bin Packing Problem is not bounded below by any constant.*

Proof. Consider any fair, deterministic on-line algorithm, $\mathbb{A}$. For the following sequence, I, defined on the basis of $\mathbb{A}$'s performance, $\mathbb{A}$ will accept all of the larger items and reject all of the small, while, for any permutation of I, OPT will be able to arrange to reject some of the larger items and accept many of the small ones.

Let $0 < \varepsilon < \frac{1}{24}$. The sequence I begins with n items of size $\frac{1}{3} + \varepsilon$, called items of type A. Suppose $\mathbb{A}$ places these items in the bins, leaving q bins empty. Then, exactly q bins have two items and $n - 2q$ have one. The sequence continues with

Type B items: $n + q$ items of size $\frac{1}{3}$
Type C items: $n - 2q$ items of size $\frac{1}{3} - \varepsilon$ Items of types A, B, C, and D are
Type D items: q items of size $\frac{1}{3} - 2\varepsilon$
Type E items: $\frac{n}{12\varepsilon} - \frac{n}{4}$ items of size ε

the "large" items, and items of type E are the small items. Since it is fair, $\mathbb{A}$ is forced to accept all of the large items, completely filling up the bins, so that it must reject all the small items. $\mathbb{A}_W(I) = 3n$.

In the worst ordering of I for OPT, OPT will be able to reject the least number of large items and thus accept the least number of small items. Consider such a worst-case ordering $I' = \sigma(I)$. Without loss of generality, we may assume that all of the large items in I' come before all of the small. In such a sequence, all items of type D are accepted since there are only $3n$ large items in all, and an item of type D will fit in any bin which only contains two large items.

In addition, we may assume that all of the items of type D come after all of the larger items. To see this, suppose there exists an item x of type D, which occurs before some larger item y. Choose these items so that x occurs immediately before y. Now consider the sequence I'', which is identical to I', except that the order of x and y is inverted. OPT accepts x in both orderings, since it is of type D. If it also accepts y in I', it will accept it in I'' too. If it rejects y in I', and also in I'', I'' is also a worst ordering. If it rejects y in I', but accepts it in I'', then its decision to reject y in I' was because that was better, so I'' is at least as bad as I'. One can continue in this manner moving all items of type D after the larger ones.

A similar argument shows that all of the items of type A can be assumed to come before all of the others.

If the items of type A are placed one per bin, then the items of type B will also have to be placed with at most one per bin, so at least q are rejected.

If the items of type A are placed so that r bins are empty, then those r bins will each get three large items, and r bins will get exactly two items of type A, plus possibly one of type D. The remaining $n - 2r$ bins can each hold at most two items of type B or C. Thus, there is space for at most $3r + 2(n - 2r) = 2n - r$ items of types B or C. This is smallest when $r = \frac{n}{2}$, its largest possible value. There are $2n - q$ items of types B and C, so $\frac{n}{2} - q$ are rejected.

OPT can choose whether to place the items of type A one per bin or two per bin, whichever leads to more rejected large items. If $q \geq \frac{n}{4}$, it will choose the first; otherwise it will choose the second. In any case, it rejects $s \geq \frac{n}{4}$ large items and accepts at least $s(\frac{1}{3} - \varepsilon)/\varepsilon$ small items.

Thus, the worst order ratio is at most

$$\mathrm{WR}_{\mathbb{A}} \leq \frac{3n}{(3n - s) + s(\frac{1}{3} - \varepsilon)/\varepsilon} = \frac{3n\varepsilon}{(3n - s)\varepsilon + s(\frac{1}{3} - \varepsilon)}$$

$$= \frac{3n\varepsilon}{(3n - 2s)\varepsilon + \frac{s}{3}} \leq \frac{3n\varepsilon}{(3n - \frac{n}{2})\varepsilon + \frac{n}{12}} = \frac{36\varepsilon}{30\varepsilon + 1} < 36\varepsilon. \qquad \square$$

Note that for the Dual Bin Packing Problem, the competitive ratio on accommodating sequences [4] can be used to get results concerning the relative worst order ratio, but the competitive ratio cannot necessarily. The problem with using the competitive ratio directly is that we are comparing to OPT which may be more able to take advantage of a fairness restriction with some orderings than with others. When the sequences are not accommodating sequences, then we may be looking at sequences where there is some order where OPT also does poorly. This cannot happen with accommodating sequences. For example, if algorithm $\mathbb{A}$ has a competitive ratio on accommodating sequences of at least p and $\mathbb{B}$ has a competitive ratio of at most $r < p$, then there is an accommodating sequence I where $\mathbb{A}_W(I) \geq p|I| > r|I| \geq \mathbb{B}_W(I)$. This can help give a result in the case where one has already shown that $\mathbb{A}$ is at least as good as $\mathbb{B}$ on all sequences.

For Dual Bin Packing, one can again show that First-Fit is the best Any-Fit algorithm, also using the proof by Johnson [6], the only difference being that now any items First-Fit rejects are concatenated to the end of the sequence created for the Any-Fit algorithm $\mathbb{A}$ and will also be rejected by $\mathbb{A}$.

Theorem 7. *For any Any-Fit algorithm* $\mathbb{A}$, $WR_{FF,\mathbb{A}} \geq 1$.

Theorem 8. $WR_{FF,BF} \geq \frac{7}{6}$.

Proof. The above theorem shows that $\mathrm{WR}_{FF,BF} \geq 1$, so it is sufficient to find a family of sequences I_n, with $\lim_{n \to \infty} \mathrm{FF_W}(I_n) = \infty$, where there exists a constant b such that for all I_n, $\mathrm{FF_W}(I_n) \geq \frac{7}{6}\mathrm{BF_W}(I_n) - b$. Let $0 < \varepsilon < \frac{1}{8n^2}$. Consider the sequence I_n starting with pairs, $(\frac{1}{2} + 2ni\varepsilon, \varepsilon)$, for $i = 0...n-1$ and followed by $\frac{1}{2} - n(2i+1)\varepsilon$, for $i = 0,..n-1$ and $n-1$ of size $n\varepsilon$. Best-Fit will reject the last $n-1$ items when they are given in this order. The worst order for First-Fit would be such that First-Fit paired together the items of size just less than $\frac{1}{2}$, so that it could only accept $\lfloor \frac{n}{2} \rfloor$ of those larger than $\frac{1}{2}$. Thus, $\mathrm{FF_W}(I_n) \geq 3n - 1 + \frac{n-1}{2} = \frac{7}{6}(3n) - \frac{3}{2} = \frac{7}{6}\mathrm{BF_W}(I_n) - \frac{3}{2}$, as required. $\square$

Recall that for the Dual Bin Packing Problem, Worst-Fit is the algorithm which places an item in one of the bins which are least full; we assume it chooses the first such bin. Worst-Fit is a fair algorithm. Its relative worst order ratio to any other fair algorithm is less than or equal to one, so it is the worst such algorithm. To prove this we consider the packing of an arbitrary sequence done by the Any-Fit algorithm under consideration and define the height of an item e to be the total size of the items packed before e in the same bins as e. If the items are given in order of non-decreasing height, Worst-Fit will produce the same packing.

Theorem 9. *For any fair algorithm* $\mathbb{A}$, $WR_{WF,\mathbb{A}} \leq 1$.

According to the relative worst order ratio, First-Fit and Best-Fit are strictly better than Worst-Fit. This is in contrast to the competitive ratio, where First-Fit and Best-Fit actually have worse ratios than Worst-Fit [4]. The relative worst order ratio corresponds more to the competitive ratio on accommodating sequences, where First-Fit and Best-Fit can be shown to perform better than Worst-Fit [4]. The result concerning the relative worst order ratio is, however, much easier to prove.

Theorem 10. $WR_{WF,FF} = WR_{WF,BF} = \frac{1}{2}$.

Proof. The proof of the above theorem shows that there is no sequence I where $\mathrm{WF_W}(I) > \mathrm{FF_W}(I)$ or $\mathrm{WF_W}(I) > \mathrm{BF_W}(I)$, so to prove the upper bound it is sufficient to find a family of sequences I_n, with $\lim_{n \to \infty} \mathrm{FF_W}(I_n) = \infty$, where there exists a constant b such that for all I_n, $\mathrm{WF_W}(I_n) \leq \frac{1}{2}\mathrm{BF_W}(I_n) + b$. Let $0 < \varepsilon \leq \frac{1}{n}$, and let I_n consist of n items of size ε, followed by $n-1$ of size 1. Worst-Fit will accept only the n items of size ε when they are given in this order. First-Fit or Best-Fit will accept all of these items, regardless of their order. This gives a ratio of $\frac{n}{2n-1}$.

Consider now the lower bound. Since Worst-Fit cannot have a lower ratio to Best-Fit than to First-Fit, it is sufficient to show that it holds for First-Fit. Consider any sequence I and the worst ordering of I for Worst-Fit. Without loss of generality, assume that all the items Worst-Fit accepts appear in I before those it rejects. Consider First-Fit's performance on this ordering of I, and suppose it accepts m items, but Worst-Fit only accepts $m' < m$.

Reorder the first m' items so that First-Fit gets them bin by bin, according to Worst-Fit's packing. Since no item will be packed in a later bin by First-Fit than by Worst-Fit, for each item Worst-Fit accepts, First-Fit will have room for it in the same bin. Thus, First-Fit will accept all the items Worst-Fit accepts. First-Fit accepts at most $n - 1$ more items than Worst-Fit, since each of the $m - m'$ items which Worst-Fit rejects must be larger than the empty space in any of Worst-Fit's bins. Thus, the total size of any n rejected items (if there are that many) would be more than the total empty space in the n bins after packing the items accepted by Worst-Fit.

Since Worst-Fit is fair, it must accept at least n items if it rejects any at all. Thus, the relative worst order ratio of Worst-Fit to either First-Fit or Best-Fit is at least $\frac{n}{2n-1}$.   □

An example of an algorithm for the Dual Bin Packing Problem which is not fair is Unfair-First-Fit [1]. It behaves as First-Fit, except that when given a request of size greater than $1/2$, it automatically rejects that item if it has already accepted at least $2/3$ of the items it has received so far. The intuition is that by rejecting some large items, it may have more room for more small items. The algorithm is defined in Figure 1.

```
Input:  S = ⟨o₁, o₂, ... , oₙ⟩
Output: A, R, and a packing for those items in A
A := {};  R := {}
while  S ≠ ⟨⟩
      o := hd(S); S := tail(S)
      if size(o) > ½  and  |A|/(|A|+|R|+1) ≥ ⅔
            R := R ∪ {o}
      else if there is space for o in some bin
                  pack o according to the First-Fit rule
                  A := A ∪ {o}
            else
                  R := R ∪ {o}
```

Fig. 1. The algorithm Unfair-First-Fit

Theorem 11. *Under the relative worst order ratio, Unfair-First-Fit is incomparable to all Any-Fit algorithms.*

Proof. It is easy to see that there exist sequences where Unfair-First-Fit does worse than any Any-Fit algorithm. Consider, for example, the request sequence containing n items of size 1. Unfair-First-Fit will only accept 2/3 of them, while any fair algorithm (and thus all Any-Fit algorithms) will accept all of them. Hence, on such a sequence, any Any-Fit algorithm accepts 3/2 times as many items as Unfair-First-Fit.

To show the other direction, it suffices to compare Unfair-First-Fit to First-Fit, the best among the Any-Fit algorithms. Since the competitive ratio on accommodating sequences for First-Fit is bounded above by $\frac{5}{8} + O(\frac{1}{\sqrt{n}})$, and the competitive ratio on accommodating sequences for Unfair-First-Fit is $\frac{2}{3} \pm \Theta(\frac{1}{n})$, for large enough n [1], there exists an accommodating sequence where Unfair-First-Fit outperforms First-Fit. Asymptotically, Unfair-First-Fit accepts $\frac{16}{15}$ times as many items as First-Fit.   □

5   Conclusion and Open Problems

This new performance measure gives the advantages that one can compare two on-line algorithms directly, that it is intuitively suitable for some natural problems where any ordering of the input is equally likely, and that it is easier to compute than the random order ratio. It is also better than the competitive ratio at distinguishing between algorithms for Classical and Dual Bin Packing. Although the competitive ratio on accommodating sequences can also be used to show that Worst-Fit is better than First-Fit for Dual Bin Packing, the proof is easier with the relative worst order ratio.

The definition of the competitive ratio has been taken rather directly from that of the approximation ratio. This seems natural in that on-line algorithms can be viewed as a special class of approximation algorithms. However, for approximation algorithms, the comparison to OPT is very natural, since one is comparing to another algorithm of the same general type, just with more computing power, while for on-line algorithms, the comparison to OPT is to a different type of algorithm.

Although the competitive ratio has been an extremely useful notion, in many cases it has appeared inadequate at differentiating between on-line algorithms. When this is the goal, doing a direct comparison between the algorithms, instead of involving an intermediate comparison to OPT, seems the obvious choice. A direct comparison on exactly the same sequences will produce the result that many algorithms are incomparable because one algorithm does well on one type of ordering, while the other does well on another type. With the relative worst order ratio, on-line algorithms are compared directly to each other on their respective worst orderings of multisets. This first study of this new measure seems very promising in that most results obtained are consistent with those obtained with the competitive ratio, but new separations are found.

Work in progress [3] shows that for the paging problem, the relative worst order ratio of LRU (FIFO) to LRU (FIFO) with lookahead l is $\min(k, l+1)$ when there are k pages in fast memory. This compares well with the competitive ratio, where these algorithms have the same competitive ratio. This result is similar to that which Koutsoupias and Papadimitriou obtained using comparative analysis

[9], and stronger than that obtained with the Max/Max ratio [2]. The relative worst order ratio should be applied to other on-line problems to see if it is also useful for those problems.

The definition presented here for the relative worst order ratio allows one to say that $\mathbb{A}$ is better than $\mathbb{B}$ even if there exist sequences where $\mathbb{B}$ does better than $\mathbb{A}$ by an additive constant. Another possible definition would require that $\mathbb{A}$ do at least as well as $\mathbb{B}$ on every sequence. In all the examples of comparable algorithms presented here, one algorithm does at least as well as the other on every sequence. It would be interesting to find an example showing that this alternative definition is different from the definition in this paper.

For the Classical Bin Packing Problem, there exist multi-sets where First-Fit's worst ordering uses one less bin than Best-Fit's worst ordering. One example of this is the following sequence: $\frac{1}{4}, \frac{1}{4}, \varepsilon, \frac{3}{4}, \varepsilon, \frac{1}{4}, \frac{1}{4}$, where Best-Fit uses three bins for its worst ordering, while First-Fit only uses two. However this seems to be hard to extend to an asymptotic result. Determining if $\mathrm{WR}_{\mathrm{FF,BF}} < 1$ is an open problem.

Acknowledgments

We would like to thank Kim Skak Larsen for helpful discussions.

References

1. Y. Azar, J. Boyar, L. Epstein, L. M. Favrholdt, K. S. Larsen, and M. N. Nielsen. Fair versus Unrestricted Bin Packing. *Algorithmica*, 34(2):181–196, 2002.
2. S. Ben-David and A. Borodin. A New Measure for the Study of On-Line Algorithms. *Algorithmica*, 11(1):73–91, 1994.
3. J. Boyar, L. M. Favrholdt, and K. S. Larsen. Work in progress.
4. J. Boyar, K. S. Larsen, and M. N. Nielsen. The Accommodating Function — a Generalization of the Competitive Ratio. *SIAM Journal of Computation*, 31(1):233–258, 2001. Also in *WADS 99*, pages 74–79.
5. R. L. Graham. Bounds for Certain Multiprocessing Anomalies. *Bell Systems Technical Journal*, 45:1563–1581, 1966.
6. D. S. Johnson. Fast Algorithms for Bin Packing. *Journal of Computer and System Sciences*, 8:272–314, 1974.
7. A. R. Karlin, M. S. Manasse, L. Rudolph, and D. D. Sleator. Competitive Snoopy Caching. *Algorithmica*, 3(1):79–119, 1988.
8. C. Kenyon. Best-Fit Bin-Packing with Random Order. In *7th Annual ACM-SIAM Symposium on Discrete Algorithms*, pages 359–364, 1996.
9. E. Koutsoupias and C. H. Papadimitriou. Beyond Competitive Analysis. In *35th Annual Symposium on Foundations of Computer Science*, pages 394–400, 1994.
10. D. D. Sleator and R. E. Tarjan. Amortized Efficiency of List Update and Paging Rules. *Communications of the ACM*, 28(2):202–208, 1985.

On-Line Stream Merging, Max Span, and Min Coverage

Wun-Tat Chan[1], Tak-Wah Lam[2*],
Hing-Fung Ting[2**], and Prudence W.H. Wong[2]

[1] Department of Computing, Hong Kong Polytechnic University, Hong Kong,
cswtchan@comp.polyu.edu.hk
[2] Department of Computer Science, University of Hong Kong, Hong Kong,
{twlam,hfting,whwong}@cs.hku.hk

Abstract. This paper introduces the notions of span and coverage for analyzing the performance of on-line algorithms for stream merging. We show that these two notions can solely determine the competitive ratio of any such algorithm. Furthermore, we devise a simple greedy algorithm that can attain the ideal span and coverage, thus giving a better performance guarantee than existing algorithms with respect to either the maximum bandwidth or the total bandwidth. The new notions also allow us to obtain a tighter analysis of existing algorithms.

1 Introduction

A typical problem encountered in video-on-demand (VOD) systems is that many requests for a particular popular video are received over a short period of time (say, Friday evening). If a dedicated video stream is used to serve each request, the total bandwidth requirement for the server is enormous. To reduce the bandwidth requirement without sacrificing the response time, a popular approach is to merge streams initiated at different times (see e.g., [11,10,3,8,9,4,5,13,15,1,12,14]).

Stream merging is based on a multicasting architecture and assumes that each client has extra bandwidth to receive data from two streams simultaneously. In such a system, a stream can run in two different states: normal state and exceptional state. A new stream X is initially in *normal* state and all its clients receive one unit of video from X in every time unit for immediate playback. Some time later, X may change to the *exceptional* state and all its clients receive and buffer an extra of $1/\lambda$ unit of video from an earlier stream Y in every time unit, where λ is an integer parameter characterizing the extra bandwidth allowed for a particular VOD system. When X's clients have buffered enough data from Y, they can synchronize with the playback of Y and all clients of X can switch to Y. At this time, X can terminate, and X is said to *merge* with Y. Note that such a merging reduces the total bandwidth requirement.

* This research was supported in part by Hong Kong RGC Grant HKU-7024/01E
** This research was supported in part by Hong Kong RGC Grant HKU-7045/02E

R. Petreschi et al. (Eds.): CIAC 2003, LNCS 2653, pp. 70–82, 2003.

To support stream merging effectively, we need an on-line algorithm to decide how streams merge with each other. The performance of such an on-line algorithm can be measured rigorously using the competitive ratio, i.e., the worst-case ratio between its total bandwidth and the total bandwidth used in an optimal schedule. The literature contains a number of on-line stream merging algorithms, e.g., the greedy algorithm [3] (also called nearest-fit), the Dynamic Fibonacci tree algorithm [3], the connector algorithm [7], and the α-dyadic algorithm [9]. The greedy heuristic is attractive because of its simplicity and ease of implementation; a stream simply merges to the nearest possible stream. Unfortunately, it has been shown to be $\Omega(n/\log n)$-competitive, where n is the total number of requests [3]. The other three algorithms provide much better performance guarantee; in particular, the connector algorithm and the α-dyadic algorithm are known to be 3-competitive [7,6]. Yet these algorithms are much more complicated than the greedy algorithm. The Dynamic Fibonacci tree algorithm is based on a data structure called Fibonacci merge tree, the connector algorithm needs to pre-compute a special reference tree to guide the on-line algorithm, and the α-dyadic algorithm is recursive in nature.

In reality, it might make more sense for a stream merging algorithm to minimize the maximum bandwidth over time instead of the total bandwidth [2]. In [6], we consider the special case when the extra bandwidth parameter λ is equal to one (i.e., a client can receive 1 unit of normal and 1 unit of extra data in one time unit) and show that with respect to the maximum bandwidth, the connector algorithm is 4-competitive and the α-dyadic algorithm is 4-competitive when α is chosen to be 1.5. Empirical studies indeed confirm that the connector algorithm and the α-dyadic algorithm do have very similar performance under different measurements [2,16].

As far as we know, the best lower bounds on the competitive ratios with respect to the maximum bandwidth and the total bandwidth are $4/3$ and 1.139, respectively. An obvious question is whether we can further improve the analysis of existing algorithms or come up with another algorithm with a better competitive ratio. With a deep thought, we want to identify the key elements in designing a good stream merging algorithm and to explain why the connector algorithm and dyadic algorithm have similar performance. In this paper we attempt to answer the above questions.

When designing a stream merging algorithm, there are two conflicting concerns in determining how long a stream should run before it merges. Obviously, we want to merge it with an earlier stream early enough so as to minimize the bandwidth requirement. Yet we also want a stream to run long enough so that more streams initiated later can merge with it. Good algorithms such as the connector algorithm and the α-dyadic algorithm must be able to balance these two concerns properly. In this paper we show how these two concerns can be measured rigorously and more importantly, can be used to determine the competitive ratio of an algorithm. More precisely, we define the notions of span factor and coverage

Table 1. Competitive ratios of different algorithms. The values enclosed are the previously best results. Unless otherwise specified, the ratios are valid for all possible $\lambda \geq 1$.

	Connector algorithm	Dyadic algorithm	Our greedy algorithm
maximum bandwidth	3 (4 with $\lambda = 1$)	2 (4 with $\lambda = 1$)	2
total bandwidth	3 (3)	2.5 (3)	2.5

factor, and show that if a stream merging algorithm has a span factor at most s and a coverage factor at least c, then the algorithm is $K_{c,s}$-competitive with respect to the maximum bandwidth and $\max(2.5, K_{c,s})$-competitive with respect to the total bandwidth, where $K_{c,s} = 1 + \max\left\{ \lceil \log_{1+c} \frac{1+2\lambda}{1+\lambda} \rceil, \lceil \log_{1+\frac{1+\lambda}{s\lambda}} \frac{1+2\lambda}{\lambda} \rceil \right\}$. Note that $K_{c,s}$ attains the smallest value of 2 when $s = 1$ and $c = \lambda/(1+\lambda)$.

Another contribution of this paper is a simple greedy algorithm that guarantees the ideal span factor and coverage factor, i.e., 1 and $\lambda/(1+\lambda)$, respectively (or in general, given any number s, guarantee a span factor at most s and a coverage factor at least $s\lambda/(1+\lambda)$). In other words, this greedy algorithm is 2-competitive with respect to the maximum bandwidth, and 2.5-competitive with respect to the total bandwidth. This result improves existing work regarding the competitiveness and generality. See Table 1 for comparison.

The notions of span and coverage factors also help us obtain a tighter analysis of the existing algorithms. For the connector algorithm, we find that the span factor is at most $\frac{1}{2}\frac{1+\lambda}{\lambda}$ and the coverage factor is at least $1/2$; thus, the connector algorithm is 3-competitive with respect to either the maximum or the total bandwidth. The α-dyadic algorithm has a better performance, its span factor is at most $(\alpha - 1)\frac{1+\lambda}{\lambda}$ and coverage factor at least $\alpha - 1$. When α is chosen as $\frac{1+2\lambda}{1+\lambda}$, the competitive ratio is exactly 2 and 2.5 with respect to the maximum and the total bandwidth, respectively.

The technique used in this paper is drastically different from the so-called "schedules-sandwiching" technique, which was used in our previous work to analyze the connector and α-dyadic algorithms. A basic step of the schedules-sandwiching technique is to "enlarge" the on-line schedule to make it more regular for comparison with an optimal schedule. The disadvantage is that the enlarged schedule may loosen the actual bandwidth required. Another major reason for not using this technique in this paper is that we have no idea how the actual on-line schedules look like (since our analysis is based on any schedule satisfying the bounds on the span factor and coverage factor). The core of our analysis is based on a many-to-one mapping between the bandwidth of the on-line schedule $\mathcal{S}$ and an optimal schedule $\mathcal{O}$. To compare the bandwidth of the two schedules at some time t, we use this mapping to map each stream X in $\mathcal{S}$ that is still running at t to a unique stream Y in $\mathcal{O}$ that is still running at t. The mapping is designed in such a way that only those streams X that start at some particular time interval can map to Y. Then, using the given bounds on the span factor and coverage factor, we prove that there are not many streams running in normal state, or running in exceptional state in this interval.

2 The Model

Let λ be a fixed positive integer. In a VOD system with $1/\lambda$ extra bandwidth there are a server and a set of clients connected through a network. A stream sends one unit of video in one time unit. A client receives one unit of video in one time unit from one stream, and has the capability to receive and buffer an extra $1/\lambda$ unit of video from another stream. From time to time, requests are received from clients for the same popular video, which is ℓ units of length. The system responses a request by multicasting a video stream of the popular video and the client making the request starts receiving data from the stream.

To model stream merging, a stream has two states: *normal* and *exceptional*. Initially, a stream, say X, is in normal state and all of its clients will receive one unit of video from X in one time unit. After some time, X may change to exceptional state. In such a state, X will be coupled with an earlier stream W, and X's clients will receive, in one time unit, one unit and $1/\lambda$ units of video from X and W, respectively. We say that X *merges* with W at time t if at this time, X terminates and all its clients switch to listen to W and become W's clients. Obviously, X merges with W only when the clients of both X and W have received the same amount of data. Followings give the formal condition.

Condition 1 *Suppose W and X are initiated at time t_W and t_X, respectively. If X merges with W, then X has to be in exceptional state for exactly $\lambda(t_X - t_W)$ time units so that its clients would receive the extra $(t_X - t_W)$ units of video played by W. Furthermore, if X is in normal state for t_N time units, then W has to be in normal state for at least $t_N + (1+\lambda)(t_X - t_W)$ time units so that it will still be in normal state at time $t_W + t_N + (1+\lambda)(t_X - t_W) = t_X + t_N + \lambda(t_X - t_W)$, the time when the merging occurs.*

The major problem in stream merging is to find a schedule to determine, for every stream multicasted by the system, the life of this stream (i.e., how long will they be in normal and exceptional states) and to which streams they merge. We say that an on-line stream merging algorithm A is c-competitive with respect to maximum bandwidth (resp. total bandwidth) if A will always produce a schedule with maximum bandwidth (resp. total bandwidth) at most c times that of an optimal schedule.

We say that a request sequence $\mathcal{R} = (t_1, t_2, \ldots, t_n)$ is *compact* if $t_n - t_1 \leq \ell/(1 + \lambda)$ (recall that ℓ is the length of the popular video). A key property about compact sequence is that except for the stream for the first request, which must be a full stream, streams for any other requests can merge with an earlier stream. Given any schedule $\mathcal{S}$, denoted by $\texttt{load}(\mathcal{S}, t)$ the *load* of $\mathcal{S}$ at time t, i.e., according to $\mathcal{S}$ how many streams are running at t. The following lemma suggests we can focus on compact request sequence.

Lemma 1. *Let c be a positive number. Suppose that A is an on-line stream merging algorithm such that given any compact sequence C, A always produces a schedule $\mathcal{S}$ for C with $\mathrm{load}(\mathcal{S}, t) \leq c\,\mathrm{load}(T, t)$ where T is any schedule for C, and t is any time. Then, we have the following:*

1. *We can construct from A a stream merging algorithm that is c-competitive with respect to maximum bandwidth for any general request sequence.*
2. *We can construct from A a stream merging algorithm that is $\max\{2.5, c\}$-competitive with respect to total bandwidth for any general request sequence.*

Proof. For Statement (1), see [6]. For Statement (2), we note that from [3,7], we know how to construct from A a $\max\{3, c\}$-competitive algorithm. A more elaborate construction can reduce 3 to 2.5. Details will be given in the full paper.

In the rest of the paper, we assume the input sequence is compact.

3   Span and Coverage

Consider any schedule $\mathcal{S}$. Let X be a stream in $\mathcal{S}$ initiated at time t_X. When we say a stream Y initiated at time $t_Y > t_X$ is *mergable* with X, we mean if Y runs in exceptional state initially, it will have enough time to merge with X; by Condition 1, it is equivalent to say that X is still in normal state at time $t_X + (1 + \lambda)(t_Y - t_X)$.

Suppose X merges with a stream initiated at time t_{parent}, and it runs in normal state and in exceptional state for τ_n and τ_e time units, respectively. Suppose the stream immediately before X is initiated at time t_{before}. Let t_{miss} be the smallest time at which there is a stream initiated and this stream is not mergable with X. We define the following characteristics of a schedule:

> The *coverage factor* of X, denoted as $\mathrm{Cr}(X)$, is the ratio $|[t_{\mathrm{before}}, t_{\mathrm{miss}}]|/|[t_{\mathrm{parent}}, t_{\mathrm{before}}]|$ where $|I|$ denote the length of the interval I, and the *span factor* of X, denoted as $\mathrm{Sr}(X)$, is the ratio τ_n/τ_e.

We call the value $\min_{X \in \mathcal{S}} \mathrm{Cr}(X)$ the minimum coverage factor and $\max_{X \in \mathcal{S}} \mathrm{Sr}(X)$ the maximum span factor of $\mathcal{S}$.

Let s be any positive number. We show below a modified greedy on-line algorithm $\mathcal{G}_s$ which, given any input sequence, returns a schedule with maximum span factor at most s, and minimum coverage factor at least $\frac{\lambda}{1+\lambda}s$. Note that this is best possible because it can be proved that for dense input sequence, there does not exist any schedule that has maximum span factor at most s, and the minimum coverage factor strictly greater than $\frac{\lambda}{1+\lambda}s$.

We say that a stream X *covers* time t if a stream initiated at time t is mergable with X. Let $\delta = \frac{1+\lambda}{1+\lambda+s\lambda}$. For any time interval $[x, y]$, we say that a time $t \in [x, y]$ is a δ-*checkpoint* for $[x, y]$ if $t = x + \delta^i(y - x)$ for some $i \geq 0$. Suppose at t_X a stream X is initiated in the system. $\mathcal{G}_s$ determines the life of this stream, as well as with which stream X merges, as follows.

- If X is the stream for the first request, X is a full stream and will not merge with any stream.
- Otherwise, find the latest stream W (i.e., the one with the largest initiation time) with which X is mergable. Suppose W is initiated at t_{parent}. Let t_{last} be the largest time covered by W. (Note that we can find W and t_{last} because we know the life of all streams before X.) Then, X will merge with W, and it will run in normal state for just long enough to cover the next δ-checkpoint for $[t_{\text{parent}}, t_{\text{last}}]$ (i.e., the smallest δ-checkpoint for $[t_{\text{parent}}, t_{\text{last}}]$ that is no less than t_X).

Theorem 1. *Let $\mathcal{S}$ be the schedule produced by $\mathcal{G}_s$ for some input sequence $\mathcal{R}$. Then we have $\text{Sr}(\mathcal{S}) \leq s$, and $\text{Cr}(\mathcal{S}) \geq \frac{\lambda}{1+\lambda}s$.*

Proof. Consider any stream X scheduled by $\mathcal{G}_s$. Suppose that X is initiated at t_X and merges with an earlier stream W initiated at time t_{parent}. We define the following notations.

- t_{last} is the largest time covered by W.
- $I = [t_{\text{parent}}, t_{\text{last}}]$.
- $t_\ell = t_{\text{parent}} + \delta^{i+1}|I|$ and $t_r = t_{\text{parent}} + \delta^i|I|$ are the two checkpoints for I that enclosed t_X, i.e. $t_\ell < t_X \leq t_r$.
- Let t_{miss} be the smallest time at where there is a stream initiated at time t_{miss} and X cannot cover t_{miss}.
- τ_n and τ_e are the duration of X in normal and exceptional state, respectively.

First, we consider the coverage factor of X. By construction, we have $t_r - t_X \leq t_{\text{miss}} - t_X$. Suppose the stream directly before X is initiated at t_{before}. Note that $t_{\text{before}} \leq t_\ell$ (otherwise, we can conclude that there is a stream later than W and X is mergable with this stream; this contradict with our choice of W), and we have $t_{\text{before}} - t_{\text{parent}} \leq t_\ell - t_{\text{parent}} = \delta^{i+1}|I|$ and

$$t_r - t_{\text{before}} \geq t_r - t_\ell = \delta^i(1 - \delta)|I| = \frac{1 - \delta}{\delta}\delta^{i+1}|I| \geq \frac{1 - \delta}{\delta}(t_{\text{before}} - t_{\text{parent}}).$$

Finally, note that $(t_{\text{miss}} - t_X) + (t_X - t_{\text{before}}) \geq (t_r - t_X) + (t_X - t_{\text{before}}) = t_r - t_{\text{before}} \geq \frac{1-\delta}{\delta}(t_{\text{before}} - t_{\text{parent}}) = \frac{\lambda}{1+\lambda}s(t_{\text{before}} - t_{\text{parent}})$ and it follows that $\text{Cr}(X) \geq \frac{\lambda}{1+\lambda}s$.

Now, we consider the span factor of X. Since X merges with W and by Condition 1, we have $\tau_e = \lambda(t_X - t_{\text{parent}}) > \lambda(t_\ell - t_{\text{parent}}) = \lambda\delta^{i+1}|I|$. According to the algorithm, X is in normal state just long enough to cover the checkpoint t_r, thus, $\tau_n = (1 + \lambda)(t_r - t_X) \leq (1 + \lambda)(t_r - t_\ell) = (1 + \lambda)(\delta^i|I| - \delta^{i+1}|I|)$. Thus,

$$\frac{\tau_n}{\tau_e} \leq \frac{(1 + \lambda)\delta^i|I|(1 - \delta)}{\lambda\delta^{i+1}|I|} = \frac{1 - \delta}{\delta}\frac{1 + \lambda}{\lambda} = s,$$

and it follows that $\text{Sr}(X) \leq s$. This completes the proof.

In the next section, we analyze schedule with minimum coverage factor at least c and maximum span factor at most s. We call such a schedule (c, s)-bounded schedule.

4   The Competitiveness of (c, s)-Bounded Schedules

In Section 4.1, we introduce a representation of a merging schedule, based on which we perform our analysis. In Section 4.2 we describe a counting argument that relates the load of a (c, s)-bounded schedule with that of an optimal schedule. Finally, in Section 4.3, we complete the analysis.

4.1   Geometric Representation of a Schedule

In [7], we introduced the following way to represent a merging schedule.

Given a schedule $\mathcal{S}$, its geometric representation comprises a set of rectilinear lines on the plane. Suppose a stream X is initiated at time t_0, run in normal state for τ_n time, and then in exceptional state for τ_e time. Then, for X, we put a crook $\mathcal{C}(X)$ (a horizontal line segment followed by a vertical line segment) on the plane as follows:

Starting from (t_0, t_0), we draw a right-going horizontal line of length $\tau_n/(1 + \lambda)$, followed by a down-going vertical line of length τ_e/λ.

The following lemma lists the important information given by $\mathcal{C}(X)$.

Lemma 2. *Suppose $\mathcal{C}(X)$ passes through the point (x, y). Then we can conclude that stream X is still running at time $t = (1 + \lambda)x - \lambda y$, and at this time t, the clients of X will have received totally $\ell = (x - y)(1 + \lambda)$ units of data.*

Proof. X is initiated at time t_0. By construction, $\mathcal{C}(X)$ has a horizontal segment from (t_0, t_0) to (x, t_0), and a vertical segment from (x, t_0) to (x, y), and thus after (t_0, t_0), X runs in normal state and exceptional state for $(1 + \lambda)(x - t_0)$ and $\lambda(t_0 - y)$ time units respectively. Then, the lemma follows directly from definition.

Recall that for a stream X to merge with some stream Y at time t, the clients of both X and Y should have received the same amount of data at t. Together with Lemma 2, it is easy to see that if X merges with Y at time t, then $\mathcal{C}(X)$ must terminate at some point (x, y) on $\mathcal{C}(Y)$ where $t = (1 + \lambda)x - \lambda y$.

Now, we introduce the notion of *timeline*. For any time $t > 0$, define the timeline for t, denoted as $\mathcal{L}_t$, to be the line $y = ((1 + \lambda)/\lambda)x - t/\lambda$. Note that for any point (x_0, y_0) on $\mathcal{L}_t$, we have $t = (1 + \lambda)x_0 - \lambda y_0$. Together with Lemma 2, we conclude that any crook that intersects $\mathcal{L}_t$ must still be running at time t

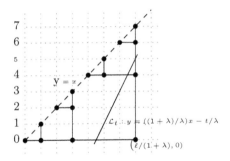

Fig. 1. A canonical schedule with the first request arriving at time 0.

and hence contributes a load at t. Let $\mathcal{S} \cap \mathcal{L}_t$ denote the set of crooks in $\mathcal{S}$ that intersects with $\mathcal{L}_t$. Then the load of $\mathcal{S}$ at t, i.e., $\text{load}(\mathcal{S}, t)$, is equal to $|\mathcal{S} \cap \mathcal{L}_t|$.

Before giving our analysis on $|\mathcal{S} \cap \mathcal{L}_t|$, we need some definitions. We say the point (t, t) is a request point if there is a request at time t, and thus there is a stream initiated at t. For any point $p = (x, y)$, let $\text{above}(p)$ and $\text{left}(p)$ denote the point (x, x) and (y, y), respectively. Note that $\text{above}(p)$ and $\text{left}(p)$ are just the points on the line $y = x$ that are directly above and to the left of p, respectively. Given any crook C, let $h(C)$ and $v(C)$ denote respectively the horizontal and vertical segment of C. We say that a crook C is *in the shadow* of another crook C' if C is lying completely above $h(C')$. A schedule $\mathcal{S}$ is called *canonical* if it satisfies the following properties:

- All the crooks in $\mathcal{S}$ form a tree in which the leaves are on the line $y = x$, and the root is at the point $(t_1 + \ell/(1 + \lambda), t_1)$. (Recall that ℓ is the length of the video.) Furthermore, the path from a leaf to the root is a monotonic rectilinear path in which every horizontal segment is right going and every vertical segment is down going.
- If crook C covers the interval $[a, b]$, then for any request point (t, t) with $t \in [a, b]$, the crook for (t, t) is in the shadow of C.
- Let p be a point on some crook C. If p is on $h(C)$, then $\text{left}(p)$ is a request point. If p is on $v(C)$, then $\text{above}(p)$ is a request point.

See Figure 1 for an example. Obviously, not every schedule is canonical. However, the following theorem suggests we can focus on canonical schedules.

Theorem 2. *Given an on-line algorithm $\mathcal{A}$, we can construct another algorithm $\mathcal{B}$ that simulates $\mathcal{A}$ such that for any input sequence R, if $\mathcal{A}$ produces a schedule $\mathcal{S}_A$ for R that is (c, s)-bounded, i.e., has minimum coverage factor c, and maximum span factor s, then $\mathcal{B}$ will construct another schedule for R that is canonical, with minimum coverage factor at least c, span cover ratio at most s, and without increasing the load at any time.*

Proof. Using the techniques described in Chan et al. [8], we can transform $\mathcal{S}_A$ to make sure every stream will merge with the nearest mergable stream, and this will ensure the schedule satisfies the first two properties of a canonical schedule. Then, using a technique of Coffman *et al.* [9], we can make sure every stream X will merge with another stream at the earliest possible moment, i.e. as soon as the last stream scheduled to merge with X has done so, X will change to exceptional state immediately; this will ensure the schedule satisfies the last property of a canonical schedule. Details will be given in the full paper.

Lemma 3. *Suppose S is (c,s)-bounded. For any crook C in S, the length of $h(C)$ is at most $\frac{\lambda}{1+\lambda}s$ times the length of $v(C)$.*

Proof. Let X be the stream C represents. Suppose X runs in normal state and exception state for τ_n and τ_e time units. Since S is (c,s)-bounded, we have $\tau_n \leq s\tau_e$. By construction, we have the length of $h(C)$ and $v(C)$ equal $\tau_n/(1+\lambda)$ and τ_e/λ. The lemma follows.

Now, we analyze the load of a (c,s)-bounded a canonical schedule S at any time t. From above discussion, it is equivalent to study $|S \cap \mathcal{L}_t|$.

4.2   A Counting Argument for Bounding $|S \cap \mathcal{L}_t|$

Suppose S is for the request sequence $\mathcal{R}$. Let $\mathcal{O}$ be an optimal canonical schedule for $\mathcal{R}$. Below, we show a technique for finding an upper bound on $|S \cap \mathcal{L}_t|$ in terms of $|\mathcal{O} \cap \mathcal{L}_t|$.

Since $\mathcal{O}$ is canonical, it is a tree with its root at $r = (t_1 + \ell/(1 + \lambda), t_1)$. For every point $p \in S \cap \mathcal{L}_t$, we associate p with a unique point $q \in \mathcal{O} \cap \mathcal{L}_t$ as follows:

Suppose p is on the crook C in S. Since S is canonical, we know either above(p) and left(p) is a request point. Let u be the request point in {above(p), left(p)} (if both above(p) and left(p) are request points, let $u = $ left(p)). Note that u and the root r are on the different sides of $\mathcal{L}_t$ (see Figure 1), and thus the path from u to r in $\mathcal{O}$ must intersect at some unique point q. By definition, $q \in \mathcal{O} \cap \mathcal{L}_t$. We say that q is the *boss* of p, and p is turned to be a *slave* of q through u.

Now, we are ready to relate $|S \cap \mathcal{L}_t|$ and $|\mathcal{O} \cap \mathcal{L}_t|$. Consider the following table $T_{S\mathcal{O}}$, in which there are $|S \cap \mathcal{L}_t|$ rows and $|\mathcal{O} \cap \mathcal{L}_t|$ columns. The rows of $T_{S\mathcal{O}}$ are labeled by the points in $S \cap \mathcal{L}_t$, and the columns by those in $\mathcal{O} \cap \mathcal{L}_t$. Every row has a single entry equal to 1, and all the other entries are equal to 0. More precisely, for a row p, we have

$$T_{S\mathcal{O}}[p,q] = \begin{cases} 1, & \text{if } q \text{ is the boss of } p; \\ 0, & \text{otherwise.} \end{cases} \tag{1}$$

Let N be the total number of 1's in $T_{S\mathcal{O}}$. Obviously $|S \cap \mathcal{L}_t| = N$. In the next section, we show that if S is (c,s)-bounded, we can derive an upper bound $K_{c,s}$ on the number of 1's in every column of $T_{S\mathcal{O}}$. Then, we have

$$|S \cap \mathcal{L}_t| = N \leq K_{c,s}|\mathcal{O} \cap \mathcal{L}_t|. \tag{2}$$

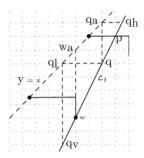

Fig. 2. The position of the points in Slave(q).

4.3 Finding the Upper Bound $K_{c,s}$

Suppose $\mathcal{S}$ is (c, s)-bounded. Consider any column q of $T_{\mathcal{SO}}$. Denoted by Slave(q) the set of all slaves of q. By definition, the total number of 1's in column q is equal to $|\mathtt{Slave}(q)|$. Below, we derive an upper bound on $|\mathtt{Slave}(q)|$.

Given any two points u and v, let $\overline{uv}$ denote the line segment joining u and v, and let $\|\overline{uv}\|$ be the length of $\overline{uv}$. For sake of simplicity, we let $q_a = \mathtt{above}(q)$ and $q_l = \mathtt{left}(q)$. Let q_h be the intersection point of the horizontal line passing through q_a and $\mathcal{L}_t$. Let q_v be the intersection point of the vertical line passing through q_l and $\mathcal{L}_t$ (See Figure 2). The following simple lemma gives the relationship about the line segments.

Lemma 4. *We have* $\|\overline{q_a q}\| = \|\overline{q_l q}\|$, $\|\overline{q_a q_h}\| = \frac{\lambda}{1+\lambda}\|\overline{q_a q}\|$ *and* $\|\overline{q_l q_v}\| = \frac{1+\lambda}{\lambda}\|\overline{q_a q}\|$.

Proof. Note that that q_a is on the line $y = x$, and q and q_h are on the line $\mathcal{L}_t : y = ((1 + \lambda)/\lambda)x - t/\lambda$. The lemma follows from simple calculation from analytic geometry.

The following theorem is useful in our analysis.

Lemma 5. *Consider any point $p \in$ Slave(q). Suppose p is turned to be a slave of q through the point u. Then u must lie on the segment $\overline{q_l q_a}$.*

Proof. By definition, the path π from u to the root in $\mathcal{O}$ passes through q. Since π is a monotonic rectilinear line, it cannot pass q if u is outside $\overline{q_l q_a}$.

Now, we are ready to estimate $|\mathtt{Slave}(q)|$. As a warm-up, we first consider a special but important case, namely when $c = \lambda/(1 + \lambda)$ and $s = 1$.

Let A and B be the points in Slave(q) that are above and below q on $\mathcal{L}_t$, respectively.

Lemma 6. *Suppose $\mathcal{S}$ is (c, s)-bounded with $c = \lambda/(1+\lambda)$ and $s = 1$. If $|A| \geq 2$ then we can conclude $|A| = 2$ and $|B| = 0$.*

Proof. Suppose $|A| \geq 2$. Let p be the highest point in A. Recall that p is some point in $S \cap L_t$, and thus must lies on the crook C in S. Note that p must lie on the horizontal segment $h(C)$ of C (otherwise, p is on $v(C)$ and by definition, p is turned to be a slave of q through $\mathtt{above}(p)$, which lies outside $\overline{q_l q_a}$; this is impossible because of Lemma 5). Thus, $\|h(C)\| \geq \|\overline{q_a q_h}\|$ (see Figure 2). From the fact that S is (c, s)-bounded, and together with Lemmas 3 and 4, we have

$$\|v(C)\| \geq \frac{1}{s}\frac{1+\lambda}{\lambda}\|h(C)\| \geq \frac{1+\lambda}{\lambda}\|\overline{q_a q_h}\| = \|q_a q\|.$$

Now consider the other point $p' \in A$ that is immediately below p on L_t. Suppose p' is on the crook C' in S. Note that C' covers $\mathtt{left}(p)$, and since S is canonical, C is in the shadow of C'. It follows that $h(C)$ and $h(C')$ must be at least $\|v(C)\| \geq \|\overline{q_a q}\|$ apart. This is only possible when $\mathtt{left}(p)$ and $\mathtt{left}(p')$ are at the two endpoints of $\overline{q_l q_a}$ (Lemma 5 asserts that they cannot lie outside $\overline{q_l q_a}$). Thus, there are no other points on $\overline{q_l q_a}$ that is in $\mathtt{Slave}(q)$, and together with Lemma 5, we conclude that $|A| = 2, |B| = 0$.

Lemma 7. *Suppose S is (c, s)-bounded with $c = \lambda/(1+\lambda)$ and $d = 1$. If $|B| \geq 2$ then we can conclude $|B| = 2$ and $|A| = 0$.*

Proof. Suppose $|B| \geq 2$. Let w be the lowest point in B, and C_w be the crook in S on which w lies. Using an argument similar to that we use in Lemma 6, we can show that $w_a = \mathtt{above}(w)$ is a request point. Let $w_a = (t_{\text{before}}, t_{\text{before}})$ (for some t_{before}). Let X be the stream in S that is initiated directly after t_{before}. Suppose X merges with a stream initiated at time t_{parent}. Note that no crooks can cross the line segments $\overline{w_a w}$; all crooks above $h(C_w)$ are in the shadow of C_w and no crooks can cross $v(C_w)$ (because S is canonical). This implies that $t_{\text{parent}} \leq t_{\text{before}} - \|\overline{w_a w}\|$. Since S has minimum coverage factor c, by definition, we conclude that the stream X covers the interval $[t_{\text{before}}, t_{\text{before}} + c(t_{\text{before}} - t_{\text{parent}})]$ and hence the interval $[t_{\text{before}}, t_{\text{before}} + c\|\overline{w_a w}\|]$.

Let $q_a = (t_a, t_a)$, and $q_l = (t_l, t_l)$. We claim that $t_{\text{before}} + c\|\overline{w_a w}\| \geq t_a$. This is obviously true for the case when $w_a = q_l$; in such case, we have $\overline{q_l q_v} = \overline{w_a w}$ and thus

$$t_{\text{before}} + c\|\overline{w_a w}\| = t_l + c\|\overline{q_l q_v}\| = t_l + \frac{\lambda}{1+\lambda}\|\overline{q_l q_v}\| = t_l + \|\overline{q_l q}\| = t_a.$$

(The third equality comes from Lemma 4.) When w_a is at some other position on $\overline{q_l q_a}$, we can prove the claim by a more complicated, but straightforward, calculation.

In conclusion, the stream X covers the interval $[t_{\text{before}}, t_a]$, and thus there are only two crooks starting in the line segments $\overline{q_l q_a}$, namely C_w and X can intersect L_t. As we assume $|B| \geq 2$, we conclude $|B| = 2$ and $|A| = 0$.

We immediately have the following theorem and corollary.

Theorem 3. *Suppose $\mathcal{S}$ is a (c, s)-bounded with $c = \lambda/(1 + \lambda)$ and $s = 1$. Then for any time t, $|\mathcal{S} \cap \mathcal{L}_t| \leq 2|\mathcal{O} \cap \mathcal{L}_t|$.*

Proof. Consider any column q of $T_{\mathcal{S}\mathcal{O}}$. The number of 1's at this column equal $|\texttt{Slave}(q)|$, which, by from Lemmas 6 and 7, is no greater than 2. Together with Inequality (2), the theorem follows.

The following corollary, which follows directly from Theorems 1 and 3, gives the first on-line algorithm for stream merging which is 2-competitive.

Corollary 1. *The on-line algorithm $\mathcal{G}_s$ with $s = 1$ is 2-competitive.*

The following theorem states our result for the general case.

Theorem 4. *Let* $K_{c,s} = \max\left\{1 + \log_{1+c}\left(1 + \frac{\lambda}{1+\lambda}\right), 1 + \log_{1 + \frac{1+\lambda}{\lambda s}}\left(1 + \frac{1+\lambda}{\lambda}\right)\right\}$. *Suppose $\mathcal{S}$ is (c, s)-bounded. Then we have for any time t, $|\mathcal{S} \cap \mathcal{L}_t| \leq K_{c,s}|\mathcal{O} \cap \mathcal{L}_t|$.*

Proof. A direct generalization of the proof of Theorem 3.

References

1. C. C. Aggarwal, J. L. Wolf, and P. S. Yu. On optimal piggyback merging policies for video-on-demand systems. In *Proc. ACM Sigmetrics*, pages 200–209, 1996.
2. A. Bar-Noy, J. Goshi, R. E. Ladner, and K. Tam. Comparison of stream merging algorithms for media-on-demand. In *Proc. Conf. on Multi. Comput. and Net. (MMCN)*, pages 115–129, 2002.
3. A. Bar-Noy and R. E. Ladner. Competitive on-line stream merging algorithms for media-on-demand. In *Proc. 12th ACM-SIAM SODA*, pages 364–373, 2001.
4. Y. Cai, K. A. Hua, and K. Vu. Optimizing patching performance. In *Proc. Conf. on Multi. Comput. and Net. (MMCN)*, pages 204–215, 1999.
5. S. W. Carter and D. D. E. Long. Improving bandwidth efficiency of video-on-demand. *Computer Networks*, 31(1-2):99–111, 1999.
6. W. T. Chan, T. W. Lam, H. F. Ting, and W. H. Wong. Competitive analysis of on-line stream merging algorithms. In *Proc. 27th MFCS*, pages 188–200, 2002.
7. W. T. Chan, T. W. Lam, H. F. Ting, and W. H. Wong. A unified analysis of hot video schedulers. In *Proc. 34th ACM STOC*, pages 179–188, 2002.
8. W. T. Chan, T. W. Lam, H. F. Ting, and W. H. Wong. On-line stream merging in a general setting. *Theoretical Computer Science*, 296(1):27–46, 2003.
9. E. Coffman, P. Jelenkovic, and P. Momcilovic. The dyadic stream merging algorithm. *Journal of Algorithms*, 43(1), 2002.
10. D. Eager, M. Vernon, and J. Zahorjan. Bandwidth skimming: A technique for cost-effective video-on-demand. In *Proc. Conf. on Multi. Comput. and Net. (MMCN)*, pages 206–215, 2000.
11. D. Eager, M. Vernon, and J. Zahorjan. Minimizing bandwidth requirements for on-demand data delivery. *IEEE Tran. on K. and Data Eng.*, 13(5):742–757, 2001.
12. L. Golubchik, J. C. S. Lui, and R. R. Muntz. Adaptive piggybacking: A novel technique for data sharing in video-on-demand storage servers. *ACM J. of Multi. Sys.*, 4(3):140–155, 1996.

13. K. A. Hua, Y. Cai, and S. Sheu. Patching: A multicast technique for true video-on-demand services. In *Proc. 6th ACM Multimedia*, pages 191–200, 1998.
14. S. W. Lau, J. C. S. Lui, and L. Golubchik. Merging video streams in a multimedia storage server: Complexity and heuristics. *ACM J. of Multi. Sys.*, 6(1):29–42, 1998.
15. S. Sen, L. Gao, J. Rexford, and D. Towsley. Optimal patching schemes for efficient multimedia streaming. In *Proc. 9th Int. W. on Net. and OS Support for Digital Audio and Video*, pages 44–55, 1999.
16. P. W. H. Wong. *On-line Scheduling of Video Streams*. PhD thesis, The University of Hong Kong.

Randomised Algorithms for Finding Small Weakly-Connected Dominating Sets of Regular Graphs

William Duckworth and Bernard Mans

Department of Computing, Macquarie University, Sydney, Australia
billy@ics.mq.edu.au, bmans@ics.mq.edu.au

Abstract. A *weakly-connected dominating set*, $\mathcal{W}$, of a graph, G, is a dominating set such that the subgraph consisting of $V(G)$ and all edges incident with vertices in $\mathcal{W}$ is connected. Finding a small weakly-connected dominating set of a graph has important applications in clustering mobile ad-hoc networks. In this paper we introduce several new randomised greedy algorithms for finding small weakly-connected dominating sets of regular graphs. These heuristics proceed by *growing* a weakly-connected dominating set in the graph. We analyse the average-case performance of the simplest such heuristic on random regular graphs using differential equations. This introduces upper bounds on the size of a smallest weakly-connected dominating set of a random regular graph. We then show that for random regular graphs, other "growing" greedy strategies have exactly the same average-case performance as the simple heuristic.

Keywords: weakly-connected, dominating sets, random regular graphs

1 Introduction

The proliferation of wireless communicating devices has created a wealth of opportunities for the field of mobile computing. However, in the extreme case of self-organising networks, such as mobile ad-hoc networks, it is challenging to guarantee efficient communications. In these infrastructure-less networks, only nodes that are sufficiently close to each other can communicate and, as mobile nodes roam at will, the network topology changes arbitrarily and rapidly. *Clustering* mobile nodes locally is an effective way to hierarchically organise the structure: one special node (the *clusterhead*) oversees the message routing within its cluster. It is straightforward to route messages between different clusters when the clusterheads form a connected component (*i.e.*, a *backbone*). However, the size of the backbone may be large in order to guarantee connectedness.

Recently, several authors [4, 6] suggested ways to reduce the size of the backbone by identifying common *gateway* nodes between clusterheads, relaxing the requirement of connectedness to weakly-connectedness. With clustering mobile ad-hoc networks in mind, we present and analyse algorithms that build small

R. Petreschi et al. (Eds.): CIAC 2003, LNCS 2653, pp. 83–95, 2003.

weakly-connected dominating sets for networks with constant regular degree (as the possible number of connections per node is a constant due to power and technology limitations).

In this paper we consider connected d-regular graphs. When discussing any d-regular graph on n vertices, dn is assumed to be even to avoid parity problems. For other basic graph theory definitions the reader is referred to Diestel [5].

A *dominating set* of a graph, G, is a set of vertices, $\mathcal{D} \subseteq V(G)$, such that every vertex of G either belongs to $\mathcal{D}$ or is incident with a vertex of $\mathcal{D}$ in G. Define the minimum cardinality of all dominating sets of G as the *domination number* of G and denote this by $\gamma(G)$. The problem of determining $\gamma(G)$ for a given graph, G, is one of the core NP-hard problems in graph theory (see, for example, Garey and Johnson [8]).

A *connected* dominating set, $\mathcal{C}$, of a graph, G, is a dominating set such that the subgraph induced by the vertices of $\mathcal{C}$ in G is connected. Define the minimum cardinality of all connected dominating sets of G as the *connected domination number* of G and denote this by $\gamma_c(G)$. The problem of determining $\gamma_c(G)$ is also NP-hard and remains so for regular and bounded degree graphs (see Haynes *et al.* [11]).

Grossman [9] introduced another NP-hard variant of the minimum dominating set problem, that being the problem of finding a minimum *weakly-connected* dominating set. A weakly-connected dominating set, $\mathcal{W}$, of a graph, G, is a dominating set such that the subgraph consisting of $V(G)$ and all edges incident with vertices in $\mathcal{W}$ is connected. Define the minimum cardinality of all weakly-connected dominating sets of G as the *weakly-connected domination number* of G and denote this by $\gamma_w(G)$.

The following section provides some known results about the problems defined above and presents our results of the average-case performance of our algorithms. In Section 3, we present our simple, yet efficient, randomised greedy algorithm. In Section 4 we describe the model we use for generating regular graphs u.a.r. (uniformly at random) and describe the notion of analysing the performance of algorithms on random graphs using systems of differential equations. The analysis of our algorithm is presented in Section 5. In the final section we consider alternative greedy heuristics and comment why any basic improvement to the simple heuristic we present will not improve the results.

2   Small Weakly-Connected Dominating Sets

Finding a small weakly-connected dominating set of a graph has applications in clustering mobile ad-hoc networks. Chen and Liestman [4] introduced worst-case $(\ln \Delta + 1)$ approximation results for $\gamma_w(G)$, when G is a graph with bounded degree Δ. Their analysis techniques are similar to those used by Guha and Khuller [10] to bound $\gamma_c(G)$ when G is a bounded degree graph. Faster distributed algorithms have been recently introduced by Dubhashi *et al.* [6].

We demonstrate the relationship between dominating sets, weakly-connected dominating sets and connected dominating sets using the small example given in

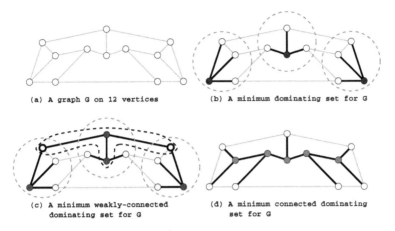

(a) A graph G on 12 vertices

(b) A minimum dominating set for G

(c) A minimum weakly-connected
dominating set for G

(d) A minimum connected dominating
set for G

Fig. 1. Comparing different types of minimum dominating set

Figure 1. The graph in Figure 1(a) is 3-regular and has twelve vertices, therefore, any dominating set must consist of at least three vertices. The darker solid (blue) vertices in Figure 1(b) denote such a set and the dotted line surrounding each such vertex denotes the vertices it dominates. Note that dominating set in Figure 1(b) is an *independent* dominating set (no edges exist between the vertices in the set). As every vertex not in the dominating set of Figure 1(b) is dominated by precisely one vertex, this implies that a weakly-connected dominating set for this graph must contain at least four vertices. The solid darker (red) vertices in Figure 1(c) form a weakly-connected dominating set of minimum size for this graph. White vertices with a dark rim are gateways.

Owing to the fact that the graph in Figure 1 is 3-regular on twelve vertices, a minimum connected dominating set for this graph must consist of at least five vertices. The solid darker (green) vertices in Figure 1(d) form a minimum connected dominating set for this graph.

The relationship between $\gamma(G)$, $\gamma_c(G)$ and $\gamma_w(G)$ has been extensively studied (see, for example, [1, 3, 7, 11]). Clearly, for any graph, G, $\gamma(G) \leq \gamma_w(G) \leq \gamma_c(G)$.

Dunbar *et al.*[7] introduced the concept of a weakly-connected *independent* dominating set, *i.e.*, a weakly-connected dominating set, $\mathcal{I}$, of a graph, G, such that no two vertices in $\mathcal{I}$ are connected by an edge of G. Define the minimum cardinality of all weakly-connected independent dominating sets of G as the *weakly-connected independent domination number* of G and denote this by $\gamma_w^i(G)$.

For arbitrary graphs, relationships amongst the parameters $\gamma(G)$, $\gamma_w(G)$, $\gamma_c(G)$ and $\gamma_w^i(G)$ are known, for example, in [7], it was shown that

$$\gamma_w(G) \leq \gamma_c(G) \leq 2\gamma_w(G) - 1 \quad \text{and} \quad \gamma(G) \leq \gamma_w(G) \leq \gamma_w^i(G) \leq 2\gamma(G) - 1.$$

As we consider random graphs, we need some notation. We say that a property, $\mathcal{B} = \mathcal{B}_n$, of a random graph on n vertices holds a.a.s. (asymptotically almost surely) if the limit of the probability that $\mathcal{B}$ holds tends to 1 as n tends to infinity.

Caro *et al.*[3] showed that, for any n-vertex graph, G, of minimum degree d, that a.a.s.

$$\gamma_c(G) = \frac{(1 + o_d(1))\ln(d+1)}{d+1}n. \tag{1}$$

This, along with a result of Alon [1] and a well-known result of Lovász [12], shows that for any n-vertex graph, G, of minimum degree d, $\gamma(G)$ and $\gamma_c(G)$ are a.a.s. the same.

In this paper we consider randomised greedy algorithms for finding small weakly-connected dominating sets of regular graphs. These heuristics proceed by *growing* a weakly-connected dominating set in the graph. Starting from a single vertex and its incident edges, the component *grows* by repeatedly deleting edges from the graph and adding vertices and edges to the component. We analyse the average-case performance of the simplest such heuristic on random regular graphs using differential equations. We then show that other "growing" greedy strategies have exactly the same average-case performance as the simple heuristic. This introduces upper bounds on $\gamma_w(G)$ when G is a random regular graph.

As the weakly-connected dominating set returned by each of the algorithms we consider is actually a weakly-connected independent dominating set, our results also give upper bounds on $\gamma_w^i(G)$. In this paper we prove the following theorem.

Theorem 1. *For a random d-regular graph, G, on n vertices, where d remains constant, the size of a minimum weakly-connected dominating set of G, $\gamma_w(G)$, asymptotically almost surely satisfies*

$$\gamma_w(G) \leq \tfrac{3\ln(3)}{8}n + o(n), \qquad\qquad\qquad\qquad \text{when } d = 3,$$

$$\gamma_w(G) \leq \tfrac{2(3-\ln(4))}{9}n + o(n), \qquad\qquad\qquad \text{when } d = 4 \text{ and}$$

$$\gamma_w(G) \leq \left(\frac{d}{2(d-1)} - \frac{d(d-3)^{\frac{d-3}{d-2}}}{2(d-2)(d-1)^{\frac{d-1}{d-2}}} \right)n + o(n), \qquad \text{when } d \geq 5.$$

3   Growing a Weakly-Connected Component

The heuristic we describe is a greedy algorithm that is based on repeatedly selecting vertices of given current degree from an ever-shrinking subgraph of the input graph. At the start of our algorithm, all vertices have degree d. Throughout the execution of our algorithm, vertices are repeatedly chosen at random from a given set. A neighbour of such a vertex may be selected for inclusion in the set under construction. In each iteration, a number of edges are deleted.

For a d-regular graph, G, the algorithm constructs a subset, $\mathcal{W}$, of the vertices of G in a series of *steps*. Each step starts by selecting a vertex u.a.r. from those vertices of a particular current degree. The first step is unique in the sense that it is the only step in which a vertex is selected u.a.r. from the vertices of degree d. We select such a vertex, u, u.a.r. from all the vertices of the input graph to add to $\mathcal{W}$ and delete its incident edges.

For each step after the first, we select a vertex, u, u.a.r. from those vertices of positive degree that is less than d. Such a vertex will always exist (after the first step and before the completion of the algorithm) as the input graph is assumed to be connected. We then select a neighbour, v, of u u.a.r. and investigate its degree. If v has degree d, we add v to $\mathcal{W}$ and delete all edges incident with v. Otherwise, (v has degree less than d) we simply delete the edge between u and v and start another step.

At any given stage of the algorithm we say that the component represents the set $\mathcal{W}$ constructed thus far, along with all edges of G that are incident with vertices in $\mathcal{W}$. Every vertex of $\mathcal{W}$ is chosen from the vertices of degree d and all edges that are deleted are either incident with a vertex in $\mathcal{W}$ or have both end-points of degree less than d. These two facts ensure that the component remains weakly-connected throughout the algorithm and once no vertices of degree d remain (at which point the algorithm terminates) $\mathcal{W}$ is a dominating set, in fact, it is a weakly-connected independent dominating set. We say that the component *grows* as edges and vertices are added to it.

The component starts out as a copy of the complete bipartite graph $K_{1,d}$ (or d-star). This is achieved by selecting the first vertex of $\mathcal{W}$ u.a.r. from all the vertices of the input graph. Pseudo-code for our algorithm, Rand_Greedy, is given in Figure 2. In the algorithm, $N(u)$ denotes the set of vertices incident to u in G.

At each iteration, the algorithm either deletes an edge from the graph that connects two vertices in the component or adds a new vertex to the component

```
Input    :  A d-regular n-vertex graph, G.
Output:     A weakly-connected dominating set W for G.

1    W ← ∅;
2    select u u.a.r. from the vertices of degree d;
3    W ← {u};
4    delete all edges incident with u;
5    while (there are vertices of degree d remaining)
     {
6        select u u.a.r. from the vertices of positive degree less than d;
             \\Random greedy selection criteria
7        select v u.a.r. from N(u);
             \\Attempt to increase the number of vertices in the component
8        delete the edge between u and v;
9        if (v has degree d − 1) { W ← W ∪ {v};
                                 delete all edges incident with v; }
     }
```

Fig. 2. Algorithm Rand_Greedy

that is incident to at least one vertex that is already in the component (along with all of its incident edges and any of its neighbours that were not already part of the component).

4   Random Graphs and Differential Equations

The analysis of the algorithm we present is carried out on random regular graphs using differential equations. We therefore introduce the model we use to generate a regular graph u.a.r. and give an overview of an established method that uses differential equations to analyse the performance of randomised algorithms. For other random graph theory definitions see Bollobás [2].

The standard model for random d-regular graphs is as follows. (See Wormald [13] for a thorough discussion of this model and the assertions made about it here, as well as other properties of random regular graphs.) Take a set of dn points in n buckets labelled $1, 2, \ldots, n$, with d points in each bucket, and choose u.a.r. a pairing, P, of the points. The resulting probability space is denoted by $\mathcal{P}_{n,d}$. Form a d-regular pseudograph on n vertices by placing an edge between vertices i and j for each pair in P having one point in bucket i and one in bucket j. This pseudograph is a simple graph (*i.e.*, has no loops or multiple edges) if no pair contains two points in the same bucket and no two pairs contain four points from just two buckets. The d-regular simple graphs on n vertices all occur with equal probabilities. With a probability that is asymptotic to $e^{(1-d^2)/4}$, the pseudograph corresponding to the random pairing in $\mathcal{P}_{n,d}$ is simple. It follows that, in order to prove that a property is a.a.s. true of a uniformly distributed random d-regular (simple) graph, it is enough to prove that it is a.a.s. true of the pseudograph corresponding to a random pairing.

As in [14], we redefine this model slightly by specifying that the pairs are chosen sequentially. The first point in a random pair may be selected using any rule whatsoever, as long as the second point in that random pair is chosen u.a.r. from all the remaining free (unpaired) points. This preserves the uniform distribution of the final pairing.

Wormald [14] gives an exposition of a method of analysing the performance of randomised algorithms. It uses a system of differential equations to express the expected changes in variables describing the state of the algorithm during its execution.

Throughout the execution of our algorithm, edges are deleted and, as this happens, the degree of a vertex may change. In what follows, we denote the set of vertices of degree i in the graph, at time t, by $V_i = V_i(t)$ and let $Y_i = Y_i(t)$ denote $|V_i|$. (For such variables, in the remainder of the paper, the parameter t will be omitted where no ambiguity arises.) We may express the state of the graph at any point during the execution of the algorithm by considering the variables Y_i where $0 \leq i \leq d$. In order to analyse a randomised algorithm for finding a small weakly-connected dominating set, $\mathcal{W}$, of regular graphs, we calculate the expected change in this state over a predefined unit of time in relation to the expected change in the size of the set $\mathcal{W}$.

Let $W = W(t)$ denote $|\mathcal{W}|$ at any stage of an algorithm (time t) and let $\mathbf{E}\Delta X$ denote the expected change in a random variable X conditional upon the history of the process (where the notation $\mathbf{E}$ denotes expectation). We then use equations representing $\mathbf{E}\Delta Y_i$ and $\mathbf{E}\Delta W$ to derive a system of differential equations. The solutions to these differential equations describe functions which represent the behaviour of the variables Y_i. Wormald [14, Theorem 6.1] describes a general result which guarantees that the solutions of the differential equations almost surely approximate the variables Y_i and W with error $o(n)$. The expected size of $\mathcal{W}$ may be deduced from these results.

5   Average-Case Analysis

Denote each iteration of the while loop in Figure 2 as one *operation*. In order to analyse the algorithm we calculate the expected change in the variables Y_i in relation to the expected change in the size of W for an operation. These equations are then used to formulate a set of differential equations.

Note that (depending on the algorithm being analysed), it *may* be necessary to calculate the expected change in the variables Y_i in relation to the expected change in W for all $0 \le i \le d$. However, as our algorithm terminates when $Y_d = 0$, computing the expected change in the variable Y_d in relation to the expected change in the size of W turns out to be sufficient. This equation may then be used to formulate a differential equation. The remainder of this section provides the proof of Theorem 1.

Proof. Let $s = s(t)$ denote the sum of the degrees of the vertices in G at a given stage (time t). Note that $s = \sum_{i=1}^{d} iY_i$. For our analysis it is convenient to assume that $s > \epsilon n$ for some arbitrarily small but fixed $\epsilon > 0$. Operations when $s \le \epsilon n$ will be discussed later.

For each operation, we select a vertex, u, u.a.r. from $V(G) \setminus \{V_0 \cup V_d\}$ and select u.a.r. a neighbour, v, of u. If v has degree d, all edges incident with v are deleted. Otherwise, just one edge incident with v is deleted (the edge between u and v). The expected change in Y_d due to decreasing the degree of v is $-dY_d/s$. Decreasing the degree of u by one has no effect on Y_d as $u \notin V_d$. In the event v has degree d, a further change in Y_d may result if any of the other neighbours of v have degree d. The expected number of neighbours of v that have degree d, given that v has degree d and $u \notin V_d$ is $d(d-1)Y_d/s$. Therefore, the expected change in Y_d when performing an operation (at time t) is $\mathbf{E}\Delta Y_d + o(1) = \mathbf{E}\Delta Y_d(t) + o(1)$ where

$$\mathbf{E}\Delta Y_d = \frac{-dY_d}{s}\left(1 + \frac{d(d-1)Y_d}{s}\right). \tag{2}$$

The expected change in the size of $\mathcal{W}$ when performing an operation (at time t) is $\mathbf{E}\Delta W + o(1) = \mathbf{E}\Delta W(t) + o(1)$ which is simply given by

$$\mathbf{E}\Delta W = \frac{dY_d}{s}. \tag{3}$$

The $o(1)$ terms in Equations (2) and (3) are due to the fact that the values of all the variables may change by a constant during the course of the operation being examined. Since $s > \epsilon n$ the error is in fact $O(1/n)$.

We use Equations (2) and (3) to formulate a differential equation. Write $Y_d(t) = nz_d(t/n)$, $W(t) = nz(t/n)$ and $s(t) = n\xi(t/n)$. From the definition of s we have $\xi = \sum_{i=1}^{d} iz_i$.

Equation (2) representing the expected change in Y_d for an operation forms the basis of a differential equation. The differential equation suggested is

$$\frac{\delta z_d}{\delta x} = \frac{-dz_d}{\xi}\left(1 + \frac{d(d-1)z_d}{\xi}\right), \tag{4}$$

where $x = t/n$ and t is the number of operations.

Equation (3) representing the expected increase in the size of $\mathcal{W}$ for an operation suggests the differential equation for z as

$$\frac{\delta z}{\delta x} = \frac{dz_d}{\xi}. \tag{5}$$

We compute the ratio $\delta z/\delta z_d$, and we have

$$\frac{\delta z}{\delta z_d} = \frac{-1}{1 + \frac{d(d-1)z_d}{\xi}}. \tag{6}$$

The solution to this differential equation represents the cardinalities of V_d and $\mathcal{W}$ (scaled by $1/n$) for given t up until $\xi = \epsilon$. After which point, the change in the variables per operation is bounded by a constant and the error in the solution is $o(1)$.

Notice that with every edge deleted, ξ decreases by 2. It follows that $\delta\xi/\delta z_d = 2\xi/dz_d$. Solving this equation with initial condition $\xi = d$ when $z_d = 1$ gives $\xi = dz_d^{2/d}$. Substituting this expression for ξ into Equation (6), we have

$$\frac{\delta z}{\delta z_d} = \frac{-1}{1 + (d-1)z_d^{(\frac{d-2}{d})}}. \tag{7}$$

The initial condition for Equation (7) is $z = 0$ when $z_d = 1$. We use the substitution $w = (d-1)^{\frac{1}{d-2}}z_d^{1/d}$. The initial condition then becomes $w = (d-1)^{\frac{1}{d-2}}$ when $z = 0$ and we wish to find the value of z when $w = 0$. So we have

$$z = d(d-1)^{\frac{-d}{d-2}}\int_0^{(d-1)^{\frac{1}{d-2}}} \frac{w^{d-1}}{1+w^{d-2}}\,\delta w$$

$$= d(d-1)^{\frac{-d}{d-2}}\int_0^{(d-1)^{\frac{1}{d-2}}} w\,\delta w - d(d-1)^{\frac{-d}{d-2}}\int_0^{(d-1)^{\frac{1}{d-2}}} \frac{w}{1+w^{d-2}}\,\delta w$$

$$= \frac{d}{2(d-1)} - d(d-1)^{\frac{-d}{d-2}}\int_0^{(d-1)^{\frac{1}{d-2}}} \frac{w}{1+w^{d-2}}\,\delta w. \tag{8}$$

Although Equation (8) has an exact solution for all $d \geq 3$, it is not of the simplest form for $d > 4$. For $d = 3$ and $d = 4$ the solution to Equation (8) does have a simple form which we provide below. We then upper bound the solution to Equation (8) when $d \geq 5$ in order to complete the proof of the theorem.

Substituting $d = 3$ into Equation (8) we have

$$z = \frac{3}{4} - \frac{3}{8} \int_0^2 \frac{w}{1+w} \delta w = \frac{3\ln(3)}{8}.$$

Thus proving that for a random 3-regular graph on n vertices $\gamma_w(G)$ is a.a.s. less than

$$\frac{3\ln(3)}{8} n + o(n).$$

Similarly, substituting $d = 4$ into Equation (8) we have

$$z = \frac{2}{3} - \frac{4}{9} \int_0^{\sqrt{3}} \frac{w}{1+w^2} \delta w = \frac{2(3 - \ln(4))}{9}.$$

Thus proving that for a random 4-regular graph on n vertices $\gamma_w(G)$ is a.a.s. less than

$$\frac{2(3 - \ln(4))}{9} n + o(n).$$

We now approximate the solution to Equation (8) for values of d larger than 4. We do so by lower bounding the function $w/(1 + w^{d-2})$ in the interval $0 \leq w \leq (d-1)^{\frac{1}{d-2}}$. Note that over this interval for w, the function $w/(1+w^{d-2})$ is always positive and in this interval, its derivative equals zero at just one point. This implies,

$$\frac{w}{1 + w^{d-2}} \leq \frac{(d-3)^{\left(\frac{d-3}{d-2}\right)}}{d-2}, \qquad 0 \leq w \leq (d-1)^{\frac{1}{d-2}}.$$

When $w/(1 + w^{d-2}) = (d-3)^{\left(\frac{d-3}{d-2}\right)}/(d-2)$ we have $w = (d-3)^{\frac{-1}{d-2}}$. Note that $0 \leq (d-3)^{\frac{-1}{d-2}} \leq (d-1)^{\frac{1}{d-2}}$ for the values of d under consideration. We compute two liner functions of w. We show that the first, in the interval $0 \leq w \leq (d-3)^{\frac{-1}{d-2}}$ is at most $w/(1+w^{d-2})$ and the other, in the interval $(d-3)^{\frac{-1}{d-2}} \leq w \leq (d-1)^{\frac{1}{d-2}}$, is also at most $w/(1 + w^{d-2})$.

Lemma 1. *For every $d > 4$ in the interval $0 \leq w \leq (d-3)^{\frac{-1}{d-2}}$,*

$$\frac{d-3}{d-2} w \leq \frac{w}{1 + w^{d-2}}.$$

Proof. Rearrange the expression above to get

$$(d-3)w(1 + w^{d-2}) \leq (d-2)w$$

$$w^{d-2} \leq \frac{1}{d-3},$$

which completes the proof as $w \leq (d-3)^{\frac{-1}{d-2}}$.   □

Lemma 2. *For every $d > 4$ in the interval $(d-3)^{\frac{-1}{d-2}} \le w \le (d-1)^{\frac{1}{d-2}}$*

$$\frac{(d-3)\left((d-1)^{\left(\frac{1}{d-2}\right)} - w\right)}{2 - d + (d-2)(d-1)^{\left(\frac{1}{d-2}\right)}(d-3)^{\left(\frac{1}{d-2}\right)}} \le \frac{w}{1 + w^{d-2}}.$$

Proof. Rearrange the expression above to get

$$(d-3)(d-1)^{\left(\frac{1}{d-2}\right)} \le (d-2)(d-1)^{\left(\frac{1}{d-2}\right)}(d-3)^{\left(\frac{1}{d-2}\right)}w + (d-3)w^{d-1}$$
$$-w - (d-3)(d-1)^{\left(\frac{1}{d-2}\right)}w^{d-2}.$$

As w and $(d-3)(d-1)^{\left(\frac{1}{d-2}\right)}w^{d-2}$ are positive:

$$(d-3)(d-1)^{\left(\frac{1}{d-2}\right)} \le (d-2)(d-1)^{\left(\frac{1}{d-2}\right)}(d-3)^{\left(\frac{1}{d-2}\right)}w + (d-3)w^{d-1}.$$

As $0 \le w \le (d-1)^{\frac{1}{d-2}}$:

$$(d-3)(d-1)^{\left(\frac{1}{d-2}\right)} \le (d-2)(d-1)^{\left(\frac{1}{d-2}\right)}(d-3)^{\left(\frac{1}{d-2}\right)}(d-1)^{\left(\frac{1}{d-2}\right)}$$
$$+(d-3)(d-1)^{\left(\frac{d-1}{d-2}\right)}.$$

As $(d-3)^{\left(\frac{1}{d-2}\right)} \le d-3$:

$$(d-3)(d-1)^{\left(\frac{1}{d-2}\right)} \le (d-2)(d-1)^{\left(\frac{1}{d-2}\right)}(d-3)(d-1)^{\left(\frac{1}{d-2}\right)}$$
$$+(d-3)(d-1)^{\left(\frac{d-1}{d-2}\right)}$$
$$-1 \le (d-1)^{\left(\frac{1}{d-2}\right)}$$

which completes the proof as $(d-1)^{\left(\frac{1}{d-2}\right)} \ge 0$. $\qquad\square$

We have

$$z = \frac{d}{2(d-1)} - d(d-1)^{\frac{-d}{d-2}} \int_0^{(d-1)^{\frac{1}{d-2}}} \frac{w}{1+w^{d-2}} \delta w$$

$$\le \frac{d}{2(d-1)} - d(d-1)^{\frac{-d}{d-2}} \int_0^{(d-3)^{\frac{-1}{d-2}}} \left(\frac{d-3}{d-2}w\right) \delta w$$

$$-d(d-1)^{\frac{-d}{d-2}} \int_{(d-3)^{\frac{-1}{d-2}}}^{(d-1)^{\frac{1}{d-2}}} \frac{(d-3)\left((d-1)^{\left(\frac{1}{d-2}\right)} - w\right)}{2 - d + (d-2)(d-1)^{\left(\frac{1}{d-2}\right)}(d-3)^{\left(\frac{1}{d-2}\right)}} \delta w.$$

Evaluating this enables us to prove that for a random d-regular graph on n vertices, $(d > 4)$, $\gamma_w(G)$ is a.a.s. less than

$$\left(\frac{d}{2(d-1)} - \frac{d(d-3)^{\frac{d-3}{d-2}}}{2(d-2)(d-1)^{\frac{d-1}{d-2}}}\right) n + o(n).$$

This completes the proof of Theorem 1.

In Table 1, we present our upper bounds on $\gamma_w(G)$, $\alpha(\gamma_w(G))$, the exact solution to Equation (6), $\beta(\gamma_w(G))$, (produced by using a Runge-Kutta method) along with the value $n \ln(d+1)/(d+1)$ from Equation (1) as a comparison to the known asymptotic results for $\gamma(G)$ and $\gamma_c(G)$.

Table 1. Bounds on $\gamma_w(G)$ for a random d-regular graph, G

d	$\alpha(\gamma_w(G))$	$\beta(\gamma_w(G))$	$\frac{n\ln(d+1)}{d+1}$	d	$\alpha(\gamma_w(G))$	$\beta(\gamma_w(G))$	$\frac{n\ln(d+1)}{d+1}$
03	0.4120n	0.4120n	0.3466n	09	0.2852n	0.2328n	0.2303n
04	0.3586n	0.3586n	0.2780n	10	0.2659n	0.2190n	0.2180n
05	0.4167n	0.3205n	0.2986n	20	0.1657n	0.1424n	0.1450n
06	0.3713n	0.2914n	0.2780n	40	0.1005n	0.0887n	0.0906n
07	0.3362n	0.2681n	0.2599n	60	0.0741n	0.0661n	0.0674n
08	0.3081n	0.2489n	0.2441n	80	0.0593n	0.0533n	0.0543n

6 Comparison with Degree-Greedy Heuristics

Having analysed what seems to be the simplest algorithm for finding a small weakly-connected dominating set of regular graphs, it is natural to consider whether different heuristics may give an improved result. So-called *degree-greedy* algorithms that are based on choosing a vertex of a particular degree at each iteration give improved results for various other problems on regular graphs.

There are several degree-greedy algorithms that one may design for finding a small weakly-connected dominating set of a regular graph. Providing all of these algorithms are based on iteratively growing a single weakly-connected component, these may only differ in two ways; namely, for each iteration (after the initial operation), the type of greedy selection criteria used and by how many vertices the set is allowed to increase per iteration. These are represented by the lines 6 and 7 in the while loop of the heuristic presented in Figure 2.

The remaining features remain the same: the initial operation (lines 1–4) must choose a vertex of degree d (deleting some or all of its incident edges) to initiate the growing component. It should also be clear that, in order to ensure only *one* component is "grown", each iteration must start by selecting a vertex of degree less than d (line 6). As vertices of degree d represent non-dominated vertices, any such algorithm may terminate once the number of vertices of degree d in the graph reaches zero.

To analyse such a heuristic, using the differential equation technique as we have, it is usually necessary to develop equations based on the expected changes in the variables Y_i, however, as the algorithm may terminate once no vertices of degree d remain (and providing all vertices of degree d encountered during an operation become part of the set and have all of their incident edges deleted), it is sufficient to track the variable Y_d as opposed to the vector $(Y_1, Y_2, \ldots, Y_d)$.

As any alternative heuristic may only modify lines 6 and 7 (and the vertex, u, selected in each iteration must have degree less than d), it is therefore immediate that any such selection may not affect the variable Y_d. The choice of which vertex of degree less than d to choose is therefore immaterial in this regard.

Once u has been selected, the only way Y_d may decrease is if a number of edges incident with u are deleted. It is well known that, for a random regular graph, the neighbourhood of a vertex, up to a constant distance, is a.a.s. acyclic, (see, for example, [13]) and as we are only interested in the degrees of vertices at distance at most 2 from u, the subgraph considered in each iteration will a.a.s. be a tree.

It may be observed that investigating the degree of one neighbour of u per iteration or investigating the degree of more than one neighbour of u per iteration will have no effect on the resulting differential equation. The latter may be seen as an algorithm that performs a sequence of operations per iteration. The first in the sequence selects a vertex u from those of a given degree and investigates the degree of one of its neighbours.

In each remaining operation in the sequence, the same vertex u is selected and one more of its neighbours has its degree investigated. (This may be achieved by a standard modification to the pairing process as outlined in Section 4.)

As soon as the required number of neighbours have had their degree investigated, the sequence terminates. Each neighbour of degree d encountered must be included in the set under construction. (Again, this comes from the assumption that vertices of degree less than d are dominated as once each of these vertices has its degree investigated, an edge incident with that vertex is deleted.) All edges incident with these vertices would then be removed from the graph.

It is not difficult to see that this would mean that the performance of algorithms that base each selection on choosing vertices of minimum (or maximum) degree and algorithms that iteratively choose one (or more than one) neighbour(s) per iteration, would be represented by Equation (6), and therefore will have the same average-case performance (as the solution to the differential equation would, of course, be the same).

References

1. Alon, N.: Transversal Numbers of Uniform Hypergraphs. Graphs and Combinatorics **6**(1) (1990) 1–4
2. Bollobás, B.: Random Graphs. Academic Press, London (1985)
3. Caro, T., West, D.B. and Yuster, R.: Connected Domination and Spanning Trees with Many Leaves. SIAM Journal on Discrete Mathematics **13**(2) (2000) 202–211
4. Chen, Y.P. and Liestman, A.L.: Approximating Minimum Size Weakly-Connected Dominating Sets for Clustering Mobile Ad-Hoc Networks. In: Proceedings of 3rd ACM International Symposium on Mobile Ad-Hoc Networking and Computing. ACM press (2002), 165–172
5. Diestel, R.: Graph Theory. Springer-Verlag (1997)
6. Dubhashi, D., Mei, A., Panconesi, A., Radhakrishnan, J. and Srinivasan, A.: Fast Distributed Algorithms for (Weakly) Connected Dominating Sets and Linear-Size Skeletons. In: Proceedings of the 14th ACM-SIAM Symposium on Discrete Algorithms (2003) 717–724
7. Dunbar, J.E., Grossman, J.W., Hattingh, J.H., Hedetniemi, S.T. and McRae, A.A.: On Weakly-Connected Domination in Graphs. Discrete Mathematics **167/168** (1997) 261–269
8. Garey, M.R. and Johnson, D.S.: Computers and Intractability: A Guide to the Theory of NP-Completeness. Freeman and Company (1979)

9. Grossman, J.W.: Dominating Sets whose Closed Stars form Spanning Trees. Discrete Mathematics **169**(1-3) (1997) 83–94
10. Guha, S. and Khuller, S.: Approximation Algorithms for Connected Dominating Sets. Algorithmica **20** (1998) 374–387
11. Haynes, T.W., Hedetniemi, S.T. and Slater, P.J.: Domination in Graphs: Advanced topics. Marcel Dekker, New York (1998)
12. Lovász, L.: On the Ratio of Optimal Integral and Fractional Covers. Discrete Mathematics **13(4)** (1975) 383–390
13. Wormald, N.C.: Models of Random Regular Graphs. In: Surveys in Combinatorics. Cambridge University Press, (1999) 239–298
14. Wormald, N.C.: The Differential Equation Method for Random Graph Processes and Greedy Algorithms. In: M. Karoński and H.-J. Prömel (editors), Lectures on Approximation and Randomized Algorithms. PWN Warsaw (1999) 73–155

Additive Spanners for k-Chordal Graphs

Victor D. Chepoi[1], Feodor F. Dragan[2], and Chenyu Yan[2]

[1] Laboratoire d'Informatique Fondamentale, Université Aix-Marseille II, France.
chepoi@lim.univ-mrs.fr
[2] Department of Computer Science, Kent State University, Kent, OH 44242, USA
dragan@cs.kent.edu, cyan1@kent.edu

Abstract. In this paper we show that every chordal graph with n vertices and m edges admits an additive 4-spanner with at most $2n-2$ edges and an additive 3-spanner with at most $O(n \cdot \log n)$ edges. This significantly improves results of Peleg and Schäffer from [Graph Spanners, J. Graph Theory, 13(1989), 99-116]. Our spanners are additive and easier to construct. An additive 4-spanner can be constructed in linear time while an additive 3-spanner is constructable in $O(m \cdot \log n)$ time. Furthermore, our method can be extended to graphs with largest induced cycles of length k. Any such graph admits an additive $(k+1)$-spanner with at most $2n-2$ edges which is constructable in $O(n \cdot k + m)$ time.

Classification: Algorithms, Sparse Graph Spanners

1 Introduction

Let $G = (V, E)$ be a connected graph with n vertices and m edges. The *length* of a path from a vertex v to a vertex u in G is the number of edges in the path. The distance $d_G(u, v)$ between vertices u and v is the length of a shortest (u, v)-path of G. We say that a graph $H = (V, E')$ is *an additive r-spanner* (*a multiplicative t-spanner*) of G, if $E' \subseteq E$ and $d_H(x, y) - d_G(x, y) \leq r$ $(d_H(x, y)/d_G(x, y) \leq t$, respectively) holds for any pair of vertices $x, y \in V$ (here $t \geq 1$ and $r \geq 0$ are real numbers). We refer to r (to t) as the *additive* (respectively, *multiplicative*) *stretch factor* of H. Clearly, every additive r-spanner of G is a multiplicative $(r+1)$-spanner of G (but not vice versa).

There are many applications of spanners in various areas; especially, in distributed systems and communication networks. In [20], close relationships were established between the quality of spanners (in terms of stretch factor and the number of spanner edges $|E'|$), and the time and communication complexities of any synchronizer for the network based on this spanner. Also sparse spanners are very useful in message routing in communication networks; in order to maintain succinct routing tables, efficient routing schemes can use only the edges of a sparse spanner [21]. Unfortunately, the problem of determining, for a given graph G and two integers $t, m \geq 1$, whether G has a t-spanner with m or fewer edges, is NP-complete (see [19]).

The sparsest spanners are tree spanners. Tree spanners occur in biology [2], and as it was shown in [18], they can be used as models for broadcast operations.

R. Petreschi et al. (Eds.): CIAC 2003, LNCS 2653, pp. 96–107, 2003.
© Springer-Verlag Berlin Heidelberg 2003

Multiplicative tree t-spanners were considered in [7]. It was shown that, for a given graph G, the problem to decide whether G has a multiplicative tree t–spanner is NP–complete for any fixed $t \geq 4$ and is linearly solvable for $t = 1, 2$ (the status of the case $t = 3$ is open for general graphs). Also, in [10], NP–completeness results were presented for tree spanners on planar graphs.

Many particular graph classes, such as cographs, complements of bipartite graphs, split graphs, regular bipartite graphs, interval graphs, permutation graphs, convex bipartite graphs, distance–hereditary graphs, directed path graphs, cocomparability graphs, AT-free graphs, strongly chordal graphs and dually chordal graphs admit additive tree r-spanners and/or multiplicative tree t-spanners for sufficiently small r and t (see [3,6,13,14,16,22,23]). We refer also to [1,5,4,6,7,15,17,18,19,24] for more background information on tree and general sparse spanners.

In this paper we are interested in finding sparse spanners with small additive stretch factors in chordal graphs and their generalizations. A graph G is *chordal* [12] if its largest induced (chordless) cycles are of length 3. A graph is k-*chordal* if its largest induced cycles are of length at most k.

The class of chordal graphs does not admit good tree spanners. As it was mentioned in [22,23], H.-O. Le and T.A. McKee have independently showed that for every fixed integer t there is a chordal graph without tree t–spanners (additive as well as multiplicative). Recently, Brandstädt et al. [4] have showed that, for any $t \geq 4$, the problem to decide whether a given chordal graph G admits a multiplicative tree t-spanner is NP-complete even when G has the diameter at most $t + 1$ (t is even), respectively, at most $t + 2$ (t is odd). Thus, the only hope for chordal graphs is to get sparse (with $O(n)$ edges) small stretch factor spanners. Peleg and Schäffer have already showed in [19] that any chordal graph admits a multiplicative 5-spanner with at most $2n - 2$ edges and a multiplicative 3-spanner with at most $O(n \cdot \log n)$ edges. Both spanners can be constructed in polynomial time.

In this paper we improve those results. We show that every chordal graph admits an additive 4-spanner with at most $2n - 2$ edges and an additive 3-spanner with at most $O(n \cdot \log n)$ edges. Our spanners are not only additive but also easier to construct. An additive 4-spanner can be constructed in linear time while an additive 3-spanner is constructable in $O(m \cdot \log n)$ time. Furthermore, our method can be extended to all k-chordal graphs. Any such graph admits an additive $(k + 1)$-spanner with at most $2n - 2$ edges which is constructable in $O(n \cdot k + m)$ time. Note that the method from [19] essentially uses the characteristic *clique trees* of chordal graphs and therefore cannot be extended (at least directly) to general k-chordal graphs for $k \geq 4$. In obtaining our results we essentially relayed on ideas developed in papers [3], [8], [9] and [19].

2   Preliminaries

All graphs occurring in this paper are connected, finite, undirected, loopless, and without multiple edges. For each integer $l \geq 0$, let $B_l(u)$ denote the *ball of* radius l centered at u: $B_l(u) = \{v \in V : d_G(u, v) \leq l\}$. Let $N_l(u)$ denote the *sphere* of radius l centered at u: $N_l(u) = \{v \in V : d_G(u, v) = l\}$. $N_l(u)$ is called also the *lth neighborhood* of u. A *layering* of G with respect to some vertex u is a partition of V into the spheres $N_l(u)$, $l = 0, 1, \ldots$. By $N(u)$ we denote the *neighborhood* of u, i.e., $N(u) = N_1(u)$. More generally, for a subset $S \subseteq V$ let $N(S) = \bigcup_{u \in S} N(u)$.

Let $\sigma = [v_1, v_2, \ldots, v_n]$ be any ordering of the vertex set of a graph G. We will write $a < b$ whenever in a given ordering σ vertex a has a smaller number than vertex b. Moreover, $\{a_1, \cdots, a_l\} < \{b_1, \cdots, b_k\}$ is an abbreviation for $a_i < b_j$ ($i = 1, \cdots, l$; $j = 1, \cdots, k$). In this paper, we will use two kind of orderings, namely, BFS-orderings and LexBFS-orderings.

In a *breadth-first search (BFS)*, started at vertex u, the vertices of a graph G with n vertices are numbered from n to 1 in decreasing order. The vertex u is numbered by n and put on an initially empty queue of vertices. Then a vertex v at the head of the queue is repeatedly removed, and neighbors of v that are still unnumbered are consequently numbered and placed onto the queue. Clearly, BFS operates by proceeding vertices in layers: the vertices closest to the start vertex are numbered first, and most distant vertices are numbered last. BFS may be seen to generate a rooted tree T with vertex u as the root. We call T the *BFS-tree* of G. A vertex v is the *father* in T of exactly those neighbors in G which are inserted into the queue when v is removed. An ordering σ generated by a BFS will be called a *BFS-ordering* of G. Denote by $f(v)$ the father of a vertex v with respect to σ. The following properties of a BFS-ordering will be used in what follows.

(P1) If $x \in N_i(u)$, $y \in N_j(u)$ and $i < j$, then $x > y$ in σ.

(P2) If $v \in N_q(u)$ ($q > 0$) then $f(v) \in N_{q-1}(u)$ and $f(v)$ is the vertex from $N(v)$ with the largest number in σ.

(P3) If $x > y$, then either $f(x) > f(y)$ or $f(x) = f(y)$.

Lexicographic breadth-first search (LexBFS), started at a vertex u, orders the vertices of a graph by assigning numbers from n to 1 in the following way. The vertex u gets the number n. Then each next available number k is assigned to a vertex v (as yet unnumbered) which has lexically largest vector $(s_n, s_{n-1}, \ldots, s_{k+1})$, where $s_i = 1$ if v is adjacent to the vertex numbered i, and $s_i = 0$ otherwise. An ordering of the vertex set of a graph generated by LexBFS we will call a *LexBFS-ordering*. Clearly any LexBFS-ordering is a BFS-ordering (but not conversely). Note also that for a given graph G, both a BFS-ordering and a LexBFS-ordering can be generated in linear time [12]. LexBFS-ordering has all the properties of the BFS-ordering. In particular, we can associate a tree T rooted at v_n with every LexBFS-ordering $\sigma = [v_1, v_2, \ldots, v_n]$ simply connecting every vertex v ($v \neq v_n$) to its neighbor $f(v)$ with the largest number in σ. We call this tree a *LexBFS-tree* of G rooted at v_n and vertex $f(v)$ *the father* of v in T.

3   Spanners for Chordal Graphs

3.1   Additive 4-Spanners with $O(n)$ Edges

For a chordal graph $G = (V, E)$ and a vertex $u \in V$, consider a BFS of G started at u and let $q = max\{d_G(u, v) : v \in V\}$. For a given k, $0 \leq k \leq q$, let S_1^k, $S_2^k, \ldots, S_{p_k}^k$ be the connected components of a subgraph of G induced by the kth neighborhood of u. In [3], there was defined a graph Γ whose vertices are the connected components S_i^k, $k = 0, 1, \ldots, q$ and $i = 1, \ldots, p_k$. Two vertices S_i^k, S_j^{k-1} are adjacent if and only if there is an edge of G with one end in S_i^k and another end in S_j^{k-1}. Before we describe our construction of the additive 4-spanner $H = (V, E')$ for a chordal graph G, first we recall two important lemmas.

Lemma 1. [3] Let G be a chordal graph. For any connected component S of the subgraph of G induced by $N_k(u)$, the set $N(S) \cap N_{k-1}(u)$ induces a complete subgraph.

Lemma 2. [3] Γ is a tree.

Now, to construct H, we choose an arbitrary vertex $u \in V$ and perform a Breadth-First-Search in G started at u. Let $\sigma = [v_1, \ldots, v_n]$ be a BFS-ordering of G. The construction of H is done according to the following algorithm (for an illustration see Figure 1).

PROCEDURE 1. Additive 4-spanners for chordal graphs

Input: A chordal graph $G = (V, E)$ with BFS-ordering σ, and connected components S_1^k, S_2^k, $\ldots$, $S_{p_k}^k$ for any k, $0 \leq k \leq q$, where $q = max\{d_G(u, v) : v \in V\}$.

Output: A spanner $H = (V, E')$ of G.

Method:
$E' = \emptyset$;
for $k = q$ **downto** 1 **do**
    for $j = 1$ **to** p_k **do**
        $M = \emptyset$;
        for each vertex $v \in S_j^k$ add edge $vf(v)$ to E' and vertex $f(v)$ to M;
        pick vertex $c \in M$ with the minimum number in σ;
        for every vertex $x \in M \setminus \{c\}$ add edge xc to E';
return $H = (V, E')$.

Lemma 3. If G has n vertices, then H contains at most $2n - 2$ edges.

Proof: The edge set of H consists of two sets E_1 and E_2, where E_1 are those edges connecting two vertices between two different layers (edges of type $vf(v)$) and E_2 are those edges which have been used to build a star for a clique M inside a layer (edges of type $cf(v)$). Obviously, E_1 has exactly $n - 1$ edges; actually,

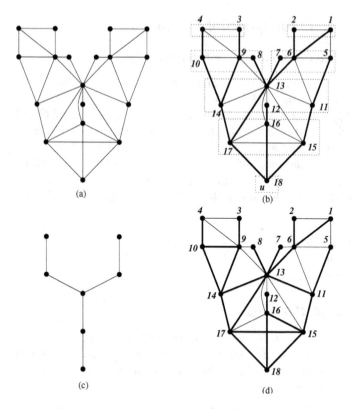

Fig. 1. (a) A chordal graph G. (b) A BFS-ordering σ, BFS-tree T associated with σ and a layering of G. (c) The tree Γ of G associated with that layering. (d) Additive 4-spanner (actually, additive 3-spanner) H of G constructed by PROCEDURE 1 (5 edges are added to the BFS-tree T).

they are the edges of the BFS-tree of G. For each connected component S_i^k of size s, we have at most s vertices in M. Therefore, while proceeding component S_i^k, at most $s-1$ edges are added to E_2. The total size of all the connected components is at most n, so E_2 contains at most $n-1$ edges. Hence, the graph H contains at most $2n-2$ edges. □

Lemma 4. H is an additive 4-spanner for G.

Proof: Consider nodes S_i^k and S_j^l of the tree Γ (rooted at $S_1^0 = \{u\}$) and their lowest common ancestor S_m^p in Γ. For any two vertices $x \in S_i^k$ and $y \in S_j^l$ of G, we have $d_G(x,y) \geq k-p+l-p$, since any path of G connecting x and y must pass S_m^p.

From our construction of H (for every vertex v of G the edge $vf(v)$ is present in H), we can easily show that there exist vertices $x',y' \in S_m^p$ such that $d_H(x,x') = k-p$, $d_H(y,y') = l-p$. Hence we only need to show that $d_H(x',y') \leq 4$. If $x' = y'$ then we are done. If vertices x' and y' are distinct, then by Lemma 1, $N(S_m^p) \cap N_{p-1}(u)$ is a clique of G. According to the Procedure 1, fathers of both vertices x' and y' are in M and they are connected in H by a path of length at most 2 via vertex c of M. Therefore, $d_H(x', y') \leq d_H(x', f(x')) + d_H(f(x'), f(y')) + d_H(f(y'), y') \leq 1 + 2 + 1 = 4$. This concludes our proof. □

We can easily show that the bounds given in Lemma 4 are tight. For a chordal graph presented in Figure 2, we have $d_G(y, b) = 1$. The spanner H of G constructed by our method is shown in bold edges. In H we have $d_H(y, b) = 5$. Therefore, $d_H(y, b) - d_G(y, b) = 4$.

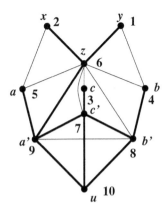

Fig. 2. A chordal graph with a BFS-ordering which shows that the bounds given in Lemma 4 are tight. We have $d_H(y, b) - d_G(y, b) = 4$ and $d_H(y, b)/d_G(y, b) = 5$.

Lemma 5. H can be constructed in linear $O(n + m)$ time.

Combining Lemmas 4-5 we get the following result.

Theorem 1. Every n-vertex chordal graph $G = (V, E)$ admits an additive 4-spanner with at most $2n - 2$ edges. Moreover, such a sparse spanner of G can be constructed in linear time.

Notice that any additive 4-spanner is a multiplicative 5-spanner. As we mentioned earlier the existence of multiplicative 5-spanners with at most $2n-2$ edges in chordal graphs was already shown in [19], but their method of constructing such spanners is more complicated than ours and can take more than linear time.

3.2 Additive 3-Spanners with $O(n \cdot \log n)$ Edges

To construct an additive 3-spanner for a chordal graph $G = (V, E)$, first we get a LexBFS-ordering σ of the vertices of G (see Figure 3). Then, we construct an additive 4-spanner $H = (V, E_1 \bigcup E_2)$ for G using the algorithm from Section 3.1. Finally, we update H by adding some more edges. In what follows, we will need the following known result.

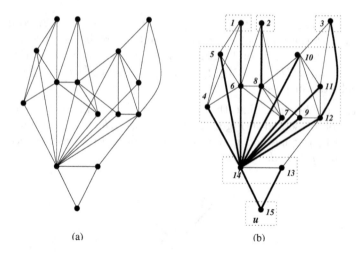

Fig. 3. (a) A chordal graph G. (b) A LexBFS-ordering σ, LexBFS-tree associated with σ and a layering of G.

Theorem 2. *[11] Every n-vertex chordal graph G contains a maximal clique C such that if the vertices in C are deleted from G, every connected component in the graph induced by any remaining vertices is of size at most $n/2$.*

An $O(n + m)$ algorithm for finding such a separating clique C is also given in [11].

As before, for a given k, $0 \le k \le q$, let S_1^k, $S_2^k, \ldots,$ $S_{p_k}^k$ be the connected components of a subgraph of G induced by the kth neighborhood of u. For each connected component S_i^k (which is obviously a chordal graph), we run the following algorithm which is similar to the algorithm in [19] (see also [18]), where a method for construction of a multiplicative 3-spanner for a chordal graph is described. The only difference is that we run that algorithm on every connected component from each layer of G instead of on the whole graph G. For the purpose of completeness, we present the algorithm here (for an example see Figure 4).

PROCEDURE 2. A balanced clique tree for a connected component S_i^k

Input: A subgraph Q of G induced by a connected component S_i^k.
Output: A balanced clique tree for Q.
Method:
        find a maximum separating clique C of the graph Q
           as prescribed in Theorem 2;
        suppose C partitions the rest of Q into connected
           components $\{Q_1, \ldots, Q_r\}$;
        for each Q_i, construct a balanced clique tree
           $T(Q_i)$ recursively;
        construct $T(Q)$ by taking C to be the root and
           connecting the root of each tree $T(Q_i)$ as a child of C.

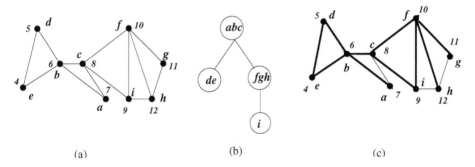

(a)   (b)   (c)

Fig. 4. (a) A chordal graph induced by set S_1^2 of the graph G presented in Figure 3, (b) its balanced clique tree and (c) edges of $E_3(S_1^2) \bigcup E_4(S_1^2)$.

The nodes of the final balanced tree for S_i^k (denote it by $T(S_i^k)$) represent a certain collection of disjoint cliques $\{C_i^k(1), \ldots C_i^k(s_i^k)\}$ that cover entire set S_i^k (see Figure 4 for an illustration). For each clique $C_i^k(j)$ $(1 \leq j \leq s_i^k)$ we build a star centered at its vertex with the minimum number in LexBFS-ordering σ. We use $E_3(i, k)$ to denote this set of star edges. Evidently, $|E_3(i, k)| \leq |S_i^k| - 1$.

Consider a clique $C_i^k(j)$ in S_i^k. For each vertex v of $C_i^k(j)$ and each clique $C_i^k(j')$ on the path of balanced clique tree $T(S_i^k)$ connecting node $C_i^k(j)$ with the root, if v has a neighbor in $C_i^k(j')$, then select one such neighbor w and put the edge vw in set $E_4(i, k)$ (initially $E_4(i, k)$ is empty). We do this for every clique $C_i^k(j)$, $j \in \{1, \ldots, s_i^k\}$. Since the depth of the tree $T(S_i^k)$ is at most $log_2|S_i^k| + 1$ (see [19], [18]), any vertex v from S_i^k may contribute at most $log_2|S_i^k|$ edges to $E_4(i, k)$. Therefore, $|E_4(i, k)| \leq |S_i^k| \cdot log_2|S_i^k|$.

Define now two sets of edges in G, namely,

$$E_3 = \bigcup_{k=1}^{q} \bigcup_{i=1}^{p_k} E_3(i, k), \qquad E_4 = \bigcup_{k=1}^{q} \bigcup_{i=1}^{p_k} E_4(i, k),$$

and consider a spanning subgraph $H^* = (V, E_1 \bigcup E_2 \bigcup E_3 \bigcup E_4)$ of G (see Figure 5). Recall that $E_1 \bigcup E_2$ is the set of edges of an additive 4-spanner H constructed for G by PROCEDURE 1 (see Section 3.1).

From what has been established above one can easily deduce that $|E_3| \leq n - 1$ and $E_4| \leq n \cdot log_2 n$, thus yielding the following result.

Lemma 6. If G has n vertices, then H^* has at most $O(n \cdot \log n)$ edges.

To prove that H^* is an additive 3-spanner for G, we will need the following auxiliary lemmas.

Lemma 7. [12] Let G be a chordal graph and σ be a LexBFS-ordering of G. Then, σ is a perfect elimination ordering of G, i.e., for any vertices a, b, c of G such that $a < \{b, c\}$ and $ab, ac \in E(G)$, vertices b and c must be adjacent.

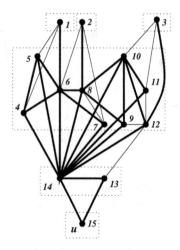

Fig. 5. An additive 3-spanner H^* of graph G presented in Figure 3

Lemma 8. *[9] Let G be an arbitrary graph and $T(G)$ be a BFS-tree of G with the root u. Let also v be a vertex of G and w ($w \neq v$) be an ancestor of v in $T(G)$ from layer $N_i(u)$. Then, for any vertex $x \in N_i(u)$ with $d_G(v, w) = d_G(v, x)$, inequality $x \leq w$ holds.*

Lemma 9. *H^* is an additive 3-spanner for G.*

Lemma 10. *If a chordal graph G has n vertices and m edges, then its additive 3-spanner H^* can be constructed in $O(m \cdot \log n)$ time.*

The main result of this subsection is the following.

Theorem 3. *Every chordal graph $G = (V, E)$ with n vertices and m edges admits an additive 3-spanner with at most $O(n \cdot \log n)$ edges. Moreover, such a sparse spanner of G can be constructed in $O(m \cdot \log n)$ time.*

In [19], it was shown that any chordal graph admits a multiplicative 3-spanner H' with at most $O(n \cdot \log n)$ edges which is constructable in $O(m \cdot \log n)$ time. It is worth to note that the spanner H' gives a better than H^* approximation of distances only for adjacent in G vertices. For pairs $x, y \in V$ at distance at least 2 in G, the stretch factor $t(x, y) = d_{H^*}(x, y)/d_G(x, y)$ given by H^* for x, y is at most 2.5 which is better than the stretch factor of 3 given by H'.

4   Spanners for k-Chordal Graphs

For each $l \geq 0$ define a graph Q^l with the lth sphere $N_l(u)$ as a vertex set. Two vertices $x, y \in N_l(u)$ ($l \geq 1$) are adjacent in Q^l if and only if they can be connected by a path outside the ball $B_{l-1}(u)$. Let $Q_1^l, \ldots, Q_{p_l}^l$ be all the connected components of Q^l.

Similar to chordal graphs and as shown in [8] we define a graph Γ whose vertex-set is the collection of all connected components of the graphs Q^l, $l = 0, 1, \ldots$, and two vertices are adjacent in Γ if and only if there is an edge of G between the corresponding components. The following lemma holds.

Lemma 11. *[8] Γ is a tree.*

Let u be an arbitrary vertex of a k-chordal graph $G = (V, E)$, σ be a BFS-ordering of G and T be the BFS-tree associated with σ. To construct our spanner H for G, we use the following procedure (for an example see Figure 6).

PROCEDURE 3. Additive $(k+1)$-spanners for k-chordal graphs

Input: A k-chordal graph $G = (V, E)$ with a BFS-ordering σ, and connected
   components $Q_1^l, Q_2^l, \ldots, Q_{p_l}^l$ for any l, $0 \le l \le q$, where $q = max\{d_G(u, v) : v \in V\}$.
Output: A spanner $H = (V, E')$ of G.
Method:
   $E' = \emptyset$;
   for $l = q$ **downto** 1 **do**
      for $j = 1$ **to** p_l **do**
         for each vertex $v \in Q_j^l$ add $vf(v)$ to E';
         pick vertex c in Q_j^l with the minimum number in σ;
         for each $v \in Q_j^l \setminus \{c\}$ **do**
            connected = FALSE;
            while connected = FALSE **do**
            /* this while loop works at most $\lfloor k/2 \rfloor$ times for each v */
               if $vc \in E(G)$ **then**
                  add vc to E';
                  connected = TRUE;
               else if $vf(c) \in E(G)$ **then**
                     add $vf(c)$ to E';
                     connected = TRUE;
                  else $v = f(v)$, $c = f(c)$
   return $H = (V, E')$.

Clearly, H contains all edges of BFS-tree T because for each $v \in V$ the edge $vf(v)$ is in H. For a vertex v of G, let P_v be the path of T connecting v with the root u. We call it *the maximum neighbor path of v in G* (evidently, P_v is a shortest path of G). Additionally to the edges of T, H contains also some bridging edges connecting vertices from different maximum neighbor paths.

Lemma 12. *Let c be vertex of Q_i^l with the minimum number in σ ($l \in \{1, \ldots, q\}$, $i \in \{1, \ldots, p_l\}$). Then, for any $a \in Q_i^l$, there is a (a, c)-path in H of length at most k consisting of a subpath $(a, \ldots, x)$ of path P_a, edge xy and a subpath $(y, \ldots, c)$ of path P_c. In particular, $d_H(a, c) \le k$. Moreover, $0 \le d_G(c, y) - d_G(a, x) \le 1$.*

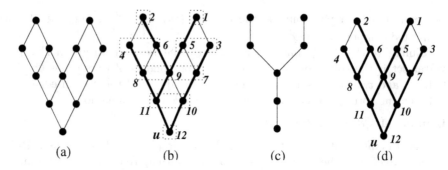

Fig. 6. (a) A 4-chordal graph G. (b) A BFS-ordering σ, BFS-tree associated with σ and a layering of G. (c) The tree Γ of G associated with that layering. (d) Additive 5-spanner (actually, additive 2-spanner) H of G constructed by PROCEDURE 3.

For any n-vertex k-chordal graph $G = (V, E)$ the following lemma holds.

Lemma 13. *H is an additive $(k + 1)$-spanner of G.*

Lemma 14. *If G has n vertices, then H has at most $2n - 2$ edges.*

Lemma 15. *If G is a k-chordal graph with n vertices and m edges, then H can be constructed in $O(n \cdot k + m)$ time.*

Summarizing, we have the following final result.

Theorem 4. *Every k-chordal graph $G = (V, E)$ with n vertices and m edges admits an additive $(k + 1)$-spanner with at most $2n - 2$ edges. Moreover, such a sparse spanner of G can be constructed in $O(n \cdot k + m)$ time.*

References

1. I. ALTHÖFER, G. DAS, D. DOBKIN, D. JOSEPH, and J. SOARES, On sparse spanners of weighted graphs, *Discrete Comput. Geom.,* 9 (1993), 81–100.
2. H.-J. BANDELT, A. DRESS, Reconstructing the shape of a tree from observed dissimilarity data, *Adv. Appl. Math.* 7 (1986) 309-343.
3. A. BRANDSTÄDT, V. CHEPOI AND F. DRAGAN. Distance Approximating Trees for Chordal and Dually Chordal Graphs, *Journal of Algorithms,* 30 (1999), 166-184.
4. A. BRANDSTÄDT, F.F. DRAGAN, H.-O. LE, AND V.B. LE, Tree Spanners on Chordal Graphs: Complexity, Algorithms, Open Problems, to appear in the *Proceedings of ISAAC 2002.*
5. U. BRANDES and D. HANDKE, NP–Completeness Results for Minimum Planar Spanners, Preprint University of Konstanz, *Konstanzer Schriften in Mathematik und Informatik,* Nr. 16, Oktober 1996.
6. L. CAI, Tree spanners: Spanning trees that approximate the distances, *Ph.D. thesis,* University of Toronto, 1992.

7. L. CAI AND D.G. CORNEIL, Tree spanners, *SIAM J. Disc. Math.*, 8 (1995), 359–387.

8. V. CHEPOI AND F. DRAGAN, A Note on Distance Approximating Trees in Graphs, *Europ. J. Combinatorics*, 21 (2000) 761-766.

9. F.F. DRAGAN, Estimating All Pairs Shortest Paths in Restricted Graph Families: A Unified Approach (extended abstract), *Proc. 27th International Workshop "Graph-Theoretic Concepts in Computer Science"(WG'01)*, June 2001, Springer, Lecture Notes in Computer Science 2204, pp. 103-116.

10. S.P. FEKETE, J. KREMER, Tree spanners in planar graphs, *Discrete Appl. Math.* 108 (2001) 85-103.

11. J.R. GILBERT, D.J. ROSE, AND A. EDENBRANDT, A separator theorem for chordal graphs, *SIAM J. Alg. Discrete Math.*, 5 (1984) 306-313.

12. M.C. GOLUMBIC, Algorithmic Graph Theory and Perfect Graphs, *Academic Press*, New York, 1980.

13. H.-O. LE, Effiziente Algorithmen für Baumspanner in chordalen Graphen, *Diploma thesis*, Dept. of mathematics, technical university of Berlin, 1994.

14. H.-O. LE, V.B. LE, Optimal tree 3-spanners in directed path graphs, *Networks* 34 (1999) 81-87.

15. A.L. LIESTMAN AND T. SHERMER, Additive graph spanners, *Networks*, 23 (1993), 343–364.

16. M.S. MADANLAL, G. VENKATESAN AND C.PANDU RANGAN, Tree 3–spanners on interval, permutation and regular bipartite graphs, *Information Processing Letters*, 59 (1996), 97–102.

17. I.E. PAPOUTSAKIS, On the union of two tree spanners of a graph, *Preprint*, 2001.

18. D. PELEG, Distributed Computing: A Locality-Sensitive Approach, *SIAM Monographs on Discrete Math. Appl.*, (SIAM, Philadelphia, 2000).

19. D. PELEG AND A. A. SCHÄFFER, Graph Spanners, *Journal of Graph Theory*, 13 (1989), 99-116.

20. D. PELEG AND J.D. ULLMAN, An optimal synchronizer for the hypercube, *in Proc. 6th ACM Symposium on Principles of Distributed Computing*, Vancouver, 1987, 77–85.

21. D.PELEG AND E.UPFAL, A tradeoff between space and efficiency for routing tables. *20th ACM Symposium on the Theory of Computing*, Chicago (1988), 43-52.

22. E. PRISNER, Distance approximating spanning trees, in *Proc. of STACS'97, Lecture Notes in Computer Science* 1200, (R. Reischuk and M. Morvan, eds.), Springer–Verlag, Berlin, New York, 1997, 499–510.

23. E. PRISNER, H.-O. LE, H. MÜLLER AND D. WAGNER, Additive tree spanners, to appear in *SIAM Journal on Discrete Mathematics*.

24. J. SOARES, Graph spanners: A survey, *Congressus Numer.* 89 (1992) 225-238.

Graph-Modeled Data Clustering: Fixed-Parameter Algorithms for Clique Generation

Jens Gramm[*], Jiong Guo[**], Falk Hüffner, and Rolf Niedermeier

Wilhelm-Schickard-Institut für Informatik, Universität Tübingen,
Sand 13, D-72076 Tübingen, Germany
{gramm,guo,hueffner,niedermr}@informatik.uni-tuebingen.de

Abstract. We present efficient fixed-parameter algorithms for the NP-complete edge modification problems CLUSTER EDITING and CLUSTER DELETION. Here, the goal is to make the fewest changes to the edge set of an input graph such that the new graph is a vertex-disjoint union of cliques. Allowing up to k edge additions and deletions (CLUSTER EDITING), we solve this problem in $O(2.27^k + |V|^3)$ time; allowing only up to k edge deletions (CLUSTER DELETION), we solve this problem in $O(1.77^k + |V|^3)$ time. The key ingredients of our algorithms are two easy to implement bounded search tree algorithms and a reduction to a problem kernel of size $O(k^3)$. This improves and complements previous work.

Keywords: NP-complete problems, edge modification problems, data clustering, fixed-parameter tractability, exact algorithms.

1 Introduction

Motivation and problem definition. There is a huge variety of clustering algorithms with applications in numerous fields (cf., e.g., [8,9]). Here, we focus on problems closely related to algorithms for clustering gene expression data (cf. [19] for a very recent survey). More precisely, Shamir et al. [17] recently studied two NP-complete problems called CLUSTER EDITING and CLUSTER DELETION[1]. These are based on the notion of a *similarity graph* whose vertices correspond to elements and in which there is an edge between two vertices iff the similarity of their corresponding elements exceeds a predefined threshold. The goal is to obtain a *cluster graph* by as few edge modifications (i.e., edge deletions and additions) as possible; a cluster graph is a graph in which each of the connected components is a clique. Thus, we arrive at the edge modification problem, CLUSTER EDITING, central to our work:

---

[*] Supported by the Deutsche Forschungsgemeinschaft (DFG), research project "OPAL" (optimal solutions for hard problems in computational biology), NI 369/2.
[**] Partially supported by the Deutsche Forschungsgemeinschaft (DFG), junior research group "PIAF" (fixed-parameter algorithms), NI 369/4.
[1] The third problem CLUSTER COMPLETION (only edge additions allowed) is easily seen to be polynomial-time solvable.

R. Petreschi et al. (Eds.): CIAC 2003, LNCS 2653, pp. 108–119, 2003.
© Springer-Verlag Berlin Heidelberg 2003

Input: An undirected graph $G = (V, E)$, and a nonnegative integer k.
Question: Can we transform G, by deleting and adding at most k edges, into a graph that consists of a disjoint union of cliques?
CLUSTER DELETION is the special case where edges can only be deleted. All these problems belong to the class of *edge modification problems,* see Natanzon et al. [14] for a recent survey.

Previous work. The most important reference point to our work is the paper of Shamir et al. [17]. Among other things, they showed that CLUSTER EDITING is NP-complete and that there exists some constant $\epsilon > 0$ such that it is NP-hard to approximate CLUSTER DELETION to within a factor of $1 + \epsilon$. In addition, Shamir et al. studied cases where the number of clusters (i.e., cliques) is fixed. Before that, Ben-Dor et al. [2] and Sharan and Shamir [18] investigated closely related clustering applications in the computational biology context, where they deal with modified versions of the CLUSTER EDITING problem together with heuristic polynomial-time solutions. From the more abstract view of graph modification problems, Leizhen Cai [3] (also cf. [5]) considered the more general problem (allowing edge deletions, edge additions, *and* vertex deletions) where the "goal graph" has a "forbidden set characterization" with respect to "hereditary graph properties" and he showed that this problem is *fixed-parameter tractable.* More precisely, Cai's result implies an $O(3^k \cdot |G|^4)$ time algorithm for both CLUSTER EDITING and CLUSTER DELETION where the forbidden set is a P_3 (i.e., a vertex-induced path consisting of three vertices).[2] Natanzon et al. [14] give a general constant-factor approximation for deletion and editing problems on bounded-degree graphs with respect to properties (such as being a cluster graph) that can be characterized by a finite set of forbidden induced subgraphs. Kaplan et al. [10] and Mahajan and Raman [13] considered other special cases of edge modification problems with particular emphasis on fixed-parameter tractability results. Khot and Raman [11] recently investigated the parameterized complexity of vertex deletion problems for finding subgraphs with hereditary properties.

New results. Following a suggestion of Natanzon et al. [14] (who note that, regarding their NP-hardness results for some edge modification problems, "... studying the parameterized complexity of the NP-hard problems is also of interest."), we present significantly improved fixed-parameter tractability results for CLUSTER EDITING and CLUSTER DELETION. More precisely, we show that CLUSTER EDITING is solvable in $O(2.27^k + |V|^3)$ worst-case time and that CLUSTER DELETION is solvable in $O(1.77^k + |V|^3)$ worst-case time. This gives simple and efficient exact algorithms for these NP-complete problems in case of reasonably small parameter values k (number of deletions and additions or number of deletions only). In particular, we present an efficient data reduction by prepro-

---

[2] Observe that a graph is a cluster graph iff it contains no P_3 as a vertex-induced subgraph. This will also be important for our work. Note that Shamir *et al.* [17] write "P_2-free," but according to the graph theory literature it should be called "P_3-free."

cessing, providing a *problem kernel* of size $O(k^3)$. Due to the lack of space, some proofs had to be omitted and are deferred to the full paper.

2   Preliminaries and Basic Notation

One of the latest approaches to attack computational intractability is to study parameterized complexity. For many hard problems, the seemingly unavoidable combinatorial explosion can be restricted to a "small part" of the input, the *parameter*, so that the problems can be solved in polynomial time when the parameter is fixed. For instance, the NP-complete VERTEX COVER problem can be solved by an algorithm with $O(1.3^k + kn)$ running time [4,16], where the parameter k is a bound on the maximum size of the vertex cover set we are looking for and where n is the number of vertices in the given graph. The parameterized problems that have algorithms of $f(k) \cdot n^{O(1)}$ time complexity are called *fixed-parameter tractable*, where f can be an arbitrary function depending only on k, and n denotes the overall input size, see [1,5,6] for details.

Our bounded search tree algorithms work recursively. The number of recursions is the number of nodes in the corresponding tree. This number is governed by homogeneous, linear recurrences with constant coefficients. It is well-known how to solve them and the asymptotic solution is determined by the roots of the characteristic polynomial (e.g., see Kullmann [12] for more details). If the algorithm solves a problem of "size" s and calls itself recursively for problems of sizes $s - d_1, \ldots, s - d_i$, then $(d_1, \ldots, d_i)$ is called the *branching vector* of this recursion. It corresponds to the recurrence $t_s = t_{s-d_1} + \cdots + t_{s-d_i}$ (to simplify matters, without any harm, we only count the number of leaves here) with the characteristic polynomial $z^d = z^{d-d_1} + \cdots + z^{d-d_i}$, where $d = \max\{d_1, \ldots, d_i\}$. If α is a root of the characteristic polynomial with maximum absolute value, then t_s is α^s up to a polynomial factor. We call $|\alpha|$ the *branching number* that corresponds to the branching vector $(d_1, \ldots, d_i)$. Moreover, if α is a single root, then $t_s = O(\alpha^s)$ and all branching numbers that will occur in this paper are single roots.

The size of the search tree is therefore $O(\alpha^k)$, where k is the parameter and α is the largest branching number that will occur; in our case, for CLUSTER EDITING, it will be shown that it is about 2.27 and it belongs to the branching vector $(1, 2, 2, 3, 3)$ which occurs in Sect. 4.

We assume familiarity with basic graph-theoretical notations. If x is a vertex in a graph $G = (V, E)$, then by $N_G(x)$ we denote the set of its neighbors. We omit the index if it is clear from the context. The whole paper only works with *simple* graphs without *self-loops*. By $|G|$ we refer to the size of graph G, which is determined by the numbers of its vertices and edges.

In our algorithms, we use Table T to store an annotation for the edges of the graph such that T has an entry for every pair of vertices $u, v \in V$:

"*permanent*": In this case, $\{u, v\} \in E$ and the algorithm is not allowed to delete $\{u, v\}$ later on; or *"forbidden"*: In this case, $\{u, v\} \notin E$ and the algorithm is not allowed to add $\{u, v\}$ later on.

Note that, whenever the algorithm deletes an edge $\{u, v\}$ from E, we set $T[u, v]$ to forbidden since it would not make sense to reintroduce previously deleted edges. In the same way, whenever the algorithm adds an edge $\{u, v\}$ to E, we set $T[u, v]$ to permanent. In the following, when adding and deleting edges, we assume that we make these adjustments even when not mentioned explicitly.

3   Problem Kernel for Cluster Editing

A *reduction rule* replaces, in polynomial time, a given CLUSTER EDITING instance (G, k) consisting of a graph G and a nonnegative integer k by a "simpler" instance (G', k') such that (G, k) has a solution iff (G', k') has a solution, i.e., G can be transformed into disjoint clusters by deleting/adding at most k edges iff G' can be transformed into disjoint clusters by deleting/adding at most k' edges. An instance to which none of a given set of reduction rules applies is called *reduced* with respect to these rules. A parameterized problem such as CLUSTER EDITING (the parameter is k) is said to have a *problem kernel* if, after the application of the reduction rules, the resulting reduced instance has size $f(k)$ for a function f depending only on k.

We present three reduction rules for CLUSTER EDITING. For each of them, we discuss its correctness and give the running time which is necessary to execute the rule. In our rules, we use table T as described in Sect. 2. Using the reduction rules, we show, at the end of this section, a problem kernel consisting of at most $2(k^2 + k)$ vertices and at most $2\binom{k+1}{2}k$ edges for CLUSTER EDITING.

Although the following reduction rules also *add* edges to the graph, we consider the resulting instances as *simplified*. The reason is that for every added edge, the parameter is decreased by one. In the following rules, it is implicitly assumed that, when edges are added or deleted, parameter k is decreased by one.

Rule 1. For every pair of vertices $u, v \in V$:
1. If u and v have more than k common neighbors, i.e.,

$$|\{ z \in V \mid \{u, z\} \in E \text{ and } \{v, z\} \in E \}| > k,$$

then $\{u, v\}$ has to belong to E and we set $T[u, v] := $ permanent. If $\{u, v\}$ is not in E, we add it to E.
2. If u and v have more than k non-common neighbors, i.e.,

$$|\{ z \in V \mid \text{either } \{u, z\} \in E \text{ or } \{v, z\} \in E \text{ but not both} \}| > k,$$

then $\{u, v\}$ may not belong to E and we set $T[u, v] := $ forbidden. If $\{u, v\}$ is in E, we delete it.
3. If u and v have both more than k common and more than k non-common neighbors, then the given instance has no solution.

Lemma 1. *Rule 1 is correct.*   □

Note that Rule 1 applies to every pair u, v of vertices for which the number of neighbors of u or v is greater than $2k$.

Rule 2. For every three vertices $u, v, w \in V$:
   1. If $T[u, v] = $ permanent and $T[u, w] = $ permanent, then $\{v, w\}$, if not already there, has to be added to E and $T[v, w] := $ permanent.
   2. If $T[u, v] = $ permanent and $T[u, w] = $ forbidden, then $\{v, w\}$, if already there, has to be deleted from E and $T[v, w] := $ forbidden.

The correctness of Rule 2 is obvious. Regarding the running time, we analyze the interleaved application of Rules 1 and 2 together.

Lemma 2. *A graph can in $O(|V|^3)$ time be transformed into a graph which is reduced with respect to Rules 1 and 2.*   □

Rule 3. Delete the connected components which are cliques from the graph.

The correctness of Rule 3 is straightforward.

Lemma 3. *Rule 3 can be executed in $O(|G|)$ time.*   □

Notably, for the problem kernel size to be shown, Rules 1 and 3 would be sufficient. Rule 2 is also taken into account since it is very easy and general and can be computed in the course of computing Rule 1. Thus, we give here a small set of general and easy reduction rules which yields a problem kernel with $O(k^2)$ vertices and $O(k^3)$ edges and which is computable in $O(|V|^3)$ time. Note that the $O(|V|^3)$ running time given here is only a worst-case bound and it is to be expected that the computation of the rules is much more efficient in practice.

Theorem 1. CLUSTER EDITING *has a problem kernel which contains at most $2(k^2 + k)$ vertices and at most $2\binom{k+1}{2}k$ edges. It can be found in $O(|V|^3)$ time.*

Proof. Let G be the reduced graph. Since Rule 3 deletes all isolated cliques from the given graph, the connected components in G are non-cliques. If the reduced instance can be solved, we need at least one edge modification for each connected component in G to transform it into a set of disjoint cliques. Let us restrict, in the following, our attention to each of the connected components separately, one of them having vertex set V' and edge set E'. We use $G[V']$ to denote the subgraph induced by V'. Let $k' \geq 1$ be the minimum number of edge modifications, namely k'_a edge additions and k'_d edge deletions, that we need to transform $G[V']$ into disjoint cliques. Under the assumption that G is reduced, i.e., none of the given reduction rules applies, we will show by contradiction that $|V'| \leq 2(k + 1) \cdot k'$ and that $|E'| \leq 2\binom{k+1}{2}k'$ as follows.

Regarding the vertex set of the connected component, assume that $|V'| > 2(k + 1) \cdot k'$. In the case that $k'_a = 0$, we have k'_d edge deletions, $1 \leq k'_d = k' \leq k$, to transform $G[V']$ into cliques. Consider the largest of these cliques. All edges connecting this clique are present ($k'_a = 0$) and at least one vertex u' of this clique is connected to a vertex v' outside the clique. In the case that v' is not connected to an other vertex in this clique, we can lowerbound the clique size by at least $|V'|/(k'_d + 1)$, i.e., more than

$$(2(k+1) \cdot k')/(k'_d + 1) \geq (2(k+1) \cdot k')/(2k'_d) = k+1.$$

Thus, u' has at least $k+1$ neighbors—all other vertices in the clique—which are not neighbors of v'. This contradicts the assumption that G is reduced since Rule 1 would apply. In the case that v' is connected to other vertices in this clique, we can lowerbound the clique size by at least $|V'|/k'_d$, i.e., more than $(2(k+1) \cdot k')/k'_d \geq 2(k+1)$. Again, u' has at least $k+1$ neighbors in this clique which are not neighbors of v', contradicting the assumption that G is reduced.

In the case $1 \leq k'_a \leq k'$, we conclude that one of the cliques to be obtained from V' has size at least $|V'|/(k'_d + 1) \geq |V'|/k' > 2(k+1)$. Denote the set of vertices of this clique by $C' \subseteq V'$. Then we have $k'_a \leq k$ many edge additions in order to insert the edges missing in C'. To recall, C' contains more than $2(k+1)$ vertices and at most k edges shall be missing. In the case that $k'_a = k'$ then, by counting arguments, there clearly has to be a pair of vertices in C' which share no edge and which have more than k common neighbors. In the case that $k'_a < k'$, C' contains a vertex u which is connected to a vertex v not in C'. Since at most $k'_a < k'$ edges in C' are missing, there are at least $k+1$ neighbors of u which are not neighbors of v. In both cases, for $k'_a = k'$ and for $1 \leq k'_a < k'$, we obtain a contradiction to the assumption that G is reduced since Rule 1 would apply.

Regarding the edge set of the connected component, we infer a contradiction from the assumption that $|E'| \geq 2\binom{k+1}{2}k'$ in an analogous way as for the vertex set.

Summing up over all connected components, the graph contains at most $2(k^2 + k)$ vertices and $2\binom{k+1}{2}k$ edges (otherwise, no solution exists).

The running time follows directly from Lemmas 2 and 3. □

4   Search Tree Algorithm for Cluster Editing

In this section, we describe a recursive algorithm for CLUSTER EDITING that follows the bounded search tree paradigm. The basic idea of the algorithm is to identify a "conflict-triple" consisting of three vertices and to branch into subcases to repair this "conflict" by adding or deleting edges between the three considered vertices. Thus, we invoke recursive calls on instances which are simplified in the sense that the value of the parameter is decreased by at least one. Before starting the algorithm and after every such branching step, we compute the problem kernel as described in Sect. 3. The running time of the algorithm is, then, mainly determined by the size of the resulting search tree. In Subsect. 4.1, we introduce a straightforward branching strategy that leads to a search tree of size $O(3^k)$; in Subsect. 4.2, we show how a more involved branching strategy leads to a search tree of worst case size $O(2.27^k)$.

Note that the more general result of Leizhen Cai [3] as discussed in the introductory section also provides an algorithm with exponential factor 3^k. By way of contrast, however, he uses a sort of enumerative approach with more computational overhead (concerning polynomial-time computations). In addition, the search tree algorithm in Subsect. 4.1 also lies the basis for a more refined search tree strategy with the improved exponential term 2.27^k. Since our mathematical analysis is purely worst-case, we expect that the search tree sizes would be usually much smaller in practical settings; this seems particularly plausible because our search tree strategy also allows numerous heuristic improvements of the running time and the search tree size without influencing the worst-case mathematical analysis.

4.1   Basic Branching Strategy

Central for the branching strategy described in this section is the following lemma which was already observed in [17].

Lemma 4. *A graph $G = (V, E)$ consists of disjoint cliques iff there are no three vertices $u, v, w \in V$ with $\{u, v\} \in E$, $\{u, w\} \in E$, but $\{v, w\} \notin E$.*   □

Lemma 4 implies that, if a given graph does not consist of disjoint cliques, then we can find a conflict-triple of vertices between which we either have to insert or to delete an edge in order to transform the graph into disjoint cliques. In the following, we describe the recursive procedure that results from this observation. Inputs are a graph $G = (V, E)$ and a nonnegative integer k, and the procedure reports, as its output, whether G can be transformed into a union of disjoint cliques by deleting and adding at most k edges.

- If the graph G is already a union of disjoint cliques, then we are done: Report the solution and return.
- Otherwise, if $k \leq 0$, then we cannot find a solution in this branch of the search tree: Return.
- Otherwise, identify $u, v, w \in V$ with $\{u, v\} \in E$, $\{u, w\} \in E$, but $\{v, w\} \notin E$ (they exist with Lemma 4). Recursively call the branching procedure on the following three instances consisting of graphs $G' = (V, E')$ with nonnegative integer k' as specified below:
- (B1) $E' := E - \{u, v\}$ and $k' := k - 1$. Set $T[u, v] :=$ forbidden.
- (B2) $E' := E - \{u, w\}$ and $k' := k - 1$. Set $T[u, v] :=$ permanent and $T[u, w] :=$ forbidden.
- (B3) $E' := E + \{v, w\}$ and $k' := k - 1$. Set $T[u, v] :=$ permanent, $T[u, w] :=$ permanent, and $T[v, w] :=$ permanent.

Proposition 1. CLUSTER EDITING *can be solved in $O(3^k \cdot k^2 + |V|^3)$ time.*

Proof. The recursive procedure suggested above is obviously correct. Concerning the running time, we observe the following. The preprocessing in the beginning to obtain the reduction to a problem kernel can be done in $O(|V|^3)$ time (Theorem 1). After that, we employ the search tree with size clearly bounded by $O(3^k)$. Hence, it remains to justify the factor k^2 which stands for the computational overhead related to every search tree node. Firstly, note that in a further preprocessing step, we can once set up a linked list of all conflict triples. This is clearly covered by the $O(|V|^3)$ term. Secondly, within every search tree node (except for the root) we deleted or added one edge and, thus, we have to update the conflict list accordingly. Due to Theorem 1 we only have $O(k^2)$ graph vertices now and with little effort, one verifies that the addition or deletion of an edge can make at most $O(k^2)$ new conflict-triples appear and it can make at most $O(k^2)$ conflict-triples disappear. Using (perfect) hashing in combination with doubly linked lists of all conflict triples, one can perform $O(k^2)$ updates of this list and the hash table in $O(k^2)$ time. □

In fact, it does "not really" matter what the polynomial factor in k is, as the interleaving technique of [15] can be applied improving Proposition 1:

Corollary 1. CLUSTER EDITING *can be solved in $O(3^k + |V|^3)$ time.*

Proof. In [15], it was shown that, in case of a polynomial size problem kernel, by doing the "kernelization" repeatedly during the course of the search tree algorithm whenever possible, the polynomial factor in parameter k can be replaced by a constant factor. □

4.2 Refining the Branching Strategy

The branching strategy from Subsect. 4.1 can be easily improved as described in the following. We still identify a conflict-triple of vertices, i.e., $u, v, w \in V$ with $\{u, v\} \in E$, $\{u, w\} \in E$, but $\{v, w\} \notin E$. Based on a case distinction, we provide for every possible situation additional branching steps. The amortized analysis of the successive branching steps, then, yields the better worst-case bound on the running time. We start with distinguishing three main situations that may apply when considering the conflict-triple:

(C1) Vertices v and w do not share a common neighbor, i.e. $\nexists x \in V, x \neq u$: $\{v, x\} \in E$ and $\{w, x\} \in E$.
(C2) Vertices v and w have a common neighbor x and $\{u, x\} \in E$.
(C3) Vertices v and w have a common neighbor x and $\{u, x\} \notin E$.

Regarding case (C1), we show in the following lemma that, here, a branching into two cases suffices.

Lemma 5. *Given a graph $G = (V, E)$, a nonnegative integer k and $u, v, w \in V$ with $\{u, v\} \in E$, $\{u, w\} \in E$, but $\{v, w\} \notin E$. If v and w do not share a common neighbor, then branching case (B3) cannot yield a better solution than both cases (B1) and (B2), and can therefore be omitted.*

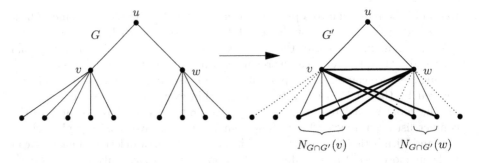

Fig. 1. In case (C1), adding edge $\{v, w\}$ does not need to be considered. Here, G is the given graph and G' is a clustering solution of G by adding edge $\{v, w\}$. The dashed lines denote the edges being deleted to transform G into G', and the bold lines denote the edges being added. Observe that the drawing only shows that parts of the graphs (in particular, edges) which are relevant for our argumentation

Proof. Consider a clustering solution G' for G where we did add $\{v, w\}$ (see Fig. 1). We use $N_{G \cap G'}(v)$ to denote the set of vertices which are neighbors of v in G and in G'. Without loss of generality, assume that $|N_{G \cap G'}(w)| \leq |N_{G \cap G'}(v)|$. We then construct a new graph G'' from G' by deleting all edges adjacent to w. It is clear that G'' is also a clustering solution for G. We compare the cost of the transformation $G \to G''$ to that of the transformation $G \to G'$:

- -1 for not adding $\{v, w\}$,
- $+1$ for deleting $\{u, w\}$,
- $-|N_{G \cap G'}(v)|$ for not adding all edges $\{w, x\}$, $x \in N_{G \cap G'}(v)$,
- $+|N_{G \cap G'}(w)|$ for deleting all edges $\{w, x\}$, $x \in N_{G \cap G'}(w)$.

Herein, we omitted possible vertices which are neighbors of w in G' but not neighbors of w in G because they would only increase the cost of transformation $G \to G'$.

In summary, the cost of $G \to G''$ is not higher than the cost of $G \to G'$, i.e., we do not need more edge additions and deletions to obtain G'' from G than to obtain G' from G.   □

Lemma 5 shows that in case (C1) a branching into two cases is sufficient, namely to recursively consider graphs $G_1 = (V, E - \{u, v\})$ and $G_2 = (V, E - \{u, w\})$, each time decreasing the parameter value by one.

For case (C2), we change the order of the basic branching. In the first branch, we add edge $\{v, w\}$. In the second and third branches, we delete edges $\{u, v\}$ and $\{u, w\}$, as illustrated by Fig. 2.

- Add $\{v, w\}$ as labeled by ② in Fig. 2. The cost of this branch is 1.
- Mark $\{v, w\}$ as forbidden and delete $\{u, v\}$, as labeled by ③. This creates the new conflict-triple (x, u, v). To resolve this conflict, we make a second branching. Since adding $\{u, v\}$ is forbidden, there are only two branches to consider:

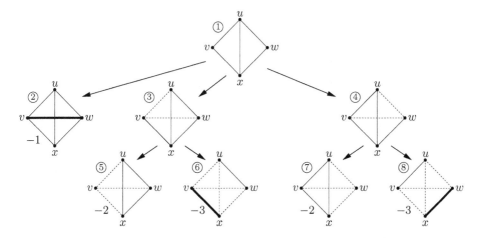

Fig. 2. Branching for case (C2). Bold lines denote permanent, dashed lines forbidden edges

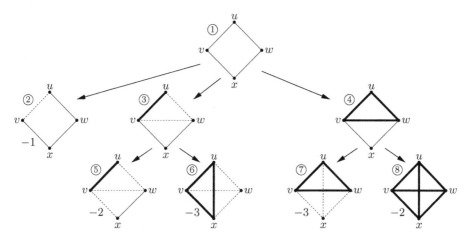

Fig. 3. Branching for case (C3)

- Delete $\{v, x\}$, as labeled by ⑤. The cost is 2.
- Mark $\{v, x\}$ as permanent and delete $\{u, x\}$. With reduction rule 2 from Sect. 3, we then delete $\{x, w\}$, too, as labeled by ⑥. The cost is 3.
- Mark $\{v, w\}$ as forbidden and delete $\{u, w\}$ (④). This case is symmetric to the previous one, so we have two branches with costs 2 and 3, respectively.

In summary, the branching vector for case (C2) is $(1, 2, 3, 2, 3)$.
For case (C3), we perform a branching as illustrated by Fig. 3:

- Delete $\{u, v\}$, as labeled by ②. The cost of this branch is 1.
- Mark $\{u, v\}$ as permanent and delete $\{u, w\}$, as labeled by ③. With Rule 2, we can additionally mark $\{v, w\}$ as forbidden. We then identify a new

conflict-triple $(v, u, x.)$ Not being allowed to delete $\{v, u\}$, we can make a 2-branching to resolve the conflict:

- Delete $\{v, x\}$, as labeled by ⑤. The cost is 2.
- Mark $\{v, x\}$ as permanent. This implies $\{x, u\}$ needs to be added and $\{x, w\}$ to be deleted, as labeled by ⑥. The cost is 3.

– Mark $\{u, v\}$ and $\{u, w\}$ as permanent and add $\{v, w\}$, as labeled by ④. Vertices (w, u, x) form a conflict-triple. To solve this conflict without deleting $\{w, u\}$, we make a 2-branching:

- Delete $\{w, x\}$ as labeled by ⑦. We then also need to delete $\{x, v\}$. The cost is 3. Additionally, we can mark $\{x, u\}$ as forbidden.
- Add $\{u, x\}$, as labeled by ⑧. The cost is 2. Additionally, we can mark $\{x, u\}$ and $\{x, v\}$ as permanent.

It follows that the branching vector for case (C3) is $(1, 2, 3, 3, 2)$.

In summary, this leads to a refinement of the branching with a worst-case branching vector of $(1, 2, 2, 3, 3)$, yielding branching number 2.27. Since the recursive algorithm stops whenever the parameter value has reached 0 or below, we obtain a search tree size of $O(2.27^k)$. This results in the following theorem.

Theorem 2. CLUSTER EDITING *can be solved in* $O(2.27^k + |V|^3)$ *time.*   □

In a similar way to Theorem 2, we can prove the following result for CLUSTER DELETION (proof deferred to the full paper).

Theorem 3. CLUSTER DELETION *can be solved in* $O(1.77^k + |V|^3)$ *time.*   □

5   Conclusion

Adopting a parameterized point of view [1,5,6], we have shed new light on the algorithmic tractability of the NP-complete problems CLUSTER EDITING and CLUSTER DELETION. We developed efficient fixed-parameter algorithms in both cases and the algorithms seem easy enough in order to allow for efficient implementations. In ongoing work, we recently succeeded in further lowering the search tree sizes for CLUSTER EDITING from 2.27^k to 1.92^k and for CLUSTER DELETION from 1.77^k to 1.62^k (further improvements seem possible). This approach, however, heavily relies on an automated generation and evaluation of numerous branching cases by computer [7] which seems "undoable" by a human being.

Shamir et al. [17] showed that so called p-CLUSTER EDITING is NP-complete for $p \geq 2$ and p-CLUSTER DELETION is NP-complete for $p \geq 3$. Herein, p denotes the number of cliques that should be generated by as few edge modifications as possible. Hence, there is no hope for fixed-parameter tractability with respect to parameter p, because fixed-parameter tractability with respect to parameter p would thus imply P = NP.

We feel that the whole field of clustering problems might benefit from more studies on parameterized complexity of the many problems related to this field. There appear to be numerous parameters that make sense in this context.

Acknowledgment

We thank Jochen Alber and Elena Prieto-Rodriguez for inspiring discussions.

References

1. J. Alber, J. Gramm, and R. Niedermeier. Faster exact solutions for hard problems: a parameterized point of view. *Discrete Mathematics*, 229:3–27, 2001.
2. A. Ben-Dor, R. Shamir, and Z. Yakhini. Clustering gene expression patterns. *Journal of Computational Biology*, 6(3/4):281–297, 1999.
3. Leizhen Cai. Fixed-parameter tractability of graph modification problems for hereditary properties. *Information Processing Letters*, 58:171–176, 1996.
4. J. Chen, I. Kanj, and W. Jia Vertex cover: further observations and further improvements. *Journal of Algorithms*, 41:280–301, 2001.
5. R. G. Downey and M. R. Fellows. *Parameterized Complexity*. Springer. 1999.
6. M. R. Fellows. Parameterized complexity: the main ideas and connections to practical computing. In *Experimental Algorithmics*, number 2547 in LNCS, pages 51–77, 2002. Springer.
7. J. Gramm, J. Guo, F. Hüffner, and R. Niedermeier. Automated generation of search tree algorithms for graph modification problems. Manuscript in preparation, March 2003.
8. P. Hansen and B. Jaumard. Cluster analysis and mathematical programming. *Mathematical Programming*, 79:191–215, 1997.
9. A. K. Jain and R. C. Dubes. Algorithms for clustering data. Prentice Hall, 1988.
10. H. Kaplan, R. Shamir, and R. E. Tarjan. Tractability of parameterized completion problems on chordal, strongly chordal, and proper interval graphs. *SIAM Journal on Computing*, 28(5):1906–1922, 1999.
11. S. Khot and V. Raman. Parameterized complexity of finding subgraphs with hereditary properties. *Theoretical Computer Science*, 289:997–1008, 2002.
12. O. Kullmann. New methods for 3-SAT decision and worst-case analysis. *Theoretical Computer Science*, 223(1-2):1–72, 1999.
13. M. Mahajan and V. Raman. Parameterizing above guaranteed values: MaxSat and MaxCut. *Journal of Algorithms*, 31:335–354, 1999.
14. A. Natanzon, R. Shamir, and R. Sharan. Complexity classification of some edge modification problems. *Discrete Applied Mathematics*, 113:109–128, 2001.
15. R. Niedermeier and P. Rossmanith. A general method to speed up fixed-parameter-tractable algorithms. *Information Processing Letters*, 73:125–129, 2000.
16. R. Niedermeier and P. Rossmanith. On efficient fixed-parameter algorithms for Weighted Vertex Cover. *Journal of Algorithms*, to appear, 2003.
17. R. Shamir, R. Sharan, and D. Tsur. Cluster graph modification problems. In *Proc. of 28th WG*, number 2573 in LNCS, pages 379–390, 2002, Springer.
18. R. Sharan and R. Shamir. CLICK: A clustering algorithm with applications to gene expression analysis. In *Proc. of 8th ISMB*, pp. 307–316, 2000. AAAI Press.
19. R. Sharan and R. Shamir. Algorithmic approaches to clustering gene expression data. In T. Jiang et al. (eds): *Current Topics in Computational Molecular Biology*, pages 269–300, The MIT Press. 2002.

Reconciling Gene Trees to a Species Tree

Paola Bonizzoni[1], Gianluca Della Vedova[2], and Riccardo Dondi[1]

[1] Dipartimento di Informatica, Sistemistica e Comunicazione
Università degli Studi di Milano-Bicocca, Milano - Italy
bonizzoni@disco.unimib.it,
riccardo.dondi@unimib.it
[2] Dipartimento di Statistica Università degli Studi di Milano-Bicocca, Milano - Italy
gianluca.dellavedova@unimib.it

Abstract. In this paper we deal with the general problem of recombining the information from evolutionary trees representing the relationships between distinct gene families. First we solve a problem from [8] regarding the construction of a minimum reconciled tree by giving an efficient algorithm. Then we show that the exemplar problem, arising from the exemplar analysis of multigene genomes [2], is NP-hard even when the number of copies of a given label is at most two. Finally we introduce two novel formulations for the problem of recombining evolutionary trees, extending the notion of the gene duplication problem studied in [8,11,9,10,6], and we give an exact algorithm (via dynamic programming) for one of the formulations given.

1 Introduction

The reconstruction and comparison of evolutionary trees is a main topic of computational biology that poses several new challenging problems to the computer science research community. An evolutionary tree or *species tree* is a rooted binary tree where each leaf is labelled by a taxon in a set of extant species of which the tree represents the ancestral relationships. The common strategy for constructing a species tree consists of two basic steps: given a gene (or a gene family) for the extant species, one constructs a gene tree representing the relationships among the sequences encoding only that gene in the different species. Indeed, a gene family is represented by homologous sequences and the initial assumption is that such genes evolve in the same way as the species. The second step consists of deriving or inferring from the gene trees the species tree. Indeed, gene trees can differ from the actual species tree because of some typical biological phenomena related to evolutionary events, such as gene divergence resulting from duplications, losses and other gene mutations. Duplications are common evolutionary events and consist of copying in multiple places a gene located along a DNA strand. Then all those copies evolve independently from each other. In a species tree the fact that an internal node has two children represents a speciation event, that is a species has evolved into two different species. In a gene tree an internal node with two children represents either a speciation event or the effect of an evolutionary event localized in a single gene.

R. Petreschi et al. (Eds.): CIAC 2003, LNCS 2653, pp. 120–131, 2003.
© Springer-Verlag Berlin Heidelberg 2003

For the above reason, gene trees for different gene families may not agree, consequently the need for summarizing different often contradictory gene trees into a unique species tree arises. A recent and biologically successful approach to this problem is the gene duplication model proposed by various authors [5,9,6] based on the idea that a possibly correct species tree can be derived from different gene trees by using a cost measure for gene duplications and mutations.

In order to relate or reconcile a gene tree to a species tree, the duplication model introduces a mapping M from nodes of a gene tree G to nodes of a species tree S. In a gene tree, an internal node g represents an ancestral gene which is associated by M to the most recent ancestral species of S (an internal node of tree S) that contains all contemporary genes (leaves of the gene tree) descending from g. The mapping M is computationally modeled by using the *least common ancestor* (lca) mapping in a species tree. When a duplication happens in g, that is the gene tree and the species tree are inconsistent because of duplication events, the lca maps a gene g and one of its children or both to the same ancestral species: a *duplication* occurs at g, or we say also that g is associated to a duplication. Biologically, this fact means that the divergence of the children of gene g in the gene tree is due to a gene duplication inside the same species, and not to a true speciation event. The number of duplications can be interpreted as a measure of the similarity between a gene tree and a species tree, while by using a parsimonious criterion the reconstruction of a species tree from a gene tree is commonly based on the idea of minimizing the duplication cost, as done recently in [8,7] where some optimization problems are introduced.

The most important difference between a species tree S and a gene tree G is given by the way leaves of the two trees are labeled; the label of S and G are taken from the same set L, but a given label occurs at most once in S.

A problem, pointed out in [5,9] as a valid biological mean to study gene phylogenies, is the one of finding the best tree that reconciles a gene tree to a species tree, more precisely in [9] a notion of reconciled tree as a smallest tree satisfying three basic properties has been defined.

Later, in [8], a recursive definition of reconciled tree used to give an algorithm to construct such a tree has been proposed. In the same paper the authors posed the question whether their construction produces a smallest reconciled tree according to the notion introduced in [9], thus leaving as an open problem to give an effective algorithm to construct a smallest reconciled tree.

In this paper we face the problem of how to reconcile a gene tree to a species tree under the duplication cost model in two cases: the species tree is given or is unknown. Both problems have a gene tree G as part of the instance, and are defined as follows: (1) if a species tree S is given, find a tree R that explains (or *reconciles*) G w.r.t. S, (2) if no species tree is given, find a species tree S that optimally reconciles G w.r.t. S, that is a tree which minimizes the number of duplications.

First we face the problem (1) by solving the open question mentioned above and posed in [8]. Indeed, we propose a recursive algorithm to build *the minimum reconciled tree*: our construction allows us also to show that such a tree is unique.

Then, we analyze the complexity of computing an optimal species tree (2) under certain restrictions.

In [8] it is shown that the reconstruction of a species tree from a gene tree, under the duplication cost is NP-hard when the number of occurrences of a label (i.e. a gene copy) in the gene tree is unbounded. Since the complexity of the problem is related to the copies of genes in a gene tree, in this paper we investigate a new approach to the reconstruction problem based on the exemplar analysis. The notion of exemplar is used in genome rearrangement to deal with multigene families when comparing two genomes, that is genomes in which a gene can occur more than once [2]: the *exemplar* of two compared genomes is a genome obtained from them by keeping only one copy of each gene, so as to minimize some rearrangement distance. In the paper, we propose to combine the phylogenetic analysis with algorithmic methods that are pertinent to genome comparison, following a new research direction recently suggested in [2]. Indeed we apply the exemplar analysis to model the reconstruction of a species tree from a gene tree by requiring that a species tree is a homomorphic subtree of the gene tree: we call this problem the *exemplar tree*. We show that even if a gene occurs in at most two copies, i.e. the occurrences of the labels are bounded by two, the exemplar tree problem is NP-hard.

We conclude the paper by suggesting a new measure for the duplication cost in reconciling a gene tree to a given species tree. We first define a variant of problem (2) in which the species tree must explain the gene tree by a general mapping that satisfies basic requirements of biological relevance, as pointed out in [10]. Then we solve efficiently, via dynamic programming the problem of reconciling a gene tree to a given species tree by using this general mapping.

2   Gene Trees and Species Trees

In this section we will present some basic terminology used in the paper. We will follow the notation used in [8] to introduce the fundamental problem of reconciling gene trees to species trees.

In the paper we focus on evolutionary trees. For this reason, unless otherwise stated, all trees are *binary*, that is each internal node x of a tree has exactly two children, the left one denoted by $a(x)$ and the right one denoted by $b(x)$. Moreover trees are *rooted*, that is there is a distinguished vertex, called *root*, that has no ancestor.

We consider two different kinds of evolutionary trees: species trees and gene trees. These trees have the property that leaves are labeled, while the internal nodes are unlabeled. Given an evolutionary tree T, that is a species tree or a gene tree, we denote with $\Lambda(T)$ its leafset and with $L(T)$ the set of labels of its leaves.

A *species* tree S has the property that its leafset is *uniquely labeled*, in the sense that no two leaves share a common label, consequently the sets $\Lambda(S)$ and $L(S)$ are isomorphic. Given a node x of a species tree S, the *cluster* of x, denoted by $\mathcal{C}(x)$, is the set of labels of all leaves of S that are descendants of x.

A different kind of tree is the *gene tree*. A gene tree is characterized by the fact that two leaves may share a label, hence, given an internal node g of a gene tree G, the labels of its descendent leaves is a multiset. Just as for species tree, we define the cluster of g, denoted by $\mathcal{C}(g)$, simply as the set of all labels of leaves in G that are descendants of g. By a slight abuse of language we will denote an internal node x of a species tree or gene tree with its associated cluster $\mathcal{C}(x)$. Given a species or gene tree T, the root of T, is denoted by $r(T)$. Moreover $\mathcal{C}(T)$ denotes the set of clusters of T.

In the following we will always use G to denote a gene tree, S to denote a species tree and T for a generic tree. Given a pair (G, S), the *lca mapping*, in short lca, is a function that associates to each node g of G the node s of S, such that $\mathcal{C}(s)$ is the smallest cluster of S containing $\mathcal{C}(g)$. Please note that the lca function is unique for a given pair (G, S).

Having defined the lca mapping we are now able to define a notion of distance between a gene tree and a species tree.

As defined in [8], whenever a node g of G and one (or both) of its children are mapped through lca in the same node x of S we will say that a *duplication* occurs in g. More formally we have a duplication in g if $\text{lca}(a(g)) = \text{lca}(g)$ or $\text{lca}(b(g)) = \text{lca}(g)$. The number of duplications induced by lca is called the *duplication distance* from G to S, that is $d(G, S)$.

Given a labeled tree T and a subset L_1 of the leaves of T, then the *homomorphic subtree* T_1 of T induced by L_1 is obtained by first removing all nodes and edges of T that are not in a path from the root of T to a leaf in L_1, and then contracting each internal node x that has only one child, that is creating an edge connecting the two neighbors of x in T, and then to remove x and all edges incident on it.

3   The Reconciled Tree Problem

A gene tree G and a species tree S are comparable if $L(G) \subseteq L(S)$. Given a pair (G, S) of comparable trees, a basic approach proposed in [9] and used in [8] for gene tree reconciliation, consists of computing a *reconciled* tree $R(G, S)$. The general notion of reconciled tree is that of a tree which contains G as a homomorphic subtree and also represents the evolutionary history of S by having exactly the clusters of the species tree S. Clearly, for any pair (G, S) there exists an infinite number of reconciled trees but, according to the parsimony principle, we are interested only in the smallest one, that is the tree with the minimum number of leaves: we will denote such tree as *minimum* reconciled tree $R(G, S)$

In [9] a formal definition of a minimum reconciled tree (Def. 1) has been given. Whenever it does not arise any ambiguities we will denote a reconciled tree R in place of $R(G, S)$.

Definition 1. *A minimum reconciled tree $R(G, S)$ of a gene tree G and a species tree S is a smallest tree satisfying the following properties:*

1. G is a homomorphic subtree of $R(G, S)$,

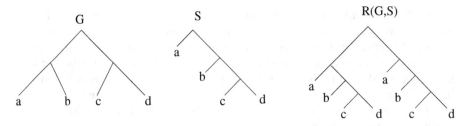

Fig. 1. A gene tree G, a species tree S and a reconciled tree of G with respect to S

2. $C(R(G, S)) = C(S)$
3. *for any internal node x of $R(G, S)$ either $a(x) \cap b(x) = \emptyset$ or $a(x) = b(x) = x$.*

Please notice that the third point of Def. 1 implies that, for each internal node x of R, the two sets $C(a(x))$ and $C(b(x))$ are either distinct or equal; this fact will be used in proving some of the following results. In [8] a recursive definition of a general reconciled tree has been introduced in order to describe an efficient algorithm for constructing such tree: this definition is reported here as Def. 2. The authors of [8] ask whether their definition and Def. 1 are equivalent. Indeed, they could not prove that their construction is optimal w.r.t. the size of the reconciled tree. We will answer affirmatively to such question by giving a recursive construction of a minimum reconciled tree (Def. 3), moreover we can prove that the optimal (smallest) tree is unique. Our construction is simpler than the one of Def. 2, and it can easily be proved to be equivalent to that one.

Our new definition uses the same operations on trees given in the definition of the reconciled tree proposed in [8]: *restriction*, *composition* and *replacement*.

- Given an internal node t of a tree T_1, the *restriction* of T_1 at t, denoted by $T_1|t$, is the subtree of T_1 rooted at t.
- Given two trees T_1, T_2 their *composition*, denoted by $T_1 \triangle T_2$, consists of the (rooted binary) tree T such that $a(r(T)) = r(T_1)$ and $b(r(T)) = r(T_2)$, $T|a(r(T)) = T_1$ and $T|b(r(T)) = T_2$. Informally T is obtained by adding a node r and connecting r to the roots of the two trees T_1 and T_2.
- Given an internal node t of a tree T_1, the *replacement of $T_1|t$ with T_2* at t, denoted by $T_1(t \rightarrow T_2)$, is the tree obtained by replacing in T_1 the restriction $T_1|t$ with T_2.

We will now give two recursive definitions of reconciled trees, successively we will prove that these definitions are equivalent. Each point of the definitions explains how the clusters of the two topmost levels of G relate to those of S. The definition of the reconciled tree given in [8] is the following:

Definition 2. *$R(G, S)$ is equal to:*

1. $R(G|a(r(G)), S) \triangle R(G|b(r(G)), S)$ *if $lca(r(G)) = r(S)$ and $lca(a(r(G)))$, $lca(b(r(G))) = r(S)$*
2. $S(lca(a(r(G))) \rightarrow R(G|a(r(G)), S)) \triangle R(G|b(r(G)), S)$ *if $lca(r(G)) = r(S)$ and $lca(a(r(G))) \subseteq a(r(S))$ and $lca(b(r(G))) = r(S)$*

3. $S(lca(a(r(G)))) \rightarrow R(G|a(r(G)), S|a(r(S))), lca(b(r(G))) \rightarrow R(G|b(r(G)),$
   $S|b(r(S))))$ if $lca(r(G)) = r(S)$ and $lca(a(r(G))) \subseteq a(r(S))$ and $lca(b(r(G)))$
   $\subseteq b(r(S))$
4. $S(a(r(S)) \rightarrow R(G, S|a(r(S)))$ if $lca(r(G)) \subseteq a(r(S))$

The case $lca(r(G)) \subseteq b(r(S))$ is symmetric to point 4. Our new definition of the reconciled tree is:

Definition 3. $R(G, S)$ is equal to G if G and S are both leaves, otherwise:

1. $R(G|a(r(G)), S) \triangle R(G|b(r(G)), S)$, if $lca(r(G)) = r(S)$, and at least one of $lca(a(r(G)))$ and $lca(b(r(G)))$ is equal to $r(S)$;
2. $R(G|a(r(G)), S|a(r(S))) \triangle R(G|b(r(G)), S|b(r(S)))$, if $lca(r(G)) = r(S)$, $lca(a(r(G)))$ and $lca(b(r(G)))$ are mapped in $s_1 \subseteq a(r(S))$ and $s_2 \subseteq b(r(S))$ respectively
3. $S(a(r(S)) \rightarrow R(G, S|a(r(S))))$ if $lca(r(G)) \subseteq a(r(S))$.

The case $lca(r(G)) \subseteq b(r(S))$ is symmetric to point 3. Notice that every point of Def. 3 modifies the two trees G and S to be reconciled so that $L(G) \subseteq L(S)$ always holds, moreover the termination condition is reached when G and S are both leaves, that is $G = S$. Also note that the conditions considered in Def. 3 are mutually exclusive and cover all possible cases. We leave to the reader the proof that the tree constructed by the algorithm induced by the definition Def. 3 satisfies the three properties of Def. 1, but it is not evident whether it is a smallest tree with such property and if this case holds whether it is unique. A classical result from graph theory (Remark 1) leads to Lemma 1, whose proof is omitted.

Remark 1. Let T_1 and T_2 be two uniquely labeled binary trees such that $\mathcal{C}(T_1) = \mathcal{C}(T_2)$. Then T_1 is isomorphic to T_2.

Lemma 1. Let T be a uniquely labeled binary tree and let T' be a binary tree (not necessarily uniquely labeled) such that $L(T) \subseteq L(T')$ and $\mathcal{C}(T) \supseteq \mathcal{C}(T')$. Then $L(T) = L(T')$ and $\mathcal{C}(T) = \mathcal{C}(T')$.

The following propositions (Prop. 1, 2 and 3) state properties that must be verified by all minimum reconciled trees for G and S, according to Def. 1. At the same time Def. 3 leads immediately to an algorithm for computing a minimum reconciliation tree, indeed such algorithm maintains Prop. 1, 2 and 3 as invariants.

Given a reconciled tree R for G, S, it must be that any homomorphic subtree T of G in R must have its root in $r(R)$ or one of such homomorphic subtrees has root in $T|a(r(R))$ or in $T|b(r(R))$.

Proposition 1. Let R be a minimum reconciled tree for (G, S), such that there exists one homomorphic subtree of G in R having the root in $R|a(r(R))$. Then $\mathcal{C}(a(r(R))) = \mathcal{C}(a(r(S)))$, $\mathcal{C}(b(r(R))) = \mathcal{C}(b(r(S)))$ and $r(G) \subseteq a(r(S))$. Moreover $R|a(r(R))$ must be a reconciled tree for G and $S|a(r(S))$.

Proof. By the property 3 of Def. 1, $\mathcal{C}(a(r(R))) \cap \mathcal{C}(b(r(R))) = \emptyset$. In fact assume to the contrary that $\mathcal{C}(a(r(R))) = \mathcal{C}(b(r(R))) = \mathcal{C}(r(R))$, then $\mathcal{C}(a(r(R))) = \mathcal{C}(b(r(R))) = \mathcal{C}(r(R)) = \mathcal{C}(r(S))$. Thus $R|a(r(R))$ is a minimum reconciled tree, since it contains a homomorphic copy of G and contains all the clusters of S, which contradicts the initial assumptions.

Since it must be $\mathcal{C}(a(r(R))) \cap \mathcal{C}(b(r(R))) = \emptyset$ it follows $a(r(R)) = a(r(S))$, $b(r(R)) = b(r(S))$ and moreover $\mathcal{C}(R|a(r(R))) = \mathcal{C}(S|a(r(S)))$, $\mathcal{C}(R|b(r(R))) = \mathcal{C}(S|b(r(S)))$ by Lemma 1.

From the property of homomorphic subtree, since by hypothesis $R|a(r(R))$ contains a homomorphic copy of G, it follows $r(G) \subseteq R|a(r(R))$ and from the fact above $r(G) \subseteq a(r(S))$.

Now consider the subtree $R|a(r(R))$. This subtree must contain a homomorphic copy of G by hypothesis and, as shown above, must have exactly the clusters of $S|a(r(S))$. Thus it is a reconciled tree for $(G, S|a(r(S)))$.

Now we will deal with the case that, given R a minimum reconciled tree for (G, S), $r(R)$ is the root of every subtree T of R homomorphic to G. Moreover we will assume that the left child of $r(T)$ is in $R|a(r(R))$ and the right child of $r(T)$ is in $R|b(r(R))$.

Proposition 2. *Let R be a minimum reconciled tree for (G, S) such that $r(R)$ is the root of all the subtrees T of R homomorphic to G, $a(r((T)) \subseteq R|a(r(R))$, $b(r(T)) \subseteq R|b(r(R))$ and $\mathcal{C}(a(r(R))) \cap \mathcal{C}(b(r(R))) = \emptyset$. Then $a(r(G)) \subseteq a(r(S))$ and $b(r(G)) \subseteq b(r(S))$. Moreover $R|a(r(R))$ must be a reconciled tree for $G|a(r(G))$ and $S|a(r(S))$ and $R|b(r(R))$ must be a reconciled tree for $G|b(r(G))$ and $S|b(r(S))$.*

Proof. Since $\mathcal{C}(R) = \mathcal{C}(S)$ and $\mathcal{C}(a(r(R))) \cap \mathcal{C}(b(r(R))) = \emptyset$, it is not restrictive to assume that $a(r(R)) = a(r(S))$ and $b(r(R)) = b(r(S))$.

Since by the first hypothesis the left child of the root of the subtree of R homomorphic to G is in $R|a(r(R))$, then $a(r(R)) \supseteq a(r(G))$, hence $a(r(G)) \subseteq a(r(S))$. Similarly $b(r(G)) \subseteq b(r(S))$.

Now we will prove that $R|a(r(R))$ must be a reconciled tree for $G|a(r(G))$ and $S|a(r(S))$. In fact $R|a(r(R))$ must contain a homomorphic copy of $G|a(r(G))$, since for hypothesis the left child of the root of any subtree of R homomorphic to G must be in $R|a(r(R))$. Moreover, since $\mathcal{C}(R|a(r(R))) \subseteq \mathcal{C}(S|a(r(S)))$ and $L(R|a(r(R))) = L(S|a(r(S)))$, by Lemma 1 follows that $\mathcal{C}(R|a(r(R))) = \mathcal{C}(S|a(r(S)))$. Clearly $R|a(r(R))$ must satisfy the third property of Def. 1, thus making $R|a(r(R))$ a reconciled tree for $G|a(r(G))$ and $S|a(r(S))$.

Similarly $R|b(r(R))$ must be a reconciled tree for $G|b(r(G))$ and $S|b(r(S))$.

Proposition 3. *Let R be a minimum reconciled tree for (G, S) such that $r(R)$ is the root of all the subtrees T of R homomorphic to G, $a(r((T)) \subseteq R|a(r(R))$, $b(r(T)) \subseteq R|b(r(R))$ and $\mathcal{C}(a(r(R))) = \mathcal{C}(b(r(R)))$. Then at least one of $lca(a(r(G)))$ and $lca(b(r(G)))$ is $r(S)$. Moreover $R|a(r(R))$ must be a reconciled tree for $G|a(r(G))$ and S and $R|b(r(R))$ must be a reconciled tree for $G|b(r(G))$ and S.*

Theorem 1. *Let R be a reconciled tree for G and S constructed as in Def. 3. Then R is a minimum reconciled tree for G and S and such a minimum tree is unique.*

The complete proof relies upon the fact that a tree built according to each case of Def. 3, satisfies Prop. 1, 2, 3 (case 1 is related to Prop. 3, case 2 is related to Prop.2, case 3 is related to Prop. 1).

4 Exemplar-Driven Reconciliation

In [8] it is proved that the phylogenetic reconstruction from gene families is an NP-hard problem (problem (2) stated in the introduction), but the proof requires that the occurrences of labels in G are unbounded. More precisely the problem can be stated as follows:

Problem 1 (Minimum Duplication Problem (MDP)). Given a gene tree G, find a species tree S such that $L(S) = L(G)$ and $d(G, S)$ is minimum.

In this section we investigate the complexity of a variant of MDP obtained by requiring that S must be a homomorphic subtree of G, that is S is obtained from G by extracting a single copy of each label, called the *exemplar*, so that the resulting tree minimizes the number of duplications in S. Given a gene tree G, we call *exemplar species tree* for G any species tree that has the same leafset of G and is a homomorphic subtree of G. The formal definition of the problem follows:

Problem 2 (Exemplar Tree (ET)). Given a gene tree G, find an exemplar species tree S so that the duplication distance $d(G, S)$ is minimized.

In the following the complexity of the Exemplar Tree problem is analyzed by bounding the number of copies of each label in a gene tree. By ET-B, with B integer, we denote the ET problem when the number of copies of a given label in a gene tree is at most B. The main result of this section is proving that ET-2 is NP-complete, that is restricting to instances of ET where each label appears at most twice does not help in designing an exact polynomial algorithm.

Theorem 2. *The decision version of the ET-2 problem is NP-complete.*

Proof. The problem is trivially in NP. To prove that ET-2 is NP-hard we use a reduction from the Vertex Cover problem on cubic graphs to the given problem [4].

Let $G = (V, E)$ be a cubic graph, with vertex set $V = \{v_1, \cdots, v_n\}$ and edge set $E = \{e_1, \cdots, e_m\}$. Then construct the gene tree T_G over set of labels $L(T_G) = V \cup E \cup L_r \cup L_l \cup \{u\}$, where $L_l = \cup_{1 \leq j \leq n} I_j$, $I_j = \{i_1^j, \cdots, i_k^j\}$, $L_r = \{r_1, \cdots, r_{2n}\}$ and $k > |V|$. Let $T_G = L[T_1, \cdots, T_n]$ denote the gene tree shown in Fig. 2.a, where each tree T_i is shown in Fig. 2.b.

The tree T_G consists of subtrees T_i, one for each vertex v_i of the graph G. Each tree T_i consists of a set of leaves I_i and the subtree T_i' which has leafset

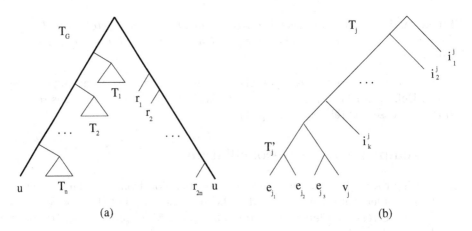

Fig. 2. The tree T_G (a), and the tree T_j (b)

$\{e_{i1}, e_{i2}, e_{i3}, v_i\}$, where e_{i1}, e_{i2}, e_{i3} are the edges that are incident on vertex v_i. We call *left line* and *right line* the paths connecting the root to respectively the leftmost and the rightmost occurrence of the label u of T_G. Both lines are represented as thick lines in Fig. 2.

The following properties relates T_G to an exemplar species tree S for T_G. Note that only labels in $E \cup \{u\}$ occur in two copies in T_G.

First note that, due to the presence of the two copies of u, the following lemma holds.

Lemma 2. *Given a species tree S that is an exemplar species tree for T_G, then either a duplication occurs in all internal nodes of T_G along the* left *line or in all internal nodes of T_G along the* right *line.*

The following lemma is immediate by the structure of the trees T_j' given in Fig. 2.b.

Lemma 3. *Given a species tree S that is an exemplar species tree for T_G, then each tree T_j' of T_G can contain at most one duplication, which is associated to the root r_j of T_j'.*

Lemma 4. *Given a species tree S that is an exemplar species tree for T_G, if some label in E of T_j is deleted, then k or $k+1$ duplications occur in T_j.*

Proof. Clearly, if any label in E of the tree T_j is deleted to obtain a subtree S of T_G, the only node of T_j' to which a duplication can be associated is the root r_j of T_j'.

Let x_s be the node of the exemplar species tree S of T_G such that $\text{lca}(r_j) = x_s$. Note that, since S is a subtree of T_G, x_s is a node along the left line of T_G and S. Now, since $c(y) = c(r_j) \cup X$ for each internal node y of T_j, where $X \subseteq I_j$ and

$c(x_s) \supseteq L(T_j)$, with $L(T_j) \supseteq X$ and $\mathrm{lca}(r_j) = x_s$, it follows that $\mathrm{lca}(y) = x_s$. But, since I_j has k elements, in T_j occur at least k-duplications and by Lemma 3, at most $k + 1$-duplications.

Proposition 4. *A cubic graph G has a node cover of size C if and only if, given the gene tree T_G, there exists an exemplar species tree S for T_G with $d(T_G, S) \leq C(k + 1) + |V|$, where $k > |V|$.*

Proof. An exemplar species tree for T_G must contain a unique copy of each label e_i and of label u, which are the only ones that occur twice in T_G. In fact, each edge e of the graph G is incident in two vertices w and v, thus e is a leaf of both trees T_w and T_v.

Note that the occurrence of u that can be deleted by inducing the least number of duplications is the one that is on the *left line* of T_G. In fact, by lemma 2 either $|V|$ duplications occur in the left line or $2|V|$ duplications occur in the right line. Now, if a given occurrence of a label e_i is deleted in a subtree T'_j of T_j in the gene tree T_G, that is e_i is an edge incident on vertex v_j, by lemma 4, at least k duplications and at most $k + 1$-duplications occur in T_j. We now show that if C vertices are enough to cover all edges in G, then at most $C(k + 1) + |V|$ duplications occur in T_G. In fact, given $N = \{v_{i,1}, \cdots, v_{i,C}\}$ the list of vertices in a node cover, then each label of an edge in E occurs in some subtree in $\{T_{i,1}, \cdots, T_{i,C}\}$. Thus, at most $C(k + 1)$ duplications are required to delete a copy of each label in E, plus $|V|$ duplications that occur along the left line.

Vice versa we show that if in an exemplar species tree for T_G at most $C(k + 1) + |V|$ duplications occur, then G has a node cover of size C. Indeed, as shown above, since by lemma 4, by deleting one or more label in E in a subtree T_j associated to the vertex v_j at least k duplications occur, if we assume that $k > C$ (this fact holds since $k \geq |V|$), it follows that labels can be deleted in at most C subtrees T_j of T_G in order to obtain the exemplar tree S such that $d(T_G, S) \leq C(k + 1) + |V|$. This fact implies that all E copies occur in at most C of such subtrees, i.e. C vertices are incident to all edges E of G, thus proving that C is the size of a node cover.

5 Mapping-Driven Reconciliation

In this section we propose a new approach to the reconciliation of a gene tree to a species tree based on the notion of a general mapping as a measure of duplication cost. Biological reasons for adopting a mapping which generalizes the lca have been proposed in [10], albeit no formal characterization of such mappings has been given.

In this direction our contribution is the definition of two new problems. The first problem generalizes the MDP, where we suggest that the mapping used to count duplications might not be restricted to the lca mapping , but in order to have biological relevance it must preserve cluster inclusion, as defined in the following definition:

Definition 4 (Inclusion-preserving mapping). *A function δ which maps each node of G to a node of S is called* inclusion-preserving *if for each $x_1, x_2 \in G$, with $L(x_1) \subseteq L(x_2)$, then $\mathcal{C}(x_1) \subseteq \mathcal{C}(\delta(x_1)) \subseteq \mathcal{C}(\delta(x_2))$.*

Now let δ be an inclusion-preserving mapping from G to S. The main biological justification for such definition is that any inclusion-preserving mapping associates to node x of a gene tree a set of species that includes all those in the subtree of G rooted at x. Please notice the lca mapping associates to such x the smallest cluster of the species tree with the desired property.

The duplication distance form G to S induced by the mapping δ, denoted by $d_\delta(G, S)$, is the number of duplications induced by δ, where analogously as lca a duplication occurs at node x if and only if $\delta(x) = \delta(a(x))$ or $\delta(x) = \delta(b(x))$.

We are now able to introduce two problems that are quite similar to the Minimum Duplication Problem.

Problem 3 (Minimum Duplication Mapped Tree Problem (MDMT)). Given a gene tree G, compute a species tree S such that $L(S) = L(G)$ and a mapping δ, from G to S, such that $d_\delta(G, S)$ is minimum.

Problem 4 (Minimum Duplication Mapping Problem (MDM)). Given a gene tree G and a species tree S such that $L(G) \subseteq L(S)$, compute a mapping δ from G to S, such that $d_\delta(G, S)$ is minimum.

Clearly the MDMT problem is more general than both the MDM and the MDP problems, more precisely the MDM is the restriction of MDMT where the species tree is given in the instance, while the MDP is the restriction of MDMT where the mapping is restricted to be the lca mapping. Moreover please notice that any solution of the MDMT and MDP problems is a feasible, but not necessarily optimal, solution of the MDM problem.

In the following we give an efficient algorithm for solving MDM via dynamic programming. The following property is central to our algorithm.

Proposition 5. *Let G and S be respectively a gene tree and a species tree, then there exists an optimal solution δ of MDM such that, for each node g of G, either both children of g are mapped to $\delta(g)$, or none of the children of g are mapped to $\delta(g)$.*

The algorithm mainly consists of filling in a bidimensional table indexed by the nodes of G and the nodes of S. In the following we will denote each cell of such table by $T[g, s]$ representing the cost of an optimal solution over the instance $(G|g, S|s)$. We order the nodes of each tree G and S according to the ancestral relation, more precisely each node x comes after all nodes (different from x) that are in the subtree rooted at x. Then the table $T[g, s]$ is filled respecting the ordering of nodes, that is starting from the leaves and going towards the root s. Clearly if S contains only one leaf the only possible solution is mapping each node g to the leaf of S, consequently, $T[g, s] = 0$ if both g and s are leaves with the same label, while $T[g, s] = \infty$ if g and s are leaves with different labels. The recurrence for $T[g, s]$ when g and s are not both leaves follows:

$$T[g,s] = \min \begin{cases} T[a(g),s]+T[b(g),s]+1 & \text{if } \mathcal{C}(g) \subseteq \mathcal{C}(s) \\ T[a(g),a(s)]+T[b(g),b(s)], \ T[a(g),b(s)]+T[b(g),a(s)] & \text{if } \mathcal{C}(g) \subseteq \mathcal{C}(s) \\ T[a(g),a(s)]+T[b(g),a(s)], \ T[a(g),b(s)]+T[b(g),b(s)] & \text{if } \mathcal{C}(g) \subseteq \mathcal{C}(s) \\ T[g,a(s)], \ T[g,b(s)] & \text{always} \end{cases}$$

The above recurrence, whose correctness follows from Prop. 5, can be implemented easily by filling in the entries according to the ordering of nodes, at the end $T[r(G), r(S)]$ contains the optimum number of duplications. Backtracking can then be used to compute the optimal solution [1].

References

1. Thomas H. Cormen, Charles E. Leiserson, Ronald L. Rivest and Clifford Stein *Introduction to Algorithms, Second Edition.* The MIT Press and McGraw-Hill Book Company, 2001.
2. N. El-Mabrouk, D. Sankoff Duplication, Rearrangement and Reconciliation. In *Comparative Genomics: Empirical and Analytical Approaches to Gene Order Dynamics, Map alignment and the Evolution of Gene Families. Computational Biology Series.* Kluwer Academic Publishers. Vol 1. pages 537-550 (2000), 2000.
3. M. R. Fellows, M. T. Hallett, U. Stege. On the Multiple Gene Duplication Problem. In *Proccedings of International Symposium on Algorithms and Computation 1998 (ISAAC 1998)*, pages 347-356, 1998.
4. M. R. Garey, D. S. Johnson. Computer and Intractability: A Guide to NP-Completeness. W. H. Freeman, 1979.
5. M. Goodman, J. Czelusniak, G.W. Moore, A.E. Romero-Herrera, and G. Matsuda. Fitting the gene lineage into its species lineage, a parsimony strategy illustrated by cladograms constructed from globin sequences. *Systematic Zoology*, (28):132–163, 1979.
6. R. Guigò, I. Muchnik, and T. Smith. Reconstruction of ancient molecular phylogeny. *Mol. Phy. and Evol.*, 6(2):189–213, 1996.
7. M. T. Hallett, and J. Lagergren. New algorithms for the duplication-loss model. In *Proceedings of the Fourth Annual International Conference on Computational Biology 2000 (RECOMB 2000)*, pages 138–146, 2000.
8. B. Ma, M. Li, and L. Zhang. From gene trees to species trees. *SIAM Journal on Computing*, 30(3):729–752, 2000.
9. R.D. M. Page. Maps between trees and cladistic analysis of historical associations among genes. *Systematic Biology*, 43:58–77, 1994.
10. R.D. M. Page and J. Cotton. Vertebrate phylogenomics: reconciled trees and gene duplications. In *Proceedings of Pacific Symposium on Biocomputing 2002 (PSB2002)*, pages 536–547, 2002.
11. U. Stege. Gene Trees and Species Trees: The Gene-Duplication Problem is Fixed-Parameter Tractable. In *Proceedings of Workshop on Algorithms And Data Structures 1999 (WADS'99)*, pages 288-293, 1999

Generating All Forest Extensions
of a Partially Ordered Set

Jayme L. Szwarcfiter*

Universidade Federal do Rio de Janeiro, Instituto de Matemática, NCE and COPPE,
Caixa Postal 2324
20001-970 Rio de Janeiro, RJ, Brasil
`jayme@nce.ufrj.br`

Abstract. Let P and P' be partially ordered sets, with ground set E, $|E| = n$, and relation sets R and R', respectively. Say that P' is an *extension* of P when $R \subseteq R'$. A partially ordered set is a *forest* when the set of ancestors of any given element forms a chain. We describe an algorithm for generating the complete set of forest extensions of an order P. The algorithm requires $O(n^2)$ time between the generation of two consecutive forests. The initialization of the algorithm requires $O(n|R|)$ time.

Keywords: algorithms, forests, forests extensions, partially ordered sets

1   Introduction

The study of extensions form an important topic in the theory of ordered sets. On the algorithmic point of view, a natural question is to find suitable algorithms for generating extensions. Algorithms finding all the linear extensions of an order have been described in [1], [5], [6], [10], [11], [12], [13]. It should be noted that [11] generates each linear extension in $O(1)$ time. In addition, [9] contains a method for finding all minimal interval extensions. Finally, [2] describes algorithms for generating all extensions of any kind.

In this article, we propose methods for generating all forests extensions of an order. Enumerative problems on forests orders have been considered before, as in [8]. In addition, [7] describes algorithms for generating all ideals in forests orders. Properties of these orders have been decribed in [3], [4]. A *partially ordered set* (or simply an *order*) P is a set of elements E, together with a reflexive, anti-symmetric and transitive binary relation R on E. Call E the *ground set* of P and R its *relation set*.

---

* Partially supported by the Conselho Nacional de Desenvolvimento Científico e Tecnológico, CNPq, and Fundação de Amparo à Pesquisa do Estado do Rio de Janeiro, FAPERJ, Brasil.

R. Petreschi et al. (Eds.): CIAC 2003, LNCS 2653, pp. 132–139, 2003.

Let $E = \{e_1, \ldots, e_n\}$. Denote $E_P^-(e_i) = \{e_j | e_j e_i \in R, i \neq j\}$ and $E_P^+(e_i) = \{e_j | e_i e_j \in R, i \neq j\}$. Also, $E_P^-[e_i] = E_P^-(e_i) \cup \{e_i\}$, $E_P^+[e_i] = E_P^+(e_i) \cup \{e_i\}$ and $E_P[e_i] = E_P^-[e_i] \cup E_P^+[e_i]$.

For $E' \subseteq E$, denote by $P(E')$ the suborder of P restricted to E'. For $e_i \in E$, write $P - e_i$ with the meaning of $P(E \setminus \{e_i\})$. If $E_P^+(e_i) = \emptyset$, say that e_i is a *maximal* element. When $E' \subseteq E$ is such that $P(E')$ has a unique maximal element, represent this element by $m_P(E')$.

Let $e_i, e_j \in E$ and $E', E'' \subseteq E$. Say that e_i and e_j are *comparable in P* when $e_i, e_j \in R$ or $e_j e_i \in R$; e_i and E' are *comparable in P* when e_i and e_k are so, for all $e_k \in EE'$; finally, E' and E'' are *comparable in P* when e_i and E'' are comparable, for all $e_i \in E'$.

Examine some special orders. Let P be an order. Say that P is a *chain* when every pair of its elements are comparable. In the case that we require only $E_P^-(e_i)$ to be a chain, for every $e_i \in E$, then P is a *forest*. A forest in which $E_P^-[e_i] \cap E_P^-[e_j] \neq \emptyset$, for all $e_i, e_j \in E$ is called a *tree*. Clearly, a forest is a union of disjoint trees.

Let F be a forest, E its ground set and $e_i \in E$. If e_i is a maximal element then call it a *leaf*, otherwise an internal element. If $E_F^-(e_i) = \emptyset$, call e_i a *root* of F. Denote by $RT(F)$ the set of roots of F and let $t(F) = |RT(F)|$. If T is a tree of F, denote by $r_F(T)$ the (unique) root of T. If $E' \subseteq E$ is entirely contained in T then there exists a unique maximal element of F comparable to E', and denoted by $m_F'(E')$. When $E_F^-(e_i) \neq \emptyset$, the element $m_F(E_F^-(e_i))$ is the *parent* of e_i, while e_i is a *child* of $m_F(E_F^-(e_i))$. Denote by $C_F(e_i)$ the set of children of $e_i \in E$. Finally, if $e_j \in E_F^-(e_i)$ then e_j is an *ancestor* of e_i, while e_i is a *descendant* of e_j.

Let P and P' be orders with ground set E and relation sets R and R', respectively. When $R \subseteq R'$ then P' is an *extension* of P. An *extension* which is a chain is called a *linear extension*.

We describe an algorithm for generating all forests extensions of an order P. The algorithm requires $O(n^2)$ time between the generation of two consecutive forests extensions. Before the generation of the first one, there is a preprocessing of $O(n|R|)$ time. The algorithm is based on a convenient decomposition of the set of all forests extensions of P.

2   The Decomposition

In this section, we prove a suitable decomposition for the set of all forests extensions of an order. The following definitions are useful.

Let P be an order, E its ground set and $e_i, e_j \in E$. Say that e_i and e_j are *P-restricted* when e_i and e_j are incomparable in P and $E_P^+(e_i) \cap E_P^+(e_j) \neq \emptyset$. In this case, write that e_i is *P-restricted* to e_j and conversely. Let P' be an extension of P. Call P' *compatible* when every pair of P-restricted elements is comparable in P'.

Let $e_1, \ldots, e_n$ be a fixed linear extension of P. Denote $E_i = \{e_1, \ldots e_i\}$ and $P_i = P(E_i)$. Let $\mathcal{F}$ be the set of forests extensions of P. Denote by $\mathcal{F}_i$ the set of compatible forests extenions of P_i. The following remark imply that we can restrict ourselves to compatible extensions.

Remark 1. : $\mathcal{F}_n = \mathcal{F}$

Proof. By definition, $\mathcal{F}_n \subseteq \mathcal{F}$. To show that $\mathcal{F} \subseteq \mathcal{F}_n$, let $F \in \mathcal{F}$. If F is not compatible then there are incomparable elements $e_i, e_j \in F$, satisfying $E_P^+(e_i) \cap E_P^+(e_j) \neq \emptyset$. However, the latter contradicts F to be a forest. Consequently, $F \in \mathcal{F}_n$, meaning that $\mathcal{F} = \mathcal{F}_n$. $\square$

Observe that there might be forest extensions of P_i which are not compatible, for $i < n$.

Let $F \in \mathcal{F}_{i-1}$ and $F' \in \mathcal{F}_i$. Say that F' is an *expansion* of F when $F = F' - e_i$. Denote by $\mathcal{F}_i(F)$ the set of expansions of F. The following remark describes a partition of the set of compatible forest extensions $\mathcal{F}_i$ of P_i.

Remark 2. : $\mathcal{F}_i = \dot{\bigcup}_{F \in \mathcal{F}_{i-1}} \mathcal{F}_i(F)$

Proof. Let $F' \in \mathcal{F}_i$ and denote $F = F' - e_i$. Because F' is compatible, so must be F. Consequently, $F \in \mathcal{F}_{i-1}$, implying that $F' \in \mathcal{F}_i(F)$. The converse follows by definition. $\square$

The following are useful properties of forests extensions.

Remark 3. : Let $F \in \mathcal{F}_{i-1}$. Then $E_P^-(e_i) = \emptyset$ or $F(E_P^-(e_i))$ is a chain.

Proof. Otherwise, assume that $E_P^-(e_i) \neq \emptyset$ and $F(E_P^-(e_i))$ is not a chain. Then there are elements $e_k, e_l \in E_P^-(e_i)$ which are incomparable in F. Consequently, e_k, e_l are P-restricted. That is, F is not compatible, which contradicts $F \in \mathcal{F}_{i-1}$. $\square$

Let $F \in \mathcal{F}_{i-1}$. Denote by $X \subseteq E_{i-1}$ the set formed by all elements of E_{i-1} which are P-restricted to e_i.

Remark 4. : Let $F \in \mathcal{F}_{i-1}$ and $F' \in \mathcal{F}_i$ an expansion of F. Then e_i and X are comparable in F'.

Proof. Otherwise, suppose that e_i and X are not comparable in F'. Then there exists $e_k \in X$, such that e_i, e_k are incomparable. Because $e_k \in X$, we know that $E_P^+(e_i) \cap E_P^+(e_k) \neq \emptyset$. These two conditions together imply that F can not be a compatible forest extension of P_{i-1}, a contradiction. $\square$

Remark 5. : Let $F \in \mathcal{F}_{i-1}$. If $E_P^-(e_i) \neq \emptyset$ then X is entirely contained in the same tree of F as $E_P^-(e_i)$.

Proof. From Remark 3, we know that $E_P^-(e_i)$ is contained in a single tree T of F. Suppose the assertion to be false. Then there exists an element $e_k \in X$ belonging in F to a tree $T' \neq T$. Since $e_k \in X$, $E_P^+(e_i) \cap E_P^+(e_k) \neq \emptyset$. Let $e_l \in E_P^-(e_i)$. By transitivity, $E_P^+(e_l) \cap E_P^+(e_k) \neq \emptyset$. However, e_l and e_k are incomparable in F. That is, F is not compatible, a contradiction. $\square$

In the sequel, we describe methods for finding the set of expansions of a given compatible forest $F \in \mathcal{F}_{i-1}$. There are two types of constructions:

ROOT EXPANSION
Let $A \subseteq RT(F)$. Denote by $F(*, A)$ the order obtained from F by adding to it the element e_i, making e_i the parent of each of the roots which form A, and adding the pairs $e_i e_k$ implied by transitivity. That is, letting $F' = F(*, A)$,

$$\begin{cases} C_{F'}(e_i) = A \\ C_{F'}(e_k) = C_F(e_k), k \neq i \end{cases}$$

INTERNAL EXPANSION
Let $e_j \in E_{i-1}$ and $A \subseteq C_F(e_j)$. Denote by $F(e_j, A)$ the order obtained from F by including in it e_i as a child of e_j, turning e_i the new parent of each of the elements of A and adding the pairs $e_i e_k$ and $e_k e_i$ implied by transitivity. That is, letting $F' = F(e_j, A)$,

$$\begin{cases} C_{F'}(e_i) = A \\ C_{F'}(e_j) = \{e_i\} \cup C_F(e_j) \setminus A \\ C_{F'}(e_k) = C_F(e_k), k \neq i, j \end{cases}$$

We associate the operators α and β to the expansions above defined, as follows. Let $F \in \mathcal{F}_{i-1}$ and $J \subseteq E_{i-1}$. Let $F_X \subseteq F$ be the subforest of F formed exactly by those trees of F containing at least one element of X. For the root expansion, let

$$\alpha(F) = \cup_{RT(F_X) \subseteq A \subseteq RT(F)} F(*, A)$$

and for the internal expansion, denote

$$\beta(F, J) = \cup_{\substack{e_j \in J \\ C_F(e_j) \cap E_F^-[X] \subseteq A \subseteq C_F(e_j)}} F(e_j, A)$$

The two following theorems lead to a description of the set $\mathcal{F}_i$ of compatible forests extensions of P_i. Denote $B = \cap_{e_k \in X} E_F^-[e_k]$.

Theorem 1. Let $F \in \mathcal{F}_{i-1}$. If $E_P^-(e_i) = \emptyset$ then

$$\mathcal{F}_i(F) = \alpha(F) \cup \begin{cases} \emptyset, & \text{if } t(F_X) > 1 \\ \beta(F, E_{i-1}), & \text{if } t(F_X) = 0 \\ \beta(F, E_F[m_F(X)]), & \text{if } F_X \text{ is a non-empty chain} \\ \beta(F, E_F^-[m_F'(B)]) & \text{otherwise} \end{cases}$$

Denote $D = E_F^+[m_F'(E_P^-(e_i))]$.

Theorem 2. Let $F \in \mathcal{F}_{i-1}$. If $E_P^-(e_i) \neq \emptyset$ then

$$\mathcal{F}_i(F) = \begin{cases} \beta(F, D), & \text{if } t(F_X) = 0 \\ \beta(F, D \cap E_F[m_F(X)]), & \text{if } F_X \text{ is a non-empty chain} \\ \beta(F, D \cap E_F^-[m_F'(B)]), & \text{otherwise} \end{cases}$$

Proof of theorem 1:

Let $F \in \mathcal{F}_{i-1}$, $E_P^-(e_i) = \emptyset$ and $X \subseteq E_{i-1}$ be the set of P-restricted elements to e_i. Denote by F_X the forest formed by those trees of F which contain some element of X. Consider the following cases:

Case 1: $t(F_X) > 1$
We have to prove that $\mathcal{F}_i(F) = \alpha(F)$. Let $F' \in \mathcal{F}_i(F)$. Then F' is a compatible forest extension of P_i. By Remark 4, e_i and X are comparable in F'. Because $t(F_X) > 1$, the only alternative for e_i is to be the root of a tree of F' containing all the elements of X. Because $E_P^-(e_i) = \emptyset$, this alternative is valid. Examine the set A of children of e_i in F'. Because $F' = F - e_i$, the roots of the trees of F_X become children of e_i in F'. That is, $RT(F_X) \subseteq A$. On the other hand, $A \subseteq RT(F)$, otherwise $F' \neq F - e_i$. Consequently, F' is precisely the expansion $F(*, A)$ of F. The latter implies $F' \in \alpha(F)$ and $\mathcal{F}_i(F) \subseteq \alpha(F)$.

To show that $\mathcal{F}_i \supseteq \alpha(F)$, let $F' \in \alpha(F)$. Then F' is an expansion of F. Because $F \in \mathcal{F}_{i-1}$ and $E_P^-(e_i) = \emptyset$ it follows that F' is a forest extension of P_i. It remains to show that it is compatible. Suppose that F' is not compatible. Then there exists a pair of elements $e_k, e_l \in E_i$ which are P-restricted and incomparable in F'.

First, suppose that e_i coincides with one of e_k, e_l, say $e_i = e_k$. Then $e_l \in X$. Because $F' \in \alpha(F)$, $F' = F(*, A)$, for some A satisfying $RT(F_X) \subseteq A \subseteq RT(F)$. By the construction of $F(*, A)$, we know that $C_{F'}(e_i) = A$. Because $RT(F_X) \subseteq A$, it follows that all P-restricted elements of X belong to the tree of F' whose root is e_i. Consequently, e_i, e_l are comparable, a contradiction. Next, examine the alternative where $e_k, e_l \neq e_i$. We are forced to conclude that e_k, e_l are again comparable in F'. Otherwise, e_k and e_l would not be comparable also in F, contradictiong F to be compatible.

Therefore F' is always comparable. That is, $F' \in \mathcal{F}_i(F)$, implying $\mathcal{F}_i = \alpha(F)$.

Case 2: $t(F_X) = 0$.
We need to prove that $\mathcal{F}_i(F) = \alpha(F) \cup \beta(F, E_{i-1})$. Let $F' \in \mathcal{F}_i(F)$. We know that F' is a compatible forest extension of P_i. Locate E_i in F'. Let $A = C_{F'}(e_i)$. Because $F = F' - e_i$, it follows that $A \subseteq RT(F)$. On the other hand, $t(F_X) = 0$, i.e $X = \emptyset$. Suppose that e_i is a root of F'. These conditions imply that $F' = F(*, A)$, that is $F' \in \alpha(F)$. In the alternative situation, consider e_i as an internal element of F'. Let $e_j \in E_{i-1}$ be the parent of e_i. Consequently, $F' = F(e_j, A)$, i.e. $F' \in \beta(F, E_{i-1})$. Therefore $F' \subseteq \alpha(F) \cup \beta(F, E_{i-1})$

Conversely, let $F' \in \alpha(F) \cup \beta(F, E_{i-1})$. By construction, F' is a forest extension of P_i. Because $t(F_X) = 0$, F' contains no elements P-restricted to e_i. On the other hand, because F is compatible, every pair of P-restricted elements $e_k, e_l \neq e_i$ must be comparable. Therefore F' is also compatible. That is, $F' \in \mathcal{F}_i(F)$ and $\mathcal{F}_i(F) = \alpha(F) \cup \beta(F, E_{i-1})$.

Case 3: F_X is a nonempty chain
From Remark 4, it follows that F_X consists of a single tree. We have to show that $\mathcal{F}_i(F) = \alpha(F) \cup \beta(F, E_F[m_F(X)])$. Let $F' \in \mathcal{F}_i(F)$ and denote $C_{F'}(e_i) = A$. Suppose that e_i is a root of F'. Because F' is compatible, it follows that $r(F_X) \in A$. Consequently, $F' \in \alpha(F)$. Alternatively, consider the situation where e_i is an internal element of F', and let e_j denote its parent. By Remark 4, e_i and X are comparable. Since F_X is a chain, we conclude that $e_i \in E[m_{F'}(X)]$. That is, $e_j \in E[m_F(X)]$. Denote by A the set of children of e_i in F'. Because $F = F' - e_i$, it follows that $A \subseteq C_F(e_j)$ and that $C_{F'}(e_j) = C_F(e_j) \setminus A$. Also, since F' is compatible, it follows that $C_F(e_j) \cap E_F^-[X] \subseteq A$. Consequently, $F' = F(e_j, A)$, implying that $F' \in \beta(F, E_F[m_F(X)])$ and $\mathcal{F}_i(F) \subseteq \alpha(F) \cup \beta(F, E_F[m_F(X)])$.

Conversely, let $F' \in \alpha(F) \cup \beta(F, E_F[m_F(X)])$. We prove that $F' \in \mathcal{F}_i(F)$. We know that F' is a forest expansion of P_i and we need only to show that it is also compatible.

By definition, F' is compatible whenever it contains no P-retricted elements. Otherwise, suppose that there exists $e_k, e_l \in E_i$ which are P-restricted. Clearly, e_i must coincide with e_k or e_l, otherwise F would not be compatible. Let $e_i = e_k$. It follows that $e_l \in X$. If $F' = \alpha(F)$ then e_i is the root of the tree containing e_l in F'. Consequently, e_i, e_l are not P-restricted. When $F' \in \beta(F(E_F[m_F(X)])$, the construction of $F(e_j, A)$ assures that e_i and e_l belong to a same chain in F'. Consequently, e_i and e_j can not be P-restricted. Therefore, F' is always compatible, meaning that $F' \in \mathcal{F}_i(F)$.

Case 4: None of the above
In the present case, $X \neq \emptyset$ and F_X is not a chain of F. Let $F' \in \mathcal{F}_i(F)$. We have to show that $F' \in \alpha(F) \cup \beta(F, B)$. By Remark 4, we know that e_i and X are comparable in F'. Consequently, e_i and X belong to a same tree of F', which is precisely F_X. The latter implies $e_i \in E_{F'}^-[m_{F'}'(B)]$. If e_i is a root of

F', again $r(F_X) \in A$, meaning that $F' \in \alpha(F)$. Alternatively, let e_i be an internal element of F' and e_j the parent of e_i. Since $e_i \in m'_{F'}(B)$, it follows that $e_j \in m'_F(B)$. By denoting A as the set of children of e_i in F' and using a similar argument as in the previous case, we conclude that $F' \in \beta(F, B)$. Consequently, $\mathcal{F}_i(F) \subseteq \alpha(F) \cup \beta(F, B)$

Conversely, let $F' \in \alpha(F) \cup \beta(F, B)$. To prove that $F' \in \mathcal{F}_i$. it suffices again to show that $\mathcal{F}_i$ is compatible. Similarly, as in Case 3, because $F' = F - e_i$ it follows that F' can not contain restricted elements. Consequently, F' is compatible and $F' \in \mathcal{F}_i(F)$. $\square$

The proof of Theorem 2 is similar, making use of Remark 5.

3   The Algorithm

Let P be an order with ground set E, $|E| = n$, and relation set R. In this section, we describe an algorithm for generating all forests extensions of P.

The algorithm is an application of Theorems 1 and 2 and can be formulated as follows:

In the initial step, compute a linear extension $e_1, \ldots, e_i$ of P and denote $E_i = \{e_1, \ldots, e_i\}$. Then find the set X of P-restricted elements to e_i, in E_{i-1}, for all i.

The general step is a recursive computation to generate all compatible forests extensions of P_i. By Remarks 1 and 2, all forests extensions of P_i are then generated.

In the recursive tree T, each node at level i corresponds to a compatible forest extension of P_i. For the computation relative to $F \in \mathcal{F}_{i-1}$, $1 < i \leq n$, apply Theorem 1 or 2, according whether $E_P^-(e_i) = \emptyset$ or not, respectively. In any situation, identify the forest F_X (recall that X has already been evaluated). According to the value of F_X, compute $m_F(X)$, $E_F[m_F(X)]$, $B = \cap_{e_k \in X} E_F^-[e_k]$, $m'_F(B)$ and $E_F^-[m'_F(B)]$. When $E_P^-(e_i) \neq \emptyset$, also find $D = E_F^+[m'_F(E_P^-(e_i))]$. Then compute the corresponding set of expansions α and β. Each of these expansions is associated to a child of the node z which corresponds to the forest extension F of P_{i-1} in T. We know that z has at least one child. The leaves of T are in correspondence with the forests extensions of P.

All the computations at each internal node of the recursion tree can be performed in $O(n)$ time, using standart techniques. The computations at the leaves require constant time. Since the leaves are at height n, we conclude thatr the elapsed time between the generation of two consecutive forest extensions is $O(n^2)$. The initialization process requires $O(n|R|)$ time.

References

1. E. R. Canfield and S. G. Williamson. A loop free algorithm for generating the linear extensions of a poset. *ORDER*, 12:1–18, 1995.
2. R. C. Correa and J. L. Szwarcfiter. On extensions, linear extensions, upsets and downsets of ordered sets. Submitted, 2000.
3. R. E. Jamison. On the average number of nodes in a subtree of a tree. *Journal of Combinatorial Theory B*, 35:207–223, 1983.
4. R. E. Jamison. Monotonicity of the mean order of subtrees. *Journal of Combinatorial Theory B*, 37:70–78, 1984.
5. A. D. Kalvin and Y. L. Varol. On the generation of topological sortings. *Journal of Algorithms*, 4:150–162, 1983.
6. D. E. Knuth and J. L. Szwarcfiter. A structured program to generate all topological sorting arrangements. *Information Processing Letters*, 2:153–157, 1974.
7. Y. Koda and F. Ruskey. A gray code for the ideals of a forest poset. *Journal of Algorithms*, 15:324–340, 1993.
8. A. Meir and J. W. Moon. On subtrees of certain families of trees. *Ars Combinatoria*, 16-B:305–318, 1983.
9. M. Morvan and L. Nourine. Generating minimal interval extensions. Technical report, Laboratoire d'Informatique Robotique et Microelectronique, Universite de Montpellier II, 1992. Rapport de Recherche 92-05.
10. G. Pruesse and F. Ruskey. Generating the linear extensions of certain posets by transpositions. *SIAM Journal on Discrete Mathematics*, 4:413–422, 1991.
11. G. Pruesse and F. Ruskey. Generating linear extensions fast. *SIAM Journal on Computing*, 23:373–386, 1994.
12. F. Ruskey. Generating linear extensions of posets. *Journal of Combinatorial Theory B*, 54:77–101, 1992.
13. Y. L. Varol and D. Rotem. An algorithm to generate al topological sorting arrangements. *Computer Journal*, 24:83–84, 1981.

Indexing Structures
for Approximate String Matching*

Alessandra Gabriele**, Filippo Mignosi,
Antonio Restivo, and Marinella Sciortino

Dipartimento di Matematica ed Applicazioni,
Università degli Studi di Palermo.
Via Archirafi 34 - 90123 Palermo - Italy
{sandra,mignosi,restivo,mari}@math.unipa.it

Abstract. In this paper we give the first, to our knowledge, structures and corresponding algorithms for approximate indexing, by considering the *Hamming distance*, having the following properties.

i) Their size is linear times a polylog of the size of the text on average.

ii) For each pattern x, the time spent by our algorithms for finding the list $occ(x)$ of all occurrences of a pattern x in the text, up to a certain distance, is proportional on average to $|x| + |occ(x)|$, under an additional but realistic hypothesis.

Keywords: Combinatorics on words, automata theory, suffix trees, DAWGs, approximate string matching, indexing.

1 Introduction

Approximate string matching concerns to find patterns in texts in presence of "mismatches" or "errors". It has several applications in data analysis and data retrieval such as, for instance, searching text under the presence of typing or spelling errors, retrieving musical passages, or finding biological sequences in presence of possibly mutations or misreads.

An index over a fixed text S is an abstract data type based on the set of all factors of S, denoted by $Fact(S)$. Such data type is equipped with some operations that allow it to answer the following query (cf. [4]): given $x \in Fact(S)$, find the list of all occurrences of x in S.

In the case of exact string matching, there exist classical data structures for indexing, such as suffix trees, suffix arrays, DAWGs, factor automata or their compacted versions (cf. [4]). The algorithms that use them run in a time that is usually independent from the size of the text or is at least substantially smaller than it. The last property is required by some authors to be essential part in the definition of an index (cf. [1]).

---

* Supported by MIUR National Project PRIN "Linguaggi Formali e Automi: teoria ed applicazioni"

** Supported by *Progetto Giovani Ricercatori anno 1999 - Comitato 01*

R. Petreschi et al. (Eds.): CIAC 2003, LNCS 2653, pp. 140–151, 2003.

In the case of approximate string matching the situation is not the same. We refer to [6,7,9,10] and to the reference therein for a *panorama* on this subject and on approximate string matching in general.

Typical approaches in this field consist on considering a percentage D of errors or fixing the number k of them. We use an hybrid approach, i.e. we allow k errors *in every window* of r symbols, where r is not necessarily constant. In the case when r is equal to the length $|x|$ of the pattern x the percentage of errors coincides with $D = \frac{k}{r}$. In the case when r is greater than or equal to the size of the pattern, the problem reduces to the case of k mismatches in x.

In our new formalism the statement of the problem is the following. Given a *text S*, a *pattern x* and two integers k and r, return all the text positions l such that x occurs in S at position l up to k errors for r symbols. In [5] we propose a first natural approach, suggested by indexing methods for exact string matching, that consists on building an automaton recognizing the language $L(S, k, r)$ of all words that occur in S up to k errors in a window of size r. NFA's arising in similar context present an exponential size after determinization and minimization. Different bounds from the classical exponential ones have been obtained by using a new parameter R called *Repetition Index* (see Section 3). In [5] we prove that the size of the automaton recognizing $L(S, k, r)$ is a function of $|S|$, R and the number of errors k' in a window of size R ($k' \sim k\frac{R}{r}$ and $r \leq R$). More precisely, the size is $O(|S| \times R^{k'})$. In the worst case, when both R and k' are $O(|S|)$, the size of the automaton is exponential again. In the average case, when $R = O(\log(|S|))$, if k' is constant, the size of the automaton is $O(|S| \times \text{polylog}(|S|))$.

The main technical contribution of our paper is algorithmic in nature and is presented in Section 4. The algorithms we propose find the list of all occurrences of a pattern x in a text S, up to k errors in every windows of r symbols, where r can be fixed or a function of the size of the text. We describe the algorithm for building the indexing data structure in two steps. The first one concerns the *"long patterns"* case and uses the automaton recognizing the language $L(S, k, r)$. The second one concerns the *"short patterns"* case and it includes a non trivial reduction to the *Document Listing Problem*, an algorithm for finding the Repetition Index and also standard filters for approximate string matching. In Section 5 we analyze the time and space required in order to implementing the indexing data structures. They are both $O(|S| \times \text{polylog}(|S|))$ on average. We also analyze the average searching time in this data structures, that turns out to be linear under an hypothesis on the distribution of $R(S, k, r)$. More precisely, we require that there exists $\beta > 1$ such that, if $\hat{R}$ is the expected value of $R(S, k, r)$ for a text S of length n, then the probability that $R(S, k, r) > \beta \hat{R}$ goes to zero faster than $\frac{1}{n}$. Further details are given in [5].

2   Basic Definitions

Let Σ be a finite set of symbols, usually called the *alphabet*. A *word* or *string* w is a finite sequence $w = a_1 a_2 \ldots a_n$ of characters taken in the alphabet Σ; its length (i.e. the number of characters in the string) is defined to be n, and is denoted by $|w|$. The set of words built on Σ is denoted by Σ^* and the empty words by ϵ. We use the notation $\|I\|$ to denote the sum of the lengths of all strings in a collection of strings I.

A word $u \in \Sigma^*$ is a *factor* of a word w if and only if there exist words $x, y \in \Sigma^*$ such that $w = xuy$. We denote by $Fact(w)$ the set of all the factors of a word w. We denote an occurrence of a factor of a string $w = a_1 a_2 \ldots a_n$ at the position i by $w(i,j) = a_i \ldots a_j$, $1 \le i \le j \le n$. The length of a substring $w(i,j)$ is the numbers of letters that compose it, i.e. $j - i + 1$. A language L is *factorial* if it contains all factors of its words. A factor u of w is a *prefix* (respectively a *suffix*) of w if $w = ux$ (respectively $w = xu$) with $x \in \Sigma^*$.

In this paper we consider the *Hamming distance* that is one of the most commonly used distance functions. The Hamming distance between two words x and y, denoted by $d(x,y)$, is the minimal number of character substitutions to transform x into y. The results and the methods showed in this paper are easily generalizable to some other distances, such as the *Weighted Hamming distance*, but not, for instance, to the *Weighted Edit distance*.

The new idea in our approach is to introduce a new parameter r and to allow at most k errors for any substring of length r of the text. We introduce the following definition.

Definition 1. *Let S be a string over the alphabet Σ, and let k, r be non negative integers such that $k \le r$. A string v occurs in S at the position l, up to k errors in a window of size r (or, simply, up to k errors for r symbols) if:*

- *in the case $|v| < r$ one has: $d(v, S(l, l + |v| - 1)) \le k$;*
- *in the case $|v| \ge r$ one has: for any i, $1 \le i \le |v| - r + 1$,*
  $d(v(i, i + r - 1), S(l + i, l + i + r - 1)) \le k$.

We denote by $L(S, k, r)$ the set of words v that satisfy the previous definition for some l, $1 \le l \le |S| - |v| + 1$. Notice that $L(S, k, r)$ is a *factorial language*, i.e. if $w \in L(S, k, r)$ then each factor (substring) of w belongs to $L(S, k, r)$. Clearly if u is a factor of S then $L(u, k, r) \subseteq L(S, k, r)$.

Definition 2. *Let u be a string over the alphabet Σ, the* neighborhood *of u (with respect to k, r) is the set*

$$V(u, k, r) = L(u, k, r) \cap \Sigma^{|u|}.$$

If u has length m, then it is easy to see that the maximal distance between a word $v \in V(u, k, r)$ and u is $k \lceil \frac{m}{r} \rceil$. Therefore, a simple combinatorial argument left to the reader shows that the number of elements in the set $V(u, k, r)$ is at most

$$\sum_{i=0}^{k \lceil m/r \rceil} \binom{m}{i} (\Sigma - 1)^i.$$

It can be proved that $L(S, k, r) = Fact(V(S, k, r))$.

Given a set I of factors of S, we denote by $I_{k,r}$ the union of the neighborhoods of the elements of I, i.e. $I_{k,r} = \bigcup_{u \in I} V(u, k, r)$.

Given a symbol $ that does not belong to Σ, we consider the sequence $S_{I,k,r}$ over the alphabet $\Sigma \cup \{\$\}$ obtained as a concatenation of all words in $I_{k,r}$ interspersed with the symbol $ (i.e. between two consecutive words there is one dollar).

Remark 1. Let $k' = \max\{d(u,v)|u \in Fact^m(S) \text{ and } v \in V(u,k,r)\}$, i.e. k' is the maximal number of mismatches between u and any word in $V(u,k,r)$. k' is obviously upper bounded by $k\lceil \frac{m}{r} \rceil$. Therefore the number of elements in the set $V(u,k,r)$, with $|u| = m$, is $\sum_{i=0}^{k'} \binom{m}{i} (|\Sigma| - 1)^i \leq \sum_{i=0}^{k\lceil \frac{m}{r} \rceil} \binom{m}{i} (|\Sigma| - 1)^i$. A classical estimate of the tail of the binomial distribution give us that $\sum_{i=0}^{m\alpha} \binom{m}{i}$ $\leq 2^{H(\alpha)m}$, for any $\alpha < \frac{1}{2}$ (cf. [3, Lemma 4.7.2]), where $H(\alpha) = -\alpha \log_2 \alpha - (1 - \alpha)\log_2(1-\alpha)$ is the entropy function.

If m is of the order of $\log_2(|S|)$ then

$$|S_{I,k,r}| \leq c\log_2(|S|) \times |S|^{1+c(H(\alpha)+\gamma\alpha)}, \tag{1}$$

where $\alpha = \frac{k\lceil \frac{m}{r} \rceil}{m}$ and $\gamma = \log_2(|\Sigma| - 1)$ (i.e.$|\Sigma| - 1 = 2^\gamma$).

If r is a function of $|S|$ then it is better to give a different estimate, based directly on the number k'.

If $k' < \frac{1}{4}m$ then $|S_{I,k,r}| \leq m \times |S| \times 2 \binom{m}{k'} (|\Sigma| - 1)^{k'} = O(|S| \times \frac{m^{k'}}{k'!} \times$ $(|\Sigma| - 1)^{k'})$. If k' is constant then

$$|S_{I,k,r}| = O(|S| \times m^{k'}). \tag{2}$$

3   Repetition Index

The basic idea of the new approach is to build an index on the text S by using only a collection I of factors (substrings) of S that satisfies a certain condition. In this way, we obtain new space and time bounds that are different from the classical exponential bounds by using a new parameter, that we call repetition index.

Definition 3. *The* Repetition Index, *denoted by* $R(S,k,r)$, *of* S *is the smallest value of* h *such that all strings of length* h *occur at most in a unique position, up to* k *errors for* r *symbols:*

$$R(S,k,r) = \min\{h \geq 1 \mid \forall \ i, j, \ 1 \leq i, j \leq |S| - h + 1,$$
$$V(S(i,i+h-1),k,r) \cap V(S(j,j+h-1),k,r) \neq \emptyset \Rightarrow i = j\}.$$

$R(S,k,r)$ is well defined because $h = |S|$ is an element of the set above described. We can prove that if $k/r \geq 1/2$ then $R(S,k,r) = |S|$.

In this section we give an algorithm for finding $R(S,k,r)$ and in Subsection 5.1 we give an evaluation of $R(S,k,r)$, on average.

Before giving the algorithm to determine $R(S, k, r)$, we state the "*Disjoint-ness Problem*": let C be a finite collection of finite languages, i.e. $C = \{X_1, X_2, ..., X_t\}$ such that for any j, $1 \leq j \leq t$, X_j is finite, non empty and composed of words having length R, with R a constant not depending on j. We say that the sets in C are mutually disjoint if for any i, j, $1 \leq i, j \leq t$, if $X_i \cap X_j \neq \emptyset \Rightarrow i = j$. This problem is easily done with an algorithm, called CHECK-DISJOINTNESS, in time linear in $|C|$.

If we consider the collection of sets $C_{h,k,r} = \{V(S(1, 1+h-1), k, r), ..., V(S(|S|-h+1, |S|), k, r)\}$, $1 \leq h \leq |S|$, then $R(S, k, r) = \min\{h \geq 1 \mid C_{h,k,r}$ is a collection of mutually disjoint sets$\}$.

Hence, we solve the disjointness problem on the collection $C_{h,k,r}$ in order to know if h is greater than $R(S, k, r)$ or not.

The following algorithm is a binary search of $R(S, k, r)$ that makes use of algorithm CHECK-DISJOINTNESS and its correctness is straightforward.

Notice that the size r of the window can be fixed, but it can also vary (for instance as a function of n, that is the length of the string S).

FIND-REPETITION-INDEX (S,k,r)
1. $h \leftarrow 1$;
2. BUILD$(C_{h,k,r})$;
3. **while** not CHECK-DISJOINTNESS$(C_{h,k,r})$ **do**
4. $h \leftarrow 2h$;
5.   BUILD$(C_{h,k,r})$;
6. $t \leftarrow \lceil h/2 \rceil$;
7. **while** $(h - t) > 1$ **do**
8.   BUILD$(C_{\lceil \frac{h+t}{2} \rceil, k, r})$;
9.   **if** CHECK-DISJOINTNESS$(C_{\lceil \frac{h+t}{2} \rceil, k, r})$
10.   **then** $h \leftarrow \lceil \frac{h+t}{2} \rceil$
11.   **else** $t \leftarrow \lceil \frac{h+t}{2} \rceil$
12. **return** h.

Procedure BUILD$(C_{h,k,r})$ uses a variation of the routine for the generation of the "*voisins fréquentables*" of the factors of a text having a fixed length ([4, Sec 7.6]), and its running time is $O(\|C_{h,k,r}\|)$.

The number of times that CHECK-DISJOINTNESS$(C_{h,k,r})$ is called in FIND-REPETITION-INDEX is at most twice $\log(R(S, k, r))$ and the greatest h considered in the algorithm is smaller than $2R(S, k, r)$. We can make minor changes to above algorithm so that it has as input just S and k. In this case we replace all occurrences of $C_{h,k,r}$ with $C_{h,k,h}$ and we obtain an algorithm that find $R(S, k, r(n))$ where $r(n)$ is a function that satisfies the equation $r(n) = R(S, k, r(n))$. Since $|C_{h,k,r}| \geq |C_{t,k,r}|$ when $h \geq t$ and $|C_{h,k,r}| = O(|S_{I,k,r}|)$, where I is the set of factors of S of length h, we can deduce the following proposition.

Proposition 1. *The worst case overall time of the algorithm* FIND-REPETI-TION-INDEX *is* $O(\log(R(S, k, r)) \times |C_{2R(S,k,r),k,r}|) = O(\log(R(S, k, r) \times |S_{I,k,r}|))$, *where I is the set of factors of S of length $2R(S, k, r)$.*

4   Indexing

We describe the algorithm for building an indexing data structures in two steps. The first one concerns the case in which the pattern x has length greater than or equal to the repetition index $R(S, k, r)$, $|x| \geq R(S, k, r)$; the second one concerns the case in which the length of the pattern is smaller than $R(S, k, r)$, $|x| < R(S, k, r)$.

In the following we suppose that $r \leq R(S, k, r)$. The case $r > R(S, k, r)$ can be reduced to the case $r(n) = R(S, k, r(n))$. Actually it is possible to construct an index for $r(n) = R(S, k, r(n))$ and then we perform queries on this new data structures. If the length of the pattern is smaller than $R(S, k, r(n))$ then the list of all occurrences of x is exactly the solution of the original problem. Otherwise the list of all occurrences has at most one element, that could be a *false positive* for the original problem and it can be checked in a time proportional to the length of x. This technique settles the case when k is fixed and there is no window size, that is equivalent to choose $r = |S|$.

4.1   Long Patterns Case

We want to find all the occurrences of x in S up to k errors every r letters, when $|x| \geq R(S, k, r)$. Indeed in this case, if x appears, it appears just once. There are several ways to solve this problem. We briefly describe here one of them.

Since $r \leq R(S, k, r)$, by using results from Formal Languages Theory concerning on minimal forbidden factors of words and their relation with the repetition index, we can build the deterministic finite automaton $\mathcal{A}$ recognizing the language $L(S, k, r)$ starting from the factor automaton of the string $S_{I,k,r}$, where I is the set of factors of S of length $R(S, k, r) + 1$ (cf. [5]). Then we add to any state of any deterministic automaton $\mathcal{A}$ recognizing the language $L(S, k, r)$ an integer that represents the length of the shortest path from that state to a state without outgoing edges. This can be easily be done with standard algorithms in time proportional to the size of $\mathcal{A}$. Once $\mathcal{A}$ is equipped in this way, it can be used to solve the problem in the standard way by "reading", with a function that we call FIND-LARGE-PATTERN(x), as long as possible the pattern string x and, if the end of x is reached, the output is the length of S minus the number associated to the arrival state minus the length of the pattern x.

Function FIND-LARGE-PATTERN(x) can be also easily implemented by using any of the other classical data structures for indexing, such as suffix trees or suffix arrays.

4.2   Short Patterns Case

Now we want to find all the occurrences of a pattern x in S up to k errors in every window of size r when $|x| < R(S, k, r)$. Let us denote by $\hat{R}$ a new variable that we initialize to be equal to $R(S, k, r)$.

For a given constant $c \geq 2$, we create a sequence of "documents" $\hat{I}$ that is in a one to one correspondence ψ with a sequence I of factors of S that has the following properties:

1) every factor of S of length $\hat{R} - 1$ is a factor of some element of the collection I, i.e. $C(I) \geq \hat{R} - 1$;

2) for every position i, $1 \leq i \leq |S|$ there are at most c factors of I of the form $S(m, n)$ such that $m \leq i \leq n$;

3) a word v of length $|v| \leq \hat{R} - 1$ appears in document $d_j \in \hat{I}$ if and only if v appears in the corresponding factor $S_{\psi(j)} \in I$ up to k errors every r characters. Moreover, the knowledge of the position of one occurrence i of v in $d_j \in \hat{I}$ allows it to find in constant time (in the constant cost model of computation) a position $\varphi(i, j)$ of one occurrence of v in S up to k errors every r characters.

There are plenty of different collections I and $\hat{I}$ that satisfy above requests. We fix here $c = 2$ and build one of such couples of collections in the following way. Let I be a collection of factors S_j of S, $1 \leq j \leq |I| = t$, where $t = \lceil \frac{|S|}{\hat{R}-2} \rceil - 1$. Each of them has length $2(\hat{R}-2)$, with the exception of v_t that has length $\hat{R}-1 \leq |v_t| \leq 2(\hat{R}-2)$. We define the words $S_1 = S(1, 2(\hat{R}-2))$, $S_2 = S((\hat{R}-2)+1, 3(\hat{R}-2))$, ..., $S_j = S((j-1)(\hat{R}-2)+1, (j+1)(\hat{R}-2))$, ..., $S_t = S((t-1)(\hat{R}-2)+1, |S|)$.

We define each document d_j of $\hat{I}$ as the concatenation of all words in $V(S(i, i+\hat{R}-2), k, r = R(S, k, r))$, interspersed by the symbol $ that is not in the alphabet of S (and that we add also at the end of document itself), where the index i ranges from $(j-1)(\hat{R}-2)+1$ to $j(\hat{R}-2)$. We define, by abuse of notation, the value V to be the length of the concatenation, interspersed by $ (added to the end too), of all words in $V(S(i, i+\hat{R}-2), k, r)$, for any fixed i. The value V is independent from i and $|d_j|$ is equal to $V \times j(\hat{R}-2) - (j-1)(\hat{R}-2)$. Readers can check that these sequences, thought as collections, satisfy above Properties 1), 2) and 3). Concerning the Property 3, if i is the position of x in the document d_j then a position of x is S is $\varphi(i, j) = [(j-1)(\hat{R}-2)+1] + \lfloor \frac{i}{V} \rfloor + [i \mod (\hat{R}-1)]$.

Now we introduce the *Document Listing Problem*. We are giving a collection D of text document $d_1, \ldots, d_k$, with each d_j being a string, $\sum_j |d_j| = n$. In the *document listing* problem, online query comprises a pattern string x of length m and our goal is to return the set of all documents that contain one or more copies of x. More precisely, the query $list(x)$ consists of a pattern x of length m, and the goal is to return the set of all documents in the library in which x is present, that is, the output is $\{j \mid d_j[i \ldots i+m-1] = x \text{ for some } i\}$. In [8] a solution of this problem is given with $O(n)$ time and spaces preprocessing, where n is the size of the collection D, and $O(m + output)$ query processing time. The solution proposed by the authors is able to return also one (the first in lexicographical order) occurrence $i(j)$ of x in every document d_j that contains it as a factor.

A first informal description of how to build indexing data structures and algorithms using them is described in the following.

First of all we compute the repetition index $R_j = R(S_j, k, r)$ for every factors S_j of the collection I that does not differ too much from the expected value given by Proposition 3. This expected value times a constant β greater than one can be used as an upper bound to stop the algorithm FIND-REPETITION-INDEX over the string S_j. In what follows we fix $\beta = 3$ and use $\frac{6 \log(n)}{\mathcal{H}(0,p)}$ as upper bound, but any other greater value could be chosen instead. Hence we use a variation of

FIND-REPETITION-INDEX that has same input S, k and r together with a new variable that we call UPPERBOUND. In this way the worst case time of this new version coincides with the average time of its previous version (cf. Proposition 4). When the repetition index of the string is greater than β times the expected value, then the output of FIND-REPETITION-INDEX is -1.

Suppose that $|x| \geq \frac{6\log(n)}{\mathcal{H}(0,p)}$ (recall that $|x| < \hat{R} = R(S,k,r)$). We can use the solution given in [8] of the *Document listing* problem on the collection $\hat{I}$ of document d_j, $1 \leq j \leq t$, for finding the list *occDocuments* of all documents d_j that contain one or more copies of x and the relative position i of one occurrence of x in d_j. By using the correspondence φ we can obtain one position of the pattern x in S. If $R_j \leq |x| \leq \hat{R} - 1$, i.e. $|x|$ is greater than or equal to the repetition index of the factor S_j, then this position is unique in S_j. Otherwise, i.e. $|x| < R_j$, the algorithm returns all the occurrences of x in S_j by using a standard procedure that works in time $O(|k'| \times |S_j|)$ where k' is defined in Remark 1 (cf. [7]).

Instead, if $|x| < \frac{6\log(n)}{\mathcal{H}(0,p)}$ we iterate the procedure above described by setting $\hat{R} = \max(k+1, \frac{6\log(n)}{\mathcal{H}(0,p)})$ until $\hat{R}$ becomes greater than or equal to the previous value of $\hat{R}$. We keep in a stack T the different values that $\hat{R}$ is assuming, putting in the bottom of the stack the value $|S|$ that is supposed to be greater than $|x|$. Together with $\hat{R}$ we keep all satellite data that we need in the sequel. More precisely, $\hat{R}$ and the satellite data corresponding to $\hat{R}$, are described by a 6-uple implemented with a record having six fields. The first (T.index) is a real number and represents the repetition index $\hat{R}$. The second (T.pointer) is a pointer to the data structure corresponding to $\hat{R}$. The third (T.documents) is a pointer to the array A of documents d_j. The fourth (T.repetitions) is a pointer to the array B that stores in position $B[j]$ the repetition index of document d_j. The last two pointers ($T.\psi$ and $T.\varphi$) represent the functions ψ and φ.

In general, given pattern x, we use the stack for finding the two consecutive values $\hat{R}_g, \hat{R}_{g+1}$ such that $\hat{R}_g > |x| \geq \hat{R}_{g+1}$. The satellite data corresponding to $\hat{R}_{g+1}$ are used for searching the occurrences of x. All above is the gist of next two algorithms CREATE-INDEXING-STRUCTURES and SEARCH-IN-INDEXING-STRUCTURE (T, x), that are described in what follows in the special case $r(n) = R(S, k, r(n))$.

Let us describe some functions used inside the algorithm CREATE-INDEXING-STRUCTURES (S, k). Function CREATE-AUTOMATON creates an automaton $\mathcal{A}$ recognizing language $L(S, r, k)$ equipped with integers, as described at the beginning of this section. The automaton $\mathcal{A}$ can be implemented following the lines described in the proof of Theorem 1. Function CREATE-DOCUMENTS builds up the collection $\hat{I}$ of documents d_j corresponding to the length $\hat{R}$ as described in this section. It returns the array A that lists them. As a side effect CREATE-DOCUMENTS generates the collection I of factors S_j and the function ψ. The information relative to I and to ψ is supposed to be also stored in the array A (for instance by linking at the end of document d_j a description of factor S_j). Function CREATE-DOCUMENT-LISTING-STRUCTURE is described in [8]. It returns a pointer to the structure.

CREATE-INDEXING-STRUCTURES (S, k)
1. $T \leftarrow$ CREATE-STACK;
2. PUSH$((|S|, \text{Null}, \text{Null}, \text{Null}, \text{Null}, \text{Null}), T)$;
3. $\hat{R} \leftarrow$ FIND-REPETITION-INDEX(S, k);
4. $P \leftarrow$ CREATE-AUTOMATON$(S, k, \hat{R})$;
5. $A = B = \psi = \varphi = \text{Null}$;
6. **repeat**
7.   PUSH$((\hat{R}, P, A, B, \psi, \varphi), T)$;
8. $A \leftarrow$ CREATE-DOCUMENTS$(S, \hat{R})$;
9. $P \leftarrow$ CREATE-DOCUMENT-LISTING-STRUCTURE(A);
10.   **for** $j \leftarrow 1$ to $\lceil \frac{|S|}{\hat{R}-2} \rceil - 1$**do**
11. $B[j] \leftarrow$ FIND-REPETITION-INDEX$(A[j], k, r, \frac{6 \log(n)}{\mathcal{H}(0,p)})$;
12. $\hat{R} \leftarrow$ MAX$(k+1, \frac{6 \log(n)}{\mathcal{H}(0,p)})$;
13. **until** $\hat{R} \geq$ TOP$(T.\text{index})$;
14. **return**(T);

Before giving the algorithm SEARCH-IN-INDEXING-STRUCTURE(T, x) we describes some functions used inside it. The Function FIND-IN-STACK finds the two consecutive values $\hat{R}_g, \hat{R}_{g+1}$ such that $\hat{R}_g > |x| \geq \hat{R}_{g+1}$ and returns the 6-uple corresponding to $\hat{R}_{g+1}$. After a call of FIND-IN-STACK, the stack remains unchanged. The Function SEARCH-IN-DOCUMENTS returns the list of all indexes j such that document d_j contains one or more copies of x. Function DUPLICATE checks whether a position in S has already appeared or not. It outputs a boolean value. It uses an auxiliary array of size $|S|$ and each call of Function DUPLICATE is done in amortized constant time. Function STANDARD-SEARCH is a standard procedure that works in time $O(|k'| \times |S_j|)$ (cf. [7]), where k' is defined in Remark 1.

SEARCH-IN-INDEXING-STRUCTURE (T, x)
1. **if** $|x| \leq k$ **then return** ("x appears everywhere");
2. **else**
3. $(\hat{R}, P, A, B, \psi, \phi) \leftarrow$ FIND-IN-STACK$(T, |x|)$;
4. $occDocuments \leftarrow$ SEARCH-IN-DOCUMENTS(P);
5.   **while** $occDocuments \neq \emptyset$ **do**
6. $(i, j) \leftarrow$ POP$(occDocuments)$;
7.   **if** $B[j] \neq -1$ **then**
8.   **if not** DUPLICATE $(\phi(i, j))$ **then**
9.   PUSH$(occ, \phi(i, j))$;
10.   **else**
11. $occ2 \leftarrow$ STANDARD-SEARCH$(x, \psi(j), k+1)$;
12. $occ \leftarrow$ CONCATENATE$(occ, occ2)$;
13.   **return** (occ);

5   Analysis of Algorithms and Data Structures

From now on in this section, let us consider an infinite sequence $\mathbf{x}$ generated by a memoryless source and let us consider the sequence of prefixes S_n of $\mathbf{x}$ of length n.

5.1   Evaluation of Repetition Index

The following theorems are a consequence of a result in (cf. [2]) and they are proved in [5]. Let $p = P(S(i,i) = S(j,j))$ be the probability that the letters in two distinct positions are equal $(1 \le i \ne j \le |S|)$. In a memoryless source with identical symbols probabilities one has that $p = \frac{1}{|\Sigma|}$.

Proposition 2. *For fixed k and r, almost surely*

1. $\lim \sup_{n \in \mathbb{N}} \frac{R(S_n, k, r)}{\log(n)} \le \frac{2}{\mathcal{H}(D,p)}$, *where $D = \frac{2k}{r} < 1 - p$,*

2. $\lim \inf_{n \in \mathbb{N}} \frac{R(S_n, k, r)}{\log(n)} \ge \frac{2}{\mathcal{H}(D,p)}$, *where $D = \frac{k}{r} < 1 - p$.*

Proposition 3. *Suppose that k is fixed and that $r(n)$ is a function that satisfies the equation $r(n) = R(S_n, k, r(n))$, then*

$$\lim_{n \to \infty} \frac{R(S_n, k, r(n))}{\log n} = \frac{2}{\mathcal{H}(0, p)}.$$

By Proposition 1 and by Remark 1 we can deduce the following proposition.

Proposition 4. *The following estimates on the average time of algorithm* FIND-REPETITION-INDEX *hold:*

- *if r and k are constant*
  $O(\log_2 \log_2(|S|) \times \log_2(|S|) \times |S|^{1+c(H(\alpha)+\gamma\alpha)})$,
  where $\gamma = \log_2(|\Sigma| - 1)$, $\alpha = \frac{k}{r} < \frac{1-p}{2}$, $D = 2\alpha$, $c = \frac{2}{\mathcal{H}(D,p)}$ and H is the classical entropy function.
- *if r is a function of S*
  $O(\log_2 \log_2(|S|) \times |S| \times \log_2(|S|)^{k'+1})$,
  where $k' = max\{d(u,v)|u \in Fact^{2R(S,k,r)}(S)$ and $v \in V(u,k,r)\}$.

The following theorem gives an estimate on the size of the DFA recognizing the language $L(S,k,r)$.

Theorem 1. *Suppose that k is fixed. If $r \le R(S,k,r)$ then the minimal deterministic finite automaton $\mathcal{A}$ recognizing the language $L(S,k,r)$ has size $O(S_{I,k,r})$, where I is the set of all factors of S of length $R(S,k,r) + 1$. More precisely it has size*

$$O\Big((R(S,k,r)+1)(|S| - R(S,k,r)) \sum_{i=0}^{k\lceil (R(S,k,r)+1)/r \rceil} \binom{R(S,k,r)+1}{i}(\Sigma - 1)^i\Big).$$

If in above theorem $r = R(S,k,r)$ then the size of $\mathcal{A}$ is $O(|S| \times (R(S,k,r)+1)^{k+1})$.

5.2   Analysis of Algorithms

It is easy to prove that the running time of FIND-LARGE-PATTERN(x), implemented as described by automaton $\mathcal{A}$, is proportional to the size $|x|$ of the pattern. This time obviously does not include the time for building the automaton. We notice that, under the unitary cost model of computation, the overall size of the automaton $\mathcal{A}$, after the addition to any state of an integer, is still bounded by the same value given in Theorem 1.

We now present the space and time analysis of algorithms CREATE-INDEXING-STRUCTURES (S, k) and SEARCH-IN-INDEXING-STRUCTURES (T, x), under the assumption that we are creating the indexing structure T over the sequence of prefixes S_n of length n, of an infinite sequence generated by a memoryless source with identical symbols probabilities, for a fixed k.

Concerning algorithm CREATE-INDEXING-STRUCTURES, the reader can easily check that the overall space is bounded by a constant times the space requirement of CREATE-AUTOMATON, that is $O(|S_{I,k,r}|)$, where I is the set of factors of S of length $R(S, k, r(n)) + 1$ (cf. Theorem 1). In fact, all the structures built up during the main cycle REPEAT have size proportional, up to a fixed constant, to the overall size of the created documents, that is, in turn, proportional to $|S_{J,k,r}|$, where J is the set of factors of S of length $\hat{R}$, and $r = \hat{R} - 1$. Since, in the average case, $\hat{R}$ is equal to a constant times the logarithm of the previous value, the series that adds up all the required spaces for each indexing data structure is convergent.

In an analogous way it is possible to prove that the overall time is bounded by a constant times the time requirement of FIND-REPETITION-INDEX. An average estimate of this time is given in Proposition 4.

Proposition 5. *Given an integer $m > 0$ the average value of the quantities $|x| + |occ(x)|$ over each possible word $x \in \Sigma^m$ is equal to*

$$m + \frac{|V(x, k, r)|(|S_n| - m + 1)}{|\Sigma|^m},$$

where the considered text is S_n.

We prove that on average the time for searching a pattern x in S up to k errors every r symbols, by using algorithm SEARCH-IN-INDEXING-STRUCTURE, is proportional to $|x| + |occ(x)|$, under an additional but realistic hypothesis. Actually, if $|x| \geq R(S_n, k, r)$ then SEARCH-IN-INDEXING-STRUCTURE is the function FIND-LARGE-PATTERN(x) and its running time is proportional to the size $|x|$ of the pattern, since $|occ(x)|$ is either 0 or 1. Now we suppose that $|x| < R(S_n, k, r)$.

Let δ_n the percentage of documents d_j such that the repetition index $R_j = R(S_j, k, r) > \frac{6 \log(n)}{\mathcal{H}(0, p)}$, where $1 \leq j \leq t = \lceil \frac{|S|}{\hat{R} - 2} \rceil - 1$. By Proposition 3, δ_n converges on average to zero as n goes up to infinite. The number of documents d_j such that $B[j] = $ FIND-REPETITION-INDEX $(A[j], k, r, \frac{6 \log(n)}{\mathcal{H}(0, p)}) = -1$ is equal to $\delta_n t$. The running time of the function STANDARD-SEARCH for a pattern x is $O(k \times |S_j|)$, where $|S_j| = 2(\hat{R} - 2)$. Therefore the total running time of the function for every pattern that matches in S_j is $O(|V(x,k,r)| \times \delta_n \times t \times k \times |S_j|^2) = O(|V(x,k,r)| \times \delta_n \times \frac{|S|}{\hat{R}} \times k \times \hat{R}^2) = O(|V(x,k,r)| \times \delta_n \times |S| \times k \times \hat{R})$. For the documents d_j such that $B[j] = $ FIND-REPETITION-INDEX $(A[j], k, \frac{6 \log(n)}{\mathcal{H}(0,p)}) > -1$ every pattern of length m that occurs in S_j up to k errors every $R(S, k, r(n))$ symbols, occurs just in one position and therefore we can have the same bound obtained in Proposition 5. Then the overall time is $O(m|\Sigma|^m + |V(x,k,r)| \times (|S| - m + 1) + |V(x,k,r)| \times \delta_n \times |S| \times k \times \hat{R})$. We conclude that:

Theorem 2. *If $\delta_n \times \hat{R} \to 0$ then the average running time of the function* SEARCH-IN-INDEXING-STRUCTURE *is*

$$O(m + \frac{|V(x,k,r)|(|S| - m + 1)}{|\Sigma|^m}).$$

Notice that large deviation theorems in similar settings state that the speed of the convergence to zero of variables that represent the average of quantities such as δ_n is quite fast. Therefore the hypothesis that $\delta_n \times \hat{R} \to 0$ is quite realistic.

References

1. A. Amir, D. Keselman, G. M. Landau, M. Lewenstein, N. Lewenstein, and M. Rodeh. Indexing and dictionary matching with one error. *LLNCS*, 1663:181–190, 1999.
2. R. Arratia and M. Waterman. The erd'os-rényi strong law for pattern matching with given proportion of mismatches. *Annals of Probability*, 4:200–225, 1989.
3. R. B. Ash. *Information Theory*. Interscience, 1965.
4. M. Crochemore, C. Hancart, and T. Lecroq. *Algorithmique du texte*. Vuibert, 2001. 347 pages.
5. A. Gabriele, F. Mignosi, A. Restivo, and M. Sciortino. Indexing structure for approximate string matching. Technical Report 169, University of Palermo, Department of Mathematics and Applications, 2002.
6. Z. Galil and R. Giancarlo. Data structures and algorithms for approximate string matching. *Journal of Complexity*, 24:33–72, 1988.
7. D. Gusfield. *Algorithms on Strings, Trees, and Sequences*. Cambridge University Press, 1997. ISBN 0 521 58519 8 hardback. 534 pages.
8. S. Muthukrishnan. Efficient algorithms for document retrieval problems. In *Proceedings of the 13th Annual ACM-SIAM Sumposium on Discrete Algorithms*, pages 657–666, 2002.
9. G. Navarro. A guided tour to approximate string matching. *ACM Computing Surveys*, 33(1):31–88, 2001.
10. G. Navarro, R. Baeza-Yates, E. Sutinen, and J. Tarhio. Indexing methods for approximate string matching. *IEEE Data Engineering Bulletin*, 24(4):19–27, 2001. Special issue on Managing Text Natively and in DBMSs. Invited paper.

Approximation Hardness for Small Occurrence Instances of NP-Hard Problems

Miroslav Chlebík[1] and Janka Chlebíková[2,*]

[1] MPI for Mathematics in the Sciences, D-04103 Leipzig
[2] CAU, Institut für Informatik und Praktische Mathematik, D-24098 Kiel
jch@informatik.uni-kiel.de

Abstract. The paper contributes to the systematic study (started by Berman and Karpinski) of explicit approximability lower bounds for small occurrence optimization problems. We present parametrized reductions for some packing and covering problems, including 3-Dimensional Matching, and prove the best known inapproximability results even for highly restricted versions of them. For example, we show that it is NP-hard to approximate MAX-3-DM within $\frac{139}{138}$ even on instances with exactly two occurrences of each element. Previous known hardness results for bounded occurence case of the problem required that the bound is at least three, and even then no explicit lower bound was known.

New structural results which improve the known bounds for 3-regular amplifiers and hence the inapproximability results for numerous small occurrence problems studied earlier by Berman and Karpinski are also presented.

1 Introduction

The research on the hardness of bounded occurrence (resp. bounded degree) optimization problems is focused on the case of very small value of the bound parameter. For many small parameter problems tight hardness results for optimization problems can be hardly achieved directly from the PCP characterization of NP. Rather, one has to use an expander/amplifier method. Considerable effort of Berman and Karpinski (see [2] and references therein) has gone into the developing of a new method of reductions for determining the inapproximability of MAXIMUM INDEPENDENT SET (MAX-IS) and MINIMUM NODE COVER (MIN-NC) in graphs of maximum degree 3 or 4.

Overview. As a starting point to our gap preserving reductions we state in Section 2 the versions of NP-hard gap results on bounded (constant) occurrence MAX-E3-LIN-2. Their weaker forms are known to experts and have been already used ([3], [7], [11], [12]). The advantage of this approach is that we need not restrict ourselves to amplifiers that can be constructed in polynomial time, to prove NP-hard gap results. Any (even nonconstructive) proof of existence of

* The author has been supported by EU-Project ARACNE, Approximation and Randomized Algorithms in Communication Networks, HPRN-CT-1999-00112.

amplifiers (or expanders) with better parameters than those currently known implies the *existence* of (deterministic, polynomial) gap-preserving reductions leading to better inapproximability result. This is our paradigm towards tighter inapproximability results inspired by the paper of Papadimitriou and Vempala on Traveling Salesman problem ([11]), that we have already used for Steiner Tree problem in [7].

We prove structural results about 3-regular amplifiers which play a crucial role in proving explicit inapproximability results for bounded occurrence optimization problems. A $(2,3)$-graph $G = (V, E)$ with nodes only of degree 2 (*Contacts*) and 3 (*Checkers*) is an *amplifier* if for very $A \subseteq V$ either $|\text{Cut } A| \geq |Contacts \cap A|$, or $|\text{Cut } A| \geq |Contacts \setminus A|$. The parameter $\tau(G) := \frac{|V|}{|Contacts|}$ measures the quality of an amplifier. We are able to prove for many bounded occurrence problems a tight correspondence between $\tau_* := \inf\{\gamma : \tau(G) < \gamma$ for infinity many amplifiers $G\}$ and inapproximability results. In this paper we slightly improve the upper bound from known $\tau_* \leq 7$ (Berman and Karpinski, [2]) to $\tau_* \leq 6.9$. This improvement is based on our structural amplifier analysis presented in Section 3. But there is still a substantial gap between the best upper and lower bounds on parameters of amplifiers and expanders. We develop our method of parametrized reductions (a parameter is a fixed amplifier) to prove inapproximability results for E3-Occ-Max-3-Lin-2 problem, and problems Max-IS and Min-NC on 3-regular graphs (Section 4). The similar method can be applied to all problems studied in [2] (with modification of amplifiers to bipartite-like for Max Cut) to improve the lower bound on approximability. Similarly, for the problem TSP with distances 1 and 2 ([3]).

We include reductions to some packing and covering problems to state the best known inapproximability results on (even highly restricted) version of Triangle Packing, 3-Set Packing, and 3-Set Covering problems (Section 4). These reductions are quite straightforward from Max-3-IS, resp. Min-3-NC and they are included as inspiration to the new reduction for 3-Dimensional Matching problem (Max-3-DM) (Section 5). APX-completeness of the problem has been well known even on instances with at most 3 occurrences of any element, but our lower bound applies to the instances with exactly 2 occurrences. We do not know about any previous hardness result on the problem with such restricted case. The best to our knowledge lower and upper approximation bounds for mentioned packing and covering problems are summarized in the following table. The upper bounds are from [5] and [6].

Problem	Param. lower bound	Approx. lower bound $(\tau_* = 6.9)$	Approx. upper bound
Max-3-DM	$1 + \frac{1}{18\tau_* + 13}$	139/138	$1.5 + \varepsilon$
Max Triangle Packing	$1 + \frac{1}{18\tau_* + 13}$	139/138	$1.5 + \varepsilon$
3-Set Packing	$1 + \frac{1}{18\tau_* + 13}$	139/138	$1.5 + \varepsilon$
3-Set Covering	$1 + \frac{1}{18\tau_* + 18}$	144/143	$1.4 + \varepsilon$

Our inapproximability result on MAX-3-DM can be applied to obtain explicit lower bounds for several problems of practical interest, e.g. scheduling problems, some (even highly restricted) cases of GENERALIZED ASSIGNMENT problem, or the other more general packing problems.

2   Inapproximability of Subproblems of Max-E3-Lin-2

In proving inapproximability results we produce new "hard gaps" from those already known using gap-preserving reductions and their compositions. We start with a restricted version of MAX-E3-LIN-2:

Definition 1. MAX-E3-LIN-2 *is the following optimization problem: Given a system I of linear equation over $\mathbb{Z}_2$, with exactly 3 (distinct) variables in each equation. The goal is to maximize, over all assignments ψ to the variables, the fraction of satisfied equations of I.*

We use the notation Ek-OCC-MAX-Ed-LIN-2 for the same maximization problem, where each equation has exactly d variables and each variable occurs exactly k times. If we drop an "E" than we have "at most d variables" and/or "at most k occurrences". Denote $Q(\varepsilon, k)$ the following restricted version of MAX-E3-LIN-2: *Given an instance of* Ek-OCC-MAX-E3-LIN-2. *The problem is to decide if the fraction of more than $(1 - \varepsilon)$ or less than $(\frac{1}{2} + \varepsilon)$ of all equations is satisfied by the optimal (i.e. maximizing) assignment.*

From Håstad [4] result one can prove NP-hard gap result also for instances of MAX-E3-LIN-2 where each variable appears bounded (or even constant) number of times (Theorem 1). For our applications the strengthening contained in Theorems 2 and 3 are more convenient. The proofs of Theorems 1–3 can be found in [8].

Theorem 1. (Håstad) *For every $\varepsilon \in \left(0, \frac{1}{4}\right)$ there is an integer $k_0(\varepsilon)$ such that the partial decision subproblem $Q(\varepsilon, k_0(\varepsilon))$ of* MAX-E3-LIN-2 *is NP-hard.*

Theorem 2. *For every $\varepsilon \in \left(0, \frac{1}{4}\right)$ there is a constant $k(\varepsilon)$ such that for every integer $k \geq k(\varepsilon)$ the partial decision subproblem $Q(\varepsilon, k)$ of* MAX-E3-LIN-2 *is NP-hard.*

To prove hard gap results for some problems using reduction from MAX-E3-LIN-2 it is sometimes useful, if all equations have the same right hand side. This can be easily enforced if we allow flipping some variables. The canonical gap versions $Q_i(\varepsilon, 2k)$ of MAX-E3-LIN-2 of this kind are as follows: *Given an instance of* MAX-E3-LIN-2 *such that all equations are of the form $x + y + z = i$ and each variable appears exactly k times negated and k times unnegated. The task is to decide if the fraction of more than $(1 - \varepsilon)$ or less than $(\frac{1}{2} + \varepsilon)$ of all equations is satisfied by the optimal (i.e. maximizing) assignment.* The corresponding hard-gap result for this restricted version reads as follows.

Theorem 3. *For every $\varepsilon \in \left(0, \frac{1}{4}\right)$ there is a constant $k(\varepsilon)$ such that for every integer $k \geq k(\varepsilon)$ the partial decision subproblems $Q_0(\varepsilon, 2k)$ and $Q_1(\varepsilon, 2k)$ of* MAX-E3-LIN-2 *are NP-hard.*

3   Amplifiers

In this section we describe our results about the structure and parameters of 3-regular amplifiers, that we use in our reductions.

Definition 2. *A graph* $G = (V, E)$ *is a* $(2, 3)$*-graph if* G *contains only nodes of degree 2 and 3. We denote* $Contacts = \{v \in V : \deg_G(v) = 2\}$, *and* $Checkers = \{v \in V : \deg_G(v) = 3\}$. *Furthermore, a* $(2, 3)$*-graph* G *is an* amplifier *(more precisely, it is a 3-regular amplifier for its contact nodes) if for every* $A \subseteq V$: $|Cut\ A| \geq |Contacts \cap A|$, *or* $|Cut\ A| \geq |Contacts \setminus A|$, *where* $Cut\ A = \{\{u, v\} \in E$: *exactly one of nodes* u *and* v *is in* $A\}$.

An amplifier G *is called a* (k, τ)*-amplifier if* $|Contacts| = k$ *and* $|V| = \tau k$. *We introduce the notation* $\tau(G) := \frac{|V|}{|Contacts|}$ *for an amplifier* G. *Let us denote* $\tau_* = \inf\{\gamma : \tau(G) < \gamma$ *for infinitely many amplifiers* $G\}$.

We have studied several probabilistic models of generating $(2, 3)$-graphs randomly. In such situation we need to estimate the probability that the random $(2, 3)$-graph G is an amplifier. It fails to be an amplifier if and only if the system of so-called *bad* sets $\mathcal{B} := \{A \subseteq V : |Cut\ A| < \min\{|Contacts \cap A|, |Contacts \setminus A|\}\}$ is nonempty. For a fixed bad set it is quite simple to estimate the probability that this candidate for a bad set doesn't occur. But the question is how to estimate the union bound over all bad sets in better way, than by adding all single probabilities. It is useful to look for a small list $\mathcal{B}_* \subseteq \mathcal{B}$, such that if $\mathcal{B} \neq \emptyset$ then $\mathcal{B}_* \neq \emptyset$ as well. In [2] the role of $\mathcal{B}_*$ play elements of $\mathcal{B}$ of the minimum size. Our analysis shows that one can produce the significantly smaller list of bad sets which is sufficient to exclude to be sure that a graph is an amplifier.

For a $(2, 3)$-graph $G = (V, E)$ we define the relation $\preceq$ on the set $\mathcal{P}(V)$ of all subsets V: $A \preceq B$ iff $|Cut\ A| \leq |Cut\ B| - |(A \triangle B) \cap Contacts|$ whenever $A, B \subseteq V$. (Here $A \triangle B$ stands for $(A \setminus B) \cup (B \setminus A)$.) Clearly, the relation $\preceq$ is reflexive and transitive. So, $\preceq$ induces a partial order on the equivalence classes $\mathcal{P}(V) / \approx$. The equivalence relation $\approx$ can be more simply characterized by $A \approx B$ iff $A \cap Contacts = B \cap Contacts$ and $|Cut\ A| = |Cut\ B|$, for $A, B \subseteq V$. Moreover, for every $A \subseteq V$, $A \preceq B$ iff $V \setminus A \preceq V \setminus B$.

Using this relation one can describe the set $\mathcal{B}$ of bad sets, as

$$\mathcal{B} := \mathcal{B}(G) = \{B \subseteq V : \text{neither } \emptyset \preceq B, \text{ nor } V \preceq B\}.$$

Clearly, for every $A, B \subseteq V$, $B \in \mathcal{B}$ and $A \preceq B$ imply $A \in \mathcal{B}$. The minimal elements of the partial order $(\mathcal{P}(V), \preceq)$ play an important role in what follows.

Further, we denote

$$\mathcal{B}_0 := \mathcal{B}_0(G) = \{B \subseteq V : B \text{ is a minimal element of } (\mathcal{B}, \preceq)\}.$$

Clearly, a set $\mathcal{B}_0$ is closed on the complementation operation $A \mapsto V \setminus A$ for any subset $A \subseteq V$.

Lemma 1. *Let* G *be a* $(2, 3)$*-graph and* $B \in \mathcal{B}_0(G)$ *be given.*

(i) For every set $Z \subseteq B$ the inequality $2 \cdot |\text{Cut } Z \cap \text{Cut } B| \leq |\text{Cut } Z| + |Z \cap \text{Contacts}|$ holds with the equality iff $B \setminus Z \approx B$. In particular, if $Z \cap \text{Contacts} \neq \emptyset$ the inequality is strict.

(ii) The set $\text{Cut } B$ is a matching in G.

For a $(2,3)$-graph $G = (V, E)$ let $Z \subseteq V$ be given. Let $G_Z = (Z, E_Z)$ stand for the subgraph of G induced by the node set Z. To see that $|\text{Cut } Z| + |Z \cap \text{Contacts}| = 3|Z| - 2|E_Z|$, we can argue as follows:

$$|\text{Cut } Z| = \sum_{v \in Z \cap Checkers} (3 - \deg_{G_Z}(v)) + \sum_{v \in Z \cap Contacts} (2 - \deg_{G_Z}(v))$$

$$= \sum_{v \in Z} (3 - \deg_{G_Z}(v)) - |Z \cap \text{Contacts}| = 3|Z| - 2|E_Z| - |Z \cap \text{Contacts}|.$$

Given $B \in \mathcal{B}_0$, $\text{Cut } B$ is a matching in G as follows from Lemma 1(ii). Let $Cutters(B)$ stand for the set of nodes in B adjacent to $\text{Cut } B$. Clearly for any $Z \subseteq B$, an edge of $\text{Cut } B$ adjacent to $v \in Cutters(B)$ belongs to $\text{Cut } Z$ if and only if $v \in Z$. Therefore $|\text{Cut } Z \cap \text{Cut } B| = |Z \cap Cutters(B)|$. Hence we can reformulate the first part of Lemma 1 as follows:

Lemma 2. Let G be a $(2,3)$-graph and $B \in \mathcal{B}_0(G)$ be given. Then for every set $Z \subseteq B$ the inequality $|Z \cap Cutters(B)| \leq \frac{3}{2}|Z| - |E_Z|$ holds with the equality iff $B \setminus Z \approx B$. In particular, if $Z \cap \text{Contacts} \neq \emptyset$ the inequality is strict.

The purpose of the following lemma is to derive some restrictions on local patterns of $\text{Cut } B$ for a general set $B \in \mathcal{B}_0$. Given $B \in \mathcal{B}_0$, we can test it with many various $Z \subseteq B$ (typically with G_Z being a small connected graph) to obtain restrictions on possible patterns of $Cutters(B)$ in B. Some of basic results of this kind are stated in the following lemma.

Lemma 3. Let G be a $(2,3)$ graph, $B \in \mathcal{B}_0(G)$ and $Z \subseteq B$ be given.

(i) If G_Z is a tree and $|Z| = 2k - 1$ $(k = 1, 2, \ldots)$ then $|Z \cap Cutters(B)| \leq k$.

(ii) If G_Z is a tree and $|Z| = 2k$ then $|Z \cap Cutters(B)| \leq k + 1$. Moreover, this inequality is strict if $Z \cap \text{Contacts} \neq \emptyset$.

(iii) If G_Z is a $(2k + 1)$-cycle then $|Z \cap Cutters(B)| \leq k$.

(iv) If G_Z is a $2k$-cycle then $|Z \cap Cutters(B)| \leq k$. Moreover, this inequality is strict if $Z \cap \text{Contacts} \neq \emptyset$.

Lemma 4. Let $G = (V, E)$ be a $(2,3)$-graph and $B \in \mathcal{B}_0(G)$ be given.

(i) If $a, b \in Cutters(B)$ and $(a, b) \in E$, then $a, b \in Checkers$ and there are 2 distinct nodes $a', b' \in B \setminus Cutters(B)$ such that $(a, a') \in E$ and $(b, b') \in E$.

(ii) If $a, c \in Cutters(B)$, $b \in B$, $(a, b) \in E$, $(b, c) \in E$, and if exactly one of nodes a, b and c belongs to Contacts, then there are 2 distinct nodes $d, e \in B \setminus Cutters(B)$, each adjacent to one of two nodes in $\{a, b, c\} \cap Checkers$.

For the purpose to provide even more restricted list we make our partial order $\preceq$ finer inside the equivalence classes $\mathcal{P}(V) / \approx$. For a $(2,3)$-graph $G = (V, E)$

let a subset F of E of "distinguished edges" be fixed. We define the following relations on the set $\mathcal{P}(V)$ of all subset V, whenever $A, B \subseteq V$: $A \overset{F}{\preceq} B$ iff either $(A \preceq B$ & $A \not\approx B)$ or $(A \approx B$ & $|F \cap \mathrm{Cut}\, A| \le |F \cap \mathrm{Cut}\, B|)$; $A \overset{F}{\preceq_*} B$ iff either $(A \overset{F}{\preceq} B$ & $A \overset{F}{\not\approx} B)$, or $(A \overset{F}{\approx} B$ & $\min\{|A \cap \mathit{Checkers}|, |\mathit{Checkers} \setminus A|\} \le \min\{|B \cap \mathit{Checkers}|, |\mathit{Checkers} \setminus B|\})$.

Denote

$$\mathcal{B}_F(G) := \{B \subseteq V : B \text{ is a minimal element of } (\mathcal{B}(G), \overset{F}{\preceq})\},$$

$$\mathcal{B}_F^*(G) := \{B \subseteq V : B \text{ is a minimal element of } (\mathcal{B}(G), \overset{F}{\preceq_*})\}.$$

The equivalence relation $\overset{F}{\approx}$ is defined by: $A \overset{F}{\approx} B$ iff $A \overset{F}{\preceq} B$ and $B \overset{F}{\preceq} A$. Clearly $\overset{F}{\preceq}$ is a partial order on equivalence classes $\mathcal{P}(V) / \overset{F}{\approx}$. The equivalence relation $\overset{F}{\approx}$ can be also characterized by $A \overset{F}{\approx} B$ iff $A \cap \mathit{Contacts} = B \cap \mathit{Contacts}$ and $|\mathrm{Cut}\, A| = |\mathrm{Cut}\, B|$ & $|F \cap \mathrm{Cut}\, A| = |F \cap \mathrm{Cut}\, B|$.

Clearly $\mathcal{B}_F^*(G) \subseteq \mathcal{B}_F(G) \subseteq \mathcal{B}_0(G)$, and $B \in \mathcal{B}_F(G)$ iff $B \in \mathcal{B}_0(G)$ & $A \approx B$ implies $|F \cap \mathrm{Cut}\, B| \le |F \cap \mathrm{Cut}\, A|$; $B \in \mathcal{B}_F^*(G)$ iff $B \in \mathcal{B}_F(G)$ & $A \overset{F}{\approx} B$ implies $\min\{|B \cap \mathit{Checkers}|, |\mathit{Checkers} \setminus B|\} \le \min\{|A \cap \mathit{Checkers}|, |\mathit{Checkers} \setminus A|\}$.

Lemma 5. *Let G be a $(2,3)$-graph and $B \in \mathcal{B}_F(G)$ be given. Then for every set $Z \subseteq B$ such that $B \setminus Z \approx B$ (equivalently, $Z \subseteq B \cap \mathit{Checkers}$ and $|Z \cap \mathit{Cutters}(B)| = \frac{3}{2}|Z| - |E_Z|)$ $|F \cap \mathrm{Cut}\, Z \cap \mathrm{Cut}\, B| \le \frac{1}{2}|F \cap \mathrm{Cut}\, Z|$ holds, with the equality if and only if $B \setminus Z \overset{F}{\approx} B$.*

Lemma 6. *Let G be a $(2,3)$-graph, $B \in \mathcal{B}_F^*(G)$ and $\emptyset \ne Z \subseteq B$ such that $B \setminus Z \approx B$ and $2 \cdot |B \cap \mathit{Checkers}| < |\mathit{Checkers}| + |Z|$. Then $|F \cap \mathrm{Cut}\, Z \cap \mathrm{Cut}\, B| < \frac{1}{2} \cdot |F \cap \mathrm{Cut}\, Z|$.*

Let us consider a $(2,3)$-graph $G = (V, E)$. For $B \subseteq V$, we denote $B_{\mathrm{red}} := B \cap \mathit{Checkers}$. Assume further that no pair of nodes in $\mathit{Contacts}$ is adjacent by an edge. We convert G to a 3-regular (multi-)graph G_{red} with a node set V_{red} equals to $\mathit{Checkers}$. Each node $v \in \mathit{Contacts}$ and two edges adjacent to v in G are replaced with an edge $e(v)$ (later called a *contact* edge) that connects the pair of nodes that were adjacent to v in G. For any $A \subseteq V_{\mathrm{red}}$ let $\mathrm{Cut}_{\mathrm{red}} A$ stand for a cut of A in G_{red}, and $\mathit{Cutters}_{\mathrm{red}}(A)$ stand for the set of nodes of A adjacent in G_{red} to an edge of $\mathrm{Cut}_{\mathrm{red}} A$.

Lemma 7. *Let G be a $(2,3)$-graph with no edge between contact nodes, and let $B \in \mathcal{B}_0(G)$. Then $|\mathrm{Cut}\, B| = |\mathrm{Cut}_{\mathrm{red}}(B_{\mathrm{red}})|$, and if any pair of nodes in $\mathit{Contacts}$ is at least at distance 3 apart, $\mathrm{Cut}_{\mathrm{red}}(B)$ is a matching in G_{red}.*

We elaborate in details on our general results in the concrete model of randomly generated (k, τ)-wheels, which generalizes slightly the notion of a wheel-amplifier used by Berman and Karpinski ([2]). A (k, τ)-wheel is a $(2,3)$-graph

$G = (V, E)$ with $|V| = \tau k$ and $|Contacts| = k$, and with the edge set E splited into two parts E_C and E_M. E_C is an edge set of several disjoint cycles in G collectively covering V. In each cycle consecutive contacts of G are separated by a chain of several (at least 2) checkers. E_M is a perfect matching for the set of checkers. We consider here the choice $F := E_C$ for the special subset of "distinguished edges" in our amplifier analysis.

Given a bad set B, we will refer to fragments of B, the connected components of B within cycles, and to reduced fragments of B_{red}, the connected components of B_{red} within corresponding reduced cycles.

The following theorem summarizes the results from Lemmas 1–7 for (k, τ)-wheel:

Proposition 1. *Let G be a (k, τ)-wheel. Then every set $B \in \mathcal{B}_0(G)$ has the following properties:*

(i) B is a bad set, i.e. $|\mathrm{Cut}\, B| < \min\{|Contacts \cap B|, |Contacts \setminus B|\}$.

(ii) $\mathrm{Cut}\, B$ is a matching in G.

(iii) $|\mathrm{Cut}_{red}(B_{red})| = |\mathrm{Cut}\, B|$, and $\mathrm{Cut}_{red}(B_{red})$ is a matching in G_{red}.

(iv) Any fragment of B contains at least 2 checkers.

(v) End nodes of any reduced fragment of B_{red} are not incident to $E_M \cap \mathrm{Cut}\, B$.

(vi) Any fragment of B consisting of 3 checkers has none of its nodes incident to $E_M \cap \mathrm{Cut}\, B$.

(vii) Any fragment of B consisting of 2 checkers and 1 contact has both its checkers matched with $B \setminus Cutters(B)$ nodes.

(viii) Any fragment of B consisting of 2 checkers has both its nodes matched with $B \setminus Cutters(B)$ nodes.

Every set $B \in \mathcal{B}_F(G)$ additionally has the following properties: (ix) Any fragment of B contains at least 3 nodes.

(x) Any fragment of B consisting of 3 checkers has all its nodes matched with $B \setminus Cutters(B)$.

(xi) Any fragment of B consisting of 4 checkers has none of its nodes incident to $E_M \cap \mathrm{Cut}\, B$.

All the above properties apply at the same time to B and $\tilde{B} := Checkers \setminus B$. The following is less symmetric, it says something more about the smaller of the sets B, $\tilde{B}$, if $B \in \mathcal{B}(G)_F^*$.

(xii) If $B \in \mathcal{B}(G)_F^*$ with $|B \cap Checkers| \leq \frac{1}{2}|Checkers|$, then no pair of checkers that are end nodes of (possibly distinct) fragments of B, are matched.

For purpose of the paper we can confine ourselves to the model with E_C consisting of 2 cycles C_1 and C_2. One consists of $(1 - \theta)k$ ($\theta \in (0, 1)$) contacts, separated by chains of checkers of length 6, and in the second one θk contacts are separated by chains of checkers of length 5. For fixed parameters θ and k consider two cycles with contacts and checkers as above and take a random perfect matching for the set of checker nodes. Then, with high probability, the produced $(k, 7 - \theta)$-wheel will be an amplifier. More precisely, for an explicit constant $\theta_0 \in (0, 1)$, for any rational $\theta \in (0, \theta_0)$, and any sufficiently large positive integer k for which θk is an even integer, $(k, 7 - \theta)$-amplifiers exist. Here τ is a rational number, $(\tau - 1)k$ is an even integer. In such model the following upper bound for τ_* can be proved

Theorem 4. $\tau_* \leq 6.9$.

The proof of this theorem is quite technical (see [8] for details and proofs of Lemmas 1–7). The further improvements of estimates on amplifier parameters of randomly generated graphs, pushing the method to its limits, is in progress.

4   Amplifier Parametrized Known Reductions

We call HYBRID a system of linear equations over $\mathbb{Z}_2$, each equation either with 2 or with 3 variables. We are interested in hard gap results for instances of HYBRID with exact 3 occurrences of each variable (a subproblem of E3-OCC-MAX-3-LIN-2). As suggested in [2], one can produce hard gaps for such restricted instances of HYBRID by gap-preserving reduction from MAX-E3-LIN-2. Our approach is simpler than in [2], since we start the reduction from the problem which is already of bounded (even constant, possibly very large) occurrence. This is a crucial point, since the number of occurrences of variables is just the value that has to be amplified using the expander or amplifier method, and in our reductions an amplifier plays a role of a constant.

Reduction from $Q(\varepsilon, k)$ to HYBRID(G). Let $\varepsilon \in (0, \frac{1}{4})$, and $k \in \mathbb{N}$ be such that $Q(\varepsilon, k)$ is NP-hard. Now we describe a gap-preserving reduction from $Q(\varepsilon, k)$ to the corresponding gap-version of HYBRID. Assume that $G = (V, E)$ is a fixed (k, τ)-amplifier with $|Contacts| = k$ and $|V| = \tau k$. Let an instance I of $Q(\varepsilon, k)$ be given, denote by $\mathcal{V}(I)$ the set of variables in I, $m := |\mathcal{V}(I)|$. Take m disjoint copies of G, one for each variable from $\mathcal{V}(I)$. Let G_x denote a copy of G that corresponds to a variable x. The contact nodes of G_x represent k occurrences of x in equations of I. Distinct occurrences of a variable x in I are now represented by distinct contact nodes of G_x. For each equation $x + y + z = i$ of I ($i \in \{0, 1\}$) we create a hyperedge of size 3, labeled by i. A hyperedge connects a triple of contact nodes, one from each G_x, G_y and G_z. The edges inside each copy G_x are labeled by 0 and any such edge (u, v) represents the equivalence equation $u + v = 0$.

The produced instance I' of HYBRID corresponds simultaneously to a system of equations and a labeled hypergraph. Clearly, nodes correspond to variables, and labeled (hyper-)edges to equations in an obvious way. The restriction of HYBRID to these instances will be called as HYBRID(G) in what follows. The most important property of a produced instance I' is that each variable occurs exactly 3 times in equations. In particular, each contact node occurs exactly in one hyperedge. If an instance I has m variables with $|I| = \frac{mk}{3}$ equations, then I' has $m\tau k$ variables, $\frac{mk}{3}$ equations with 3 variables, and $\frac{mk}{2}(3\tau - 1)$ equations with 2 variables. Hence $|I'| = \frac{mk}{6}(9\tau - 1)$ equations in total.

Clearly, any assignment to variables from $\mathcal{V}(I)$ generates so called *standard* assignment to variables of I': the value of a variable x is assigned to all variables of G_x. To show that the optimum $\text{OPT}(I')$ is achieved on standard assignments is easy. But for a standard assignment the number of unsatisfied equations for I' is the same as for I. Consequently, $\text{OPT}(I')$ depends affinely on $\text{OPT}(I)$, namely $(1 - \text{OPT}(I'))|I'| = (1 - \text{OPT}(I))|I|$. Now we see that $\text{OPT}(I) > 1 - \varepsilon$ implies $\text{OPT}(I') > 1 - \frac{2\varepsilon}{9\tau - 1}$, and $\text{OPT}(I) < \frac{1}{2} + \varepsilon$ implies $\text{OPT}(I') < \frac{9\tau - 2}{9\tau - 1} + \frac{2\varepsilon}{9\tau - 1}$. This proves that it is NP-hard to decide whether an instance of $\text{HYBRID}(G)$ with $|I|$ equations has the maximum number of satisfied equations above $(1 - \frac{2\varepsilon}{9\tau - 1})|I|$ or below $(\frac{9\tau - 2}{9\tau - 1} + \frac{2\varepsilon}{9\tau - 1})|I|$. Hence, we have just proved the following:

Theorem 5. *Assume that $\varepsilon \in (0, \frac{1}{4})$, let k be an integer such that $Q(\varepsilon, k)$ is NP-hard, and G be a (k, τ)-amplifier. Then it is NP-hard to decide whether an instance of $\text{HYBRID}(G)$ with $|I|$ equations has the maximum number of satisfied equations above $(1 - \frac{2\varepsilon}{9\tau - 1})|I|$ or below $(\frac{9\tau - 2}{9\tau - 1} + \frac{2\varepsilon}{9\tau - 1})|I|$.*

Corollary 1. *It is NP-hard to approximate the solution of* E3-Occ-Max-3-Lin-2 *within any constant smaller than $1 + \frac{1}{9\tau_* - 2}$.*

Reductions from HYBRID(G) to other problems. We refer to [2] where Berman and Karpinski provide gadgets for reductions from HYBRID to small bounded instances of Maximum Independent Set and Minimum Node Cover. We can use exactly the same gadgets in our context, but instead of their wheel-amplifier we use a general (k, τ)-amplifier. The proofs from [2] apply in our context as well.

Theorem 6. *Let $\varepsilon \in (0, \frac{1}{4})$, $k \in \mathbb{N}$ be such that $Q(\varepsilon, k)$ is NP-hard, and τ be such that a (k, τ)-amplifier exists. It is NP-hard to decide whether an instance of* Max-3-IS *with n nodes has the maximum size of an independent set above $\frac{18\tau + 14 - 2\varepsilon}{4(9\tau + 8)}n$, or below $\frac{18\tau + 13 + 2\varepsilon}{4(9\tau + 8)}n$. Consequently, it is NP-hard to approximate the solution of* Max-3-IS *within any constant smaller than $1 + \frac{1}{18\tau_* + 13}$. Similarly, it is NP-hard to decide whether an instance of* Min-3-NC *with n nodes has the minimum size of a node cover above $\frac{18\tau + 19 - 2\varepsilon}{4(9\tau + 8)}n$, or below $\frac{18\tau + 18 + 2\varepsilon}{4(9\tau + 8)}n$. Consequently, it is NP-hard to approximate the solution of* Min-3-NC *within any constant smaller than $1 + \frac{1}{18\tau_* + 18}$. The same hard-gap and inapproximability results apply to 3-regular triangle-free graphs.*

In the following we present the inapproximability results for three similar APX-complete problems. From L-reductions used in the proofs of Max-SNP completeness (see [9] or [10]) some lower bounds can be computed but they would be worse as the lower bounds presented here.

Maximum Triangle Packing problem. A triangle packing for a graph $G = (V, E)$ is a collection $\{V_i\}$ of disjoint 3-sets of V, such that every V_i induces a 3-clique in G. The goal is to find cardinality of maximum triangle packing. The problem is APX-complete even for graphs with maximum degree 4 ([9]).

Maximum 3-Set Packing problem. Given a collection C of sets, the cardinality of each set in C is at most 3. A set packing is a collection of disjoint sets $C' \subseteq C$. The goal is to find cardinality of maximum set packing. If the number of occurrences of any element in C is bounded by a constant K, $K \geq 2$, the problem is still APX-complete ([1]).

Minimum 3-Set Covering problem. Given a collection C of subsets of a finite set S, the cardinality of each set in C is at most 3. The goal is to find cardinality of minimum subset $C' \subseteq C$ such that every element in S belongs to at least one member of C'. If the number of occurrences of any element in sets of C is bounded by a constant $K \geq 2$, the problem is still APX-complete [10].

Theorem 7. *Assume that $\varepsilon \in (0, \frac{1}{4})$, $k \in \mathbb{N}$ is such that $Q(\varepsilon, k)$ is NP-hard, and τ is such that there is a (k, τ)-amplifier.*

(i) *It is NP-hard to decide whether an instance of* TRIANGLE PACKING *with n nodes has the maximum size of a triangle packing above $\frac{18\tau + 14 - 2\varepsilon}{6(9\tau + 8)} n$, or below $\frac{18\tau + 13 + 2\varepsilon}{6(9\tau + 8)} n$. Consequently, it is NP-hard to approximate the solution of* MAXIMUM TRIANGLE PACKING *problem (even on 4-regular graphs) within any constant smaller than $1 + \frac{1}{18\tau_* + 13}$.*

(ii) *It is NP-hard to decide whether an instance of* 3-SET PACKING *with n triples and the occurrence of each element exactly in two triples has the maximum size of a packing above $\frac{18\tau + 14 - 2\varepsilon}{4(9\tau + 8)} n$, or below $\frac{18\tau + 13 + 2\varepsilon}{4(9\tau + 8)} n$. Consequently, it is NP-hard to approximate the solution of* 3-SET PACKING *with exactly two occurrences of each element within any constant smaller than $1 + \frac{1}{18\tau_* + 13}$.*

(iii) *It is NP-hard to decide whether an instance of* 3-SET COVERING *with n triples and the occurrence of each element exactly in two triples has the minimum size of a covering above $\frac{18\tau + 19 - 2\varepsilon}{4(9\tau + 8)} n$, or below $\frac{18\tau + 18 + 2\varepsilon}{4(9\tau + 8)} n$. Consequently, it is NP-hard to approximate the solution of* 3-SET COVERING *with exactly two occurrences of each element within any constant smaller than $1 + \frac{1}{18\tau_* + 18}$.*

Proof. Consider a 3-regular triangle-free graph G as an instance of MAX-3-IS from Theorem 6. (i) Take a line-graph $L(G)$ of G. Nodes of G are transformed to triangles in $L(G)$ and this is one-to-one correspondence, as G was triangle-free. Clearly, independent sets of nodes in G are in one-to-one correspondence with triangle packings in $L(G)$, so the conclusion easily follows from Theorem 6. (ii) Create an instance of 3-SET PACKING that uses for 3-sets exactly triples of edges of G adjacent to each node of G. Clearly, independent sets of nodes in G are in one-to-one correspondence with packings of triples in the corresponding instance. Now the conclusion easily follows from the hard-gap for MAX-3-IS problem. (iii) Now a graph G from Theorem 6 is viewed as an instance of MIN-3-NC. Using the same collection of 3-sets as in the part (ii) we see that node covers in G are in one-to-one correspondence with coverings by triples in the new instance. The conclusion follows from the hard-gap result for MIN-3-NC from Theorem 6.

5   New Reduction for 3-Dimensional Matching

Definition and known results. Given the disjoint sets A, B, and C and a set $T \subseteq A \times B \times C$. A matching for T is a subset $T' \subseteq T$ such that no elements in T'

agree in any coordinate. The goal of the MAXIMUM 3-DIMENSIONAL MATCHING problem (shortly, MAX-3-DM) is to find cardinality of a maximum matching. The problem is APX-complete even in case if the number of occurrences of any element in A, B or C is bounded by a constant K ($K \geq 3$) [9].

Recall that usually the hardness of MAX-3-DM is proved by reduction from bounded instances of MAX-3-SAT. The L reduction given in [9] implies lower bound $(1+\varepsilon)$ for some small ε. In what follows we present the new transformation from HYBRID to edge 3-colored instances of MAX-3-IS. To the best of our knowledge we provide the first explicit lower bound on approximation of MAX-3-DM.

Idea: If we have hardness result for MAX-3-IS on 3-regular edge-3-colored graphs, it is at the same time the result for MAX-3-DM due to the following transformation. Suppose that edges of graph $G = (V, E)$ are properly colored with three colors a, b, c. Now define the sets $A = \{$all edges of color $a\}$, $B = \{$all edges of color $b\}$, $C = \{$all edges of color $c\}$ and a set $T \subseteq A \times B \times C$ as $T = \{(e_a(v), e_b(v), e_c(v)), \text{ for all } v \in V\}$, where $e_i(v)$ denotes an edge of color i incident to the node v. It is easy to see that independent sets of nodes in G are in one-to-one correspondence with matchings of an instance obtained by the reduction above. So, the hardness result for MAX-3-DM will immediately follow from the hardness result for MAX-3-IS on edge-3-colored graphs.

Theorem 8. *Given $\varepsilon \in \left(0, \frac{1}{4}\right)$ and let k be an integer such that $Q(\varepsilon, k)$ is NP-hard. Assume τ is such that there is a (k, τ)-amplifier. Then it is NP-hard to decide whether an instance of MAX-3-DM with n-triples, each element occurring in exactly two triples, has the maximum size of a matching above $\frac{18\tau+14-2\varepsilon}{4(9\tau+8)}n$, or below $\frac{18\tau+13+2\varepsilon}{4(9\tau+8)}n$. Hence it is NP-hard to approximate MAX-3-DM within any constant smaller than $1 + \frac{1}{18\tau_*+13}$ even on instances with exactly two occurrences of each element.*

Proof. Let $\varepsilon \in \left(0, \frac{1}{4}\right)$, $k \in \mathbb{N}$ be such that $Q(\varepsilon, k)$ is NP-hard, and $G = (V, E)$ be a fixed (k, τ)-amplifier. We use the same reduction of an instance of HYBRID(G) to an instance G' of MAX-3-IS as in Theorem 6. Each variable x of I is replaced with a gadget A_x. The gadget of a checker is a hexagon H_x in which nodes with labels 0 and 1 alternate. A gadget of a contact is a hexagon H_x augmented with a trapezoid T_x, a cycle of 6 nodes that shares one edge with a hexagon H_x. Again, labels 0 and 1 of nodes in those cycles alternate. If two variables x, y are connected by an equation, $x = y$, we connect their hexagons with a pair of edges, so called *connections*, to form a rectangle in which the nodes with label 0 and 1 alternate. The rectangle thus formed is a gadget of an equation with two variable (Fig. 1).

If three variables are connected by an equation (i.e. an hyperedge), say, $x + y + z = 0$, the trapezoids T_x, T_y and T_z (6-cycles) are coupled with the set S_{xyz} of four *special* nodes (Fig. 2).

It is easy to see that G' is 3-edge-colored. As follows from Fig. 2, the edges of the equation gadget can be colored in such way that all edges adjacent to contacts are colored by one fixed color, say A (dotted lines). All *connections* are

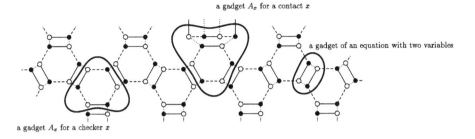

Fig. 1. Example of gadgets for checkers, contacts and equations with two variables.

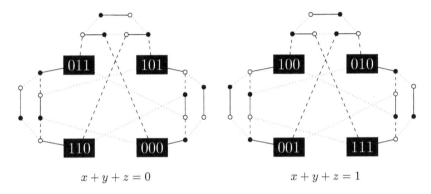

$$x + y + z = 0 \qquad x + y + z = 1$$

Fig. 2. Equation gadgets with three variables.

of the same color A, which alternates on rectangles with color B (full lines), see Fig. 1. The hard gap of MAX-3-IS from Theorem 6 implies the hard gap of MAX-3-DM.

Conclusion

There is still substantial gap between the lower and upper approximation bounds for small occurrence combinatorial optimization problems. The method of parametrized amplifiers shows better the quality of used reductions and the possibilities for further improvement of lower bounds. But it is quite possible that the upper bounds can be improved more significantly.

References

1. P. Berman and T. Fujito: *Approximating independent sets in degree 3 graphs*, Proc. 4th Workshop on Algorithms and Data Structures, 1995, Springer-Verlag, Berlin, LNCS **955**, 449–460.
2. P. Berman and M. Karpinski: *On some tighter inapproximability results, further improvements*, ECCC, Report No. **65**, 1998.

3. L. Engebretsen and M. Karpinski: *Approximation hardness of TSP with bounded metrics*, Proceedings of 28th ICALP, 2001, LNCS **2076**, 201–212.
4. J. Håstad: *Some optimal inapproximability results*, Journal of ACM **48** (2001), 798–859.
5. M. M. Halldórsson: *Approximating k-set cover and complementary graph coloring*, Proc. 5th International Conference on Integer Programming and Combinatorial Optimization, 1996, Springer-Verlag, Berlin, LNCS **1084**, 118–131.
6. C. A. J. Hurkens and A. Schrijver: *On the size of systems of sets every t of which have an SDR, with an application to the worst-case ratio of heuristics for packing problems*, SIAM J. Discrete Mathematics **2** (1989), 68–72.
7. M. Chlebík and J. Chlebíková: *Approximation Hardness of the Steiner Tree Problem on Graphs*, Proceedings of the 8th Scandinavian Workshop on Algorithm Theory, SWAT 2002, Springer, LNCS **2368**, 170–179.
8. M. Chlebík and J. Chlebíková: *Approximation Hardness for Small Occurrence Instances of NP-Hard Problems*, ECCC, Report No. **73**, 2002.
9. V. Kann: *Maximum bounded 3-dimensional matching is MAX SNP-complete*, Information Processing Letters **37** (1991), 27–35.
10. C. H. Papadimitriou and M. Yannakakis: *Optimization, approximation, and complexity classes*, J. Computer and System Sciences **43** (1991), 425–440.
11. C. H. Papadimitriou and S. Vempala: *On the Approximability of the Traveling Salesman Problem*, Proceedings of the 32nd ACM Symposium on the theory of computing, Portland, 2000.
12. M. Thimm: *On the Approximability of the Steiner Tree Problem*, Proceedings of the 26th International Symposium, MFCS 2001, Springer, LNCS **2136**, 678–689.

Fast Approximation of Minimum Multicast Congestion – Implementation versus Theory

Extended Abstract

Andreas Baltz and Anand Srivastav

Mathematisches Seminar der Christian-Albrechts-Universität zu Kiel,
Ludewig-Meyn-Str. 4, D-24098 Kiel, Germany
aba@numerik.uni-kiel.de
asr@numerik.uni-kiel.de

Abstract. The problem of minimizing the maximum edge congestion in a multicast communication network generalizes the well-known NP-hard standard routing problem. We present the presently best theoretical approximation results as well as efficient implementations.

Keywords: Approximation algorithms; Multicast routing; Integer programming

1 Introduction

A communication network can be modeled as an undirected graph $G = (V, E)$, where each node represents a computer that is able to receive, copy and send packets of data. In *multicast* traffic, several nodes have a simultaneous demand to receive a copy of a single packet. These nodes together with the packet's source specify a *multicast request* $S \subseteq V$. Meeting the request means to establish a connection represented by a *Steiner tree* with *terminal set* S, i.e. a subtree of G containing S having no leaf in $V \setminus S$. Given G and a set of multicast requests it is natural to ask about Steiner trees such that the maximum *edge congestion*, i.e. the maximum number of trees sharing an edge is as small as possible. Solving this minimization problem is NP-hard, since it contains as a special case the well-known NP-hard standard routing problem of finding (unsplittable) paths with minimum congestion. The MAX-SNP-hard minimum Steiner tree problem is also closely related.

1.1 Previous Work

Vempala and Vöcking [17] gave a randomized algorithm for approximating the minimum multicast congestion problem within a factor of $O(\log n)$. Their approach is based on applying randomized rounding to an integer linear program relaxation (LP). As an additional obstacle, the considered LP contains an *exponential* number of constraints corresponding to the exponential number of

R. Petreschi et al. (Eds.): CIAC 2003, LNCS 2653, pp. 165–177, 2003.

possible Steiner trees. Vempala and Vöcking overcame this difficulty by considering a multicommodity flow relaxation for which they could devise a polynomial separation oracle. By rounding the fractional paths thus obtained in an $O(\log n)$-stage process they proved an $O(\log n)$ approximation. Carr and Vempala [2] gave an algorithm for approximating the minimum multicast congestion within a constant factor *plus* $O(\log n)$. They showed that an r-approximate solution to an LP-relaxation can be written as a convex combination of Steiner trees. Randomized rounding yields a solution not exceeding a congestion of $\max\{2\exp(1) \cdot r\mathrm{OPT}, 2(\varepsilon + 2)\log n\}$ with probability at least $1 - n^{-\varepsilon}$, where r is the approximation factor of the network Steiner problem. Both algorithms heavily rely on the ellipsoid method with separation oracle and thus are of mere theoretical value. A closely related line of research is concerned with combinatorial approximation algorithms for multicommodity flow and – more general – fractional packing and covering problems. Matula and Shahrokhi [11] were the first who developed a combinatorial strongly polynomial approximation algorithm for the uniform concurrent flow problem. Their method was subsequently generalized and improved by Goldberg [4], Leighton et al. [10], Klein et. al. [9], Plotkin et al. [13], and Radzik [14]. A fast version that is particularly simple to analyze is due to Garg and Könemann [3]. Independently, similar results were given by Grigoriadis and Khachiyan [6]. In two recent papers Jansen and Zhang extended the latter approach and discussed an application to the minimum multicast congestion problem [7],[8]. In particular, they presented a randomized algorithm for approximating the minimum multicast congestion within $O(\mathrm{OPT} + \log n)$ in time $O(m(\ln m + \varepsilon^{-2}\ln \varepsilon)(k\beta + m\ln\ln(m/\varepsilon))$, where β is the running time of an approximate block solver. The online version of the problem was considered by Aspnes et al. [1].

1.2   Our Results

In section 2 we present three algorithms which solve the *fractional* multicast congestion problem up to a relative error of $r(1 + \varepsilon)$. The fastest of these takes time $O(k(\beta + m)\varepsilon^{-2}\ln k \ln m)$ (where β bounds the time for computing an r-approximate minimum Steiner tree), and thus improves over the previously best running time bound of Jansen and Zhang. The three algorithms are based on the approaches of Plotkin, Shmoys and Tardos [13], Radzik [14], and Garg and Könemann [3] to path-packing and general packing problems. In experiments it turned out that straightforward implementations of the above (theoretically fast) combinatorial algorithms are quite slow. However, simple modifications lead to a very fast new algorithm with much better approximation results than hinted at by the worst case bounds (section 3). The central method is to repeatedly improve badly served requests via approximate minimum Steiner where the length function punishes highly congested edges more heavily than proposed in the theory. However, a proof confirming the performance shown in experiments remains at the moment a challenging open problem.

1.3   Notations

By $G = (V, E)$ we will denote the given graph on n nodes and m edges. We consider k multicast requests, $S_1, \ldots, S_k \subseteq V$. The set of all Steiner trees with terminal nodes given by S_i is denoted by $\mathcal{T}_i$, while $\hat{\mathcal{T}}_i$ is the set of Steiner trees for S_i computed in the course of the considered algorithm. As to the (possibly fractional) congestion, we distinguish between $c_i(T)$, the congestion caused by tree T for request S_i, $c_i(e) := \sum_{\substack{T \in \mathcal{T}_i \\ e \in E(T)}} c_i(T)$, the congestion of edge $e \in E$ due to request S_i, $c(e) := \sum_{i=1}^{k} c_i(e)$, the total congestion of edge $e \in E$ and $c_{\max} := \max_{e \in E} c(e)$, the maximum edge congestion. We will define a nonnegative length function l on the edges and abbreviate the sum of edge lengths of a subgraph U of G by $l(U) := \sum_{e \in E(U)} l(e)$. The time for computing an r-approximate optimum Steiner tree will be denoted by β. Throughout the paper, $r \geq 1.55$ and $\varepsilon \in (0, 1]$ will be constant parameters describing the guaranteed ratio of our minimum Steiner tree approximation [18] and determining the target approximation of the fractional optimization problem, respectively. $\mathrm{MST}_l(S_i)$ and $\widetilde{\mathrm{MST}}_l(S_i)$ are used to denote a minimum Steiner tree and an approximate minimum Steiner tree for S_i with respect to l (l is usually omitted); we assume $l(\widetilde{\mathrm{MST}}(S_i)) \leq r \cdot l(\mathrm{MST}(S_i))$. Generally, a variable indexed by a subscript i refers to the ith multicast request S_i, while a superscript (i) refers to the end of the ith iteration of a `while`- or `for`-loop in an algorithm.

2   Approximation Algorithms

2.1   LP Formulation

The minimum multicast congestion problem can be translated into an integer linear program. We study a natural LP relaxation and its dual:

$$(\mathrm{LP}) \quad \min z$$

s. t.

$$\sum_{T \in \mathcal{T}_i} c_i(T) \geq 1 \qquad \text{for all } i \in \{1, \ldots, k\}$$

$$c(e) = \sum_{i=1}^{k} \sum_{\substack{T \in \mathcal{T}_i \\ e \in E(T)}} c_i(T) \leq z \quad \text{for all } e \in E$$

$$c_i(T) \geq 0 \qquad \text{for all } i \text{ and all } T \in \mathcal{T}_i$$

$$(\mathrm{LP}^*) \quad \max \sum_{i=1}^{k} Y_i$$

s. t.

$$\sum_{e \in E} l(e) \leq 1$$

$$\sum_{e \in E(T)} l(e) \geq Y_i \qquad \text{for all } i \text{ and all } T \in \mathcal{T}_i$$

$$l(e) \geq 0 \qquad \text{for all } e \in E$$

$$Y_i \geq 0 \qquad \text{for all } i \in \{1, \ldots, k\}$$

The proof of the following lemma is straightforward.

Lemma 1. $(Y_1^*, \ldots, Y_k^*, l^* : E \to \mathbb{R}_{\geq 0})$ *is an optimal dual solution if and only if* $\sum_{i=1}^{k} Y_i^* = \sum_{i=1}^{k} l^*(\mathrm{MST}(S_i)) = \max \left\{ \frac{\sum_{i=1}^{k} l(\mathrm{MST}(S_i))}{l(G)} \mid l : E \to \mathbb{R}_{\geq 0} \right\}.$

Corollary 1. $OPT \geq \frac{\sum_{i=1}^{k} l(MST(S_i))}{l(G)}$ *for all* $l : E \to \mathbb{R}_{\geq 0}$.

By complementary slackness, feasible solutions to (LP) and (LP*) are optimal if and only if

$$l(e)(c(e) - z) = 0 \text{ for all } e \in E \qquad (1)$$

$$c_i(T)(l(T) - Y_i) = 0 \text{ for all } i \in \{1, \ldots, k\} \text{ and all } T \in \mathcal{T}_i \qquad (2)$$

$$Y_i \left(\sum_{T \in \mathcal{T}_i} c_i(T) - 1 \right) = 0 \text{ for all } i \in \{1, \ldots, k\}. \qquad (3)$$

To guarantee optimality, it is sufficient to replace (1) and (2) by

$$l(e)(c(e) - c_{\max}) \geq 0 \text{ for all } e \in E, \qquad (1')$$

$$c_i(T)(l(T) - l(MST(S_i))) \leq 0 \text{ for all } i \in \{1, \ldots, k\}. \qquad (2')$$

Summing (1') over all edges and (2') over all $i \in \{1, \ldots, k\}$ and all $T \in \mathcal{T}_i$ yields

$$\sum_{e \in E} c(e)l(e) \geq l(G)c_{\max} \qquad (1'')$$

$$\sum_{i=1}^{k} l(MST(S_i)) \geq \sum_{i=1}^{k} \sum_{T \in \mathcal{T}_i} c_i(T)l(T) = \sum_{e \in E} c(e)l(e). \qquad (2'')$$

Any approximately optimal solution to (LP) that is found in polynomial time determines for each multicast request S_i a (polynomial) set of trees with fractional congestion. The task of selecting one of these trees for each S_i such that the maximum congestion is minimized constitutes a vector selection problem. Raghavan [15] bounds the quality of a solution by analyzing a randomized rounding procedure. For our problem his analysis gives the following result.

Theorem 1. *There exists a solution to the minimum multicast congestion problem with congestion bounded by*

$$\begin{cases} (1 + \varepsilon)r\,OPT + (1 + \varepsilon)(\exp(1) - 1)\sqrt{r\,OPT \cdot \ln m} & \text{if } r\,OPT \geq \ln m \\ (1 + \varepsilon)r\,OPT + \frac{(1+\varepsilon)\exp(1)\ln m}{1 + \ln(\frac{\ln m}{r\,OPT})} & \text{otherwise.} \end{cases}$$

2.2   Approximating the Fractional Optimum

The Plotkin-Shmoys-Tardos Approach. Plotkin, Shmoys and Tardos specify conditions that are sufficient to guarantee relaxed optimality. Adapted to the multicast problem, these conditions are the following relaxations of (1'') and (2'').

$$(1 - \varepsilon)c_{\max}l(G) \leq \sum_{e \in E} c(e)l(e) \qquad (R1)$$

$$(1 - \varepsilon) \sum_{e \in E} c(e)l(e) \leq \sum_{i=1}^{k} l(\tilde{\mathrm{MST}}(S_i)) + \varepsilon c_{\max} l(G). \tag{R2}$$

Their idea is to choose l such that (R1) is automatically satisfied, and to gradually satisfy (R2) by repeated calls to the following routine.

Algorithm IMPROVECONGESTION(1)

$\lambda := c_{\max}/2, \quad \alpha := \frac{4 \ln(2m\varepsilon^{-1})}{\varepsilon \cdot c_{\max}}, \quad \sigma = 5\varepsilon/(9\alpha k)$
while (R2) is not satisfied and $c_{\max} > \lambda$
   { for $e \in E$: $l(e) := \exp(\alpha \cdot c(e))$
     for $i = 1$ to k { $T_i := \tilde{\mathrm{MST}}(S_i)$, for $e \in E(T_i)$: $c(e) := c(e) + \sigma/(1 - \sigma)$ }
     for $e \in E$: $c(e) := c(e) \cdot (1 - \sigma)$ }

Lemma 2. *a) (R1),(R2)* $\Rightarrow$ $c_{\max} \leq (1 + 6\varepsilon)r \cdot OPT$ *for* $\varepsilon \leq 1/6$.
b) $l(e) := \exp(\alpha \cdot c(e))$ *for all* $e \in E$ *and* $\alpha \geq \frac{2 \ln(2m\varepsilon^{-1})}{c_{\max}\varepsilon}$ $\Rightarrow$ *(R1)*.

The following lemma serves to bound the number of iterations until (R2) is satisfied. For a proof, we can follow the general line of [13].

Lemma 3. *Let* $\varepsilon \leq \frac{1}{6}$ *and consider a feasible solution* $(c_i(T), c_{\max})$ *to (LP), such that for* $\alpha \geq \frac{2 \ln(2m\varepsilon^{-1})}{c_{\max}\varepsilon}$ *and* $l(e) := \exp(\alpha \cdot c(e))$ *(for all* $e \in E$*) (R2) is not satisfied. Define* $\sigma := 5\varepsilon/(9\alpha k)$ *and let* $T_i := \tilde{\mathrm{MST}}(S_i)$ *be an approximate minimum Steiner tree for* $i \in \{1, \ldots, k\}$. *Then* $\tilde{c}_i(T) := \begin{cases} (1 - \sigma)c_i(T), & \text{if } T \neq T_i \\ (1 - \sigma)c_i(T) + \sigma, & \text{if } T = T_i, \end{cases}$

and $\tilde{l}(e) := \exp(\alpha \cdot \tilde{c}(e)) := \exp\left(\alpha \sum_{i=1}^{k} \sum_{\substack{T \in \mathcal{T}_i \\ e \in E(T)}} \tilde{c}_i(T)\right)$ *satisfy*

$$l(G) - \tilde{l}(G) \geq \frac{5\varepsilon^2 c_{\max}}{9k} l(G).$$

We thus see that in one iteration of the while loop of IMPROVECONGESTION(1) $l(G)$ decreases by a factor of at least $(1 - 5\varepsilon^2 c_{\max}/(9k)) \leq \exp(-5\varepsilon^2 c_{\max}/(9k))$. The initial value of $l(G)$ is at most $m \cdot \exp(\alpha c_{\max})$. Surely, $r(1 + 6\varepsilon)$-optimality is achieved, if $l(G) \leq \exp(r(1 + 6\varepsilon)\alpha OPT)$. We will use this criterium to bound the required time complexity. Consider the following two phase process: in phase one, the overall length is reduced to at most $\exp(2rOPT)$; phase two successively reaches $r(1 + 6\varepsilon)$-optimality. Each phase consists of a number of subphases, corresponding to calls to IMPROVECONGESTION(1). Let us look at the phases in detail. During the whole of phase 1 – i.e. for each subphase i – we fix $\varepsilon^{(i)}$ at $1/6$. The congestion is halved in each subphase; hence the number of subphases of phase 1 is $O(\ln c_{\max}) = O(\ln k)$, and by Lemma 3, we can bound the number z of iterations the while-loop of IMPROVECONGESTION(1) needs to terminate for the ith subphase of phase 1 by $O(k \ln m/c_{\max}^{(i)})$. As $c_{\max}$ is halved in each subphase, the number of iterations in the last subphase dominates the overall number. Consequently the running-time of phase 1 is $O(k^2(\beta + m) \ln m) = \tilde{O}(k^2 m)$ for the

Mehlhorn-Steiner [12] approximation. (Remember, we assume $c_{\max} \geq \text{OPT} \geq 1$.) The ith subphase of phase 2 starts with $c_{\max}^{(i-1)} \leq r(1+6\varepsilon^{(i)})\text{OPT}$ and terminates when the congestion is at most $r(1 + 3\varepsilon^{(i)})\text{OPT}$. In order to reach $r(1 + 6\varepsilon)$-optimality we thus need $O(\ln \varepsilon^{-1})$ subphases in phase 2, and by Lemma 3, the ith subphase terminates within z iterations, where z satisfies

$$m \exp(\alpha^{(i)} c_{\max}^{(i-1)}) \cdot \exp(-z \cdot 5 c_{\max}^{(i-1)} \varepsilon^{(i)^2}/(9k)) = \exp\left(\alpha^{(i)} c_{\max}^{(i-1)} \cdot \frac{1+3\varepsilon^{(i)}}{1+6\varepsilon^{(i)}}\right), \text{ i.e.}$$

$z = O(k\varepsilon^{(i)^{-2}} \ln(m/\varepsilon^{(i)}))$. Hence the running-time of the ith subphase of phase 2 is $O(k^2 \varepsilon^{(i)^{-2}} \ln(m/\varepsilon^{(i)})(\beta + m))$. Since $\varepsilon^{(i+1)} = \varepsilon^{(i)}/2$, the running-time of the last subphase again is dominating, resulting in an overall estimate of $O(k^2\varepsilon^{-2} \cdot \ln(m/\varepsilon)(\beta + m)) = \tilde{O}(k^2\varepsilon^{-2}m)$ for the Mehlhorn-Steiner approximation. This proves the following theorem.

Theorem 2. *For $\varepsilon \leq \frac{1}{6}$ and assuming $OPT \geq 1$, the fractional minimum multicast congestion problem can be approximated within $r(1 + 6\varepsilon)OPT$ in time $O((k^2\varepsilon^{-2} \ln(m/\varepsilon)(\beta + m))$. Using the Mehlhorn-Steiner approximation, we obtain a $2(1 + 6\varepsilon)$-approximation in time $\tilde{O}(k^2\varepsilon^{-2}m)$.*

Here is a suggestion of how to implement the described algorithm. Instead of testing for (R2), it seems favorable to use $\texttt{maxdual} := \sum_{i=1}^{k} l(\tilde{\text{MST}}(S_i))/l(G)$ as a lower bound on $r\text{OPT}$ and terminate the while-loop as soon as $c_{\max}$ falls below $(1 + 6\varepsilon) \cdot \texttt{maxdual}$. Deviating slightly from our previous notations we use ε_0 to denote the target approximation quality. For reasons of efficiency, $c_i(T)$ is updated to hold the fractional congestion caused by tree T for request S_i only in the end of the algorithm; in previous iterations $c_i(T)$ deviates from the actual definition by a factor called $\texttt{factor}$. We scale the length function by a factor of $\exp(-c_{\max})$ in order to stay within the computers floating point precision.

Algorithm APPROXIMATECONGESTION(1)

```
Compute a start solution, initialize c(e), c_max, maxdual accordingly.
ε = 1/3,  factor = 1
while ε > ε_0/6
   { ε := ε/2
       while c_max > max{1, (1 + 6ε)maxdual}
          { α = (2 ln(2m/ε))/(ε·c_max), σ = 5ε/(9αk), for e ∈ E : l(e) = exp(α(c(e) − c_max))
             for i = 1 to k
             { T_i = M̃ST(S_i), c(T_i) := c(T_i) + σ/(1−σ)
                 for e ∈ E(T_i): c(e) := c(e) + σ/(1−σ)
             }
             maxdual := max{maxdual, (∑_{i=1}^{k} ∑_{e∈E(T_i)} l(e))/(∑_{e∈E} l(e))}
             c_max = 0
             for e ∈ E : c(e) := c(e) · (1 − σ), if c(e) > c_max then c_max = c(e)
             factor := factor · (1 − σ)
          }
   }
for T ∈ ∪_{i=1}^{k} T̂_i : c(T) := c(T)·factor
```

Goldberg, Oldham, Plotkin, and Stein [5] point out that the practical performance of the described algorithm significantly depends on the choice of α and σ.

Instead of using the theoretical value for α, they suggest to choose α such that the ratio $\frac{(l(G)c_{\max}/\sum_{e\in E}c(e)l(e))-1}{(\sum_{e\in E}c(e)l(e)/\sum_{i=1}^{k}l(\tilde{MST}(S_i)))-1}$ remains balanced, i.e. they increase α by a factor of $(\sqrt{5}+1)/2$ if the ratio is larger than 0.5 and otherwise decrease it by $2/(\sqrt{5}+1)$. Since our aim at choosing α is to satisfy (R1), one could alternatively think of taking α as $\alpha := \min\left\{\alpha \in \mathbb{R}_{>0} \mid \frac{l(G)c_{\max}}{\sum_{e\in E}c(e)l(e)} \leq \frac{1}{1-\varepsilon}\right\}$. Moreover, they emphasize the necessity of choosing σ dynamically such as to maximize $l(G) - \tilde{l}(G)$ in each iteration. Since $\tilde{l}$ is the sum of exponential functions, we can easily compute $\frac{d\tilde{l}(\sigma)}{d\sigma} = \sum_{e\in E}\alpha(c_0(e) - c(e))\exp(\alpha c(e) + \alpha\sigma(c_0(e) - c(e)))$, where $c_0(e) := |\{T \mid \exists i \in \{1,\dots,k\}$ such that $T = \tilde{MST}(S_i)$ and $e \in E(T)\}|$ abbreviates the congestion of edge e caused by choosing for each i an approximate minimum Steiner tree. The second derivative is positive, so $\tilde{l}$ has a unique minimum in $[\sigma_{\text{theor.}}, 1]$. Goldberg, Oldham, Plotkin, and Stein suggest to determine this minimum with the Newton-Raphson method. One could also try to use binary search.

The Radzik Approach. Adapting the algorithm of Radzik [14] to our problem, yields a theoretical speedup by a factor of k. The key idea is to update the length function after each single approximate Steiner tree computation, if the length of the new tree is sufficiently small. Radzik uses relaxed optimality conditions different from (R1) and (R2).

Theorem 3. Let $\lambda \geq OPT$ and $\varepsilon \leq 1/3$. If $c_{\max} \leq (1 + \varepsilon/3)\lambda$ and $\sum_{i=1}^{k}l(\tilde{MST}(S_i)) \geq (1 - \varepsilon/2)\lambda l(G)$ then $c_{\max} \leq r(1 + \varepsilon)OPT$.

The following routine IMPROVECONGESTION(2) is the heart of Radzik's algorithm. As the input it takes sets of trees $\hat{\mathcal{T}}_i^{(0)}$ with congestion $c_i^{(0)}$ and a parameter ε determining the approximation quality.

Algorithm IMPROVECONGESTION(2)

$\lambda := c_{\max}, \ \alpha := \frac{3(1+\varepsilon)\ln(m\varepsilon^{-1})}{\lambda\varepsilon}, \ \sigma := \frac{\varepsilon}{4\alpha\lambda}, \ \text{for } e \in E : l(e) := \exp(\alpha\cdot c(e))$
do $\{$ $LG := l(G)$
    for $i = 1$ to k
      $\{$ $T_i := \tilde{MST}(S_i)$
        $Li\_neu := l(T_i), \ Li\_alt := \sum_{e \in E}c_i(e)l(e)$
        if $Li\_alt - Li\_neu \geq \varepsilon\cdot Li\_alt$ then
          $\{$ for $e \in E$
            $\{$ $c(e) := c(e) - c_i(e), \ c_i(e) := c_i(e)\cdot(1 - \sigma)$
              $c(e) := c(e) + c_i(e), \ c_{\max} := \max\{c_{\max}, c(e)\}$
              $l(e) := \exp(\alpha\cdot c(e))$ $\}$
           for $e \in E(T_i)$
            $\{$ $c_i(e) := c_i(e) + \sigma, \ c(e) := c(e) + \sigma, \ l(e) := \exp(\alpha\cdot c(e))$ $\}$
           $\hat{\mathcal{T}}_i := \hat{\mathcal{T}}_i \cup \{T_i\}$
          $\}$
      $\}$
    $\}$
while $c_{\max} > \left(1 - \frac{\varepsilon}{3}\right)\lambda$ and $LG - l(G) > \frac{\varepsilon^2}{8}LG$

Theorem 4. *Let* $\varepsilon \leq \frac{1}{24}$. *Algorithm* IMPROVECONGESTION(2) *terminates with maximum congestion* $c_{\max}^{(q)} \leq \left(1 - \frac{\varepsilon}{3}\right) c_{\max}^{(0)}$ *or* $c_{\max}^{(q)} \leq r(1 + 8\varepsilon)OPT$.

Theorem 5. *Let* $\varepsilon \leq 1$. IMPROVECONGESTION(2) *takes* $O(\varepsilon^{-2} \ln n)$ *iterations of the* while-*loop to terminate.*

Corollary 2. *a)* $O\left(\ln k + \ln \frac{1}{\varepsilon}\right)$ *calls to* IMPROVECONGESTION(2) *suffice to produce an* $r(1 + \varepsilon)$-*optimal solution. b) The overall running-time required is* $O\left(\varepsilon^{-2}\right.$
$k \ln n \ln k(m + \beta)) = \tilde{O}(km\varepsilon^{-2})$ *for the Mehlhorn-Steiner approximation.*

The Garg-Könemann Approach. The algorithm of Garg and Könemann [3] differs from the previous algorithms in the fact that there is no transfer of congestion from $\tilde{\mathrm{MST}}(S_i)^{(j)}$ to $\tilde{\mathrm{MST}}(S_i)^{(j+1)}$. Instead, the functions describing the congestion are built up from scratch. A feasible solution is obtained by scaling the computed quantities in the very end.

Algorithm IMPROVECONGESTION(3)

```
for  e ∈  E:  l(e) := δ
while  Σₑ∈ₑ l(e) < ξ
    { for  i = 1 to  k
        {  T := M̃ST(Sᵢ),  cᵢ(T) := cᵢ(T) + 1, for  e  ∈  T: l(e) := l(e) · (1 + ε/u)  }
    }
scale  c̄ᵢ(T) := cᵢ(T)/#iterations
```

Lemma 4. *Suppose that* $l(e)$ *is initialized to some constant* δ *for all* $e \in E$. *Let* q *be the number of iterations of the* while-*loop* IMPROVECONGESTION(3) *needs to terminate. Then* $\bar{c}_{\max}^{(q)} < \frac{u \cdot \ln(\xi/\delta)}{(q-1) \cdot \ln(1+\varepsilon)}$ *and* $q \leq \left\lceil \frac{u}{OPT} \cdot \frac{\ln(\xi/\delta)}{\ln(1+\varepsilon)} \right\rceil$

Proof. Whenever $c(e)$ increases by u, $l(e)$ grows by a factor of at least $(1 + \varepsilon)$. Let z count the number of times this happens. Since $l^{(q-1)}(e) < \xi$, we have $\delta(1 + \varepsilon)^z < \xi$, so $z < \frac{\ln(\xi/\delta)}{\ln(1+\varepsilon)}$. After $q - 1$ iterations, $\sum_{T \in \hat{\mathcal{T}}_i} c_i(T) = q - 1$ for all $i \in \{1, \ldots, k\}$. Thus $\left(\bar{c}_{\max}, (\bar{c}_i(T)_{\substack{i \in \{1,\ldots,k\} \\ T \in \mathcal{T}_i}}\right) := \left(\frac{c_{\max}}{q-1}, \left(\frac{c_i(T)}{q-1}\right)_{\substack{i \in \{1,\ldots,k\} \\ T \in \mathcal{T}_i}}\right)$ is a feasible solution satisfying $\bar{c}_{\max} \leq \frac{u \cdot z}{q-1} < \frac{u \cdot \ln(\xi/\delta)}{(q-1)\ln(1+\varepsilon)}$. The second estimate follows since $1 \leq \frac{\bar{c}_{\max}}{OPT} < \frac{u \ln(\xi/\delta)}{(q-1)\ln(1+\varepsilon)OPT}$. $\square$

To achieve a target approximation guarantee of $r(1+\varepsilon)OPT$, the free parameters have to be chosen appropriately.

Lemma 5. *Let* $\varepsilon \leq \frac{1}{36}$. *For* $u < 2OPT$ *and* $\xi := \delta \cdot \left(\frac{m}{1-2\varepsilon r}\right)^{\frac{1+\varepsilon}{\varepsilon}}$ *the algorithm terminates with* $\bar{c}_{\max} < r(1 + 6\varepsilon)OPT$.

Theorem 6. *An $r(1+6\varepsilon)OPT$-approximate solution to the fractional minimum multicast problem can be obtained in time $O((\beta+m)\varepsilon^{-2}k\cdot\ln k\ln m) = \tilde{O}(\varepsilon^{-2}km)$ using the Mehlhorn-Steiner approximation.*

Proof. We repeatedly call IMPROVECONGESTION(3) in the following u-scaling process. Throughout the process we maintain the condition $u < 2OPT$ to make sure that the algorithm terminates with the claimed guarantee. So all we have to worry about is the number of iterations. The first call is performed with $u := \frac{k}{2}$. By Lemma 5, if $u \leq OPT$ the algorithm terminates within $\left\lceil\frac{\ln(\xi/\delta)}{\ln(1+\varepsilon)}\right\rceil$ iterations. Otherwise, we know that $OPT < \frac{k}{2}$ and restart the algorithm with $u := \frac{k}{4}$. Again, IMPROVECONGESTION(3) either terminates in $\left\lceil\frac{\ln(\xi/\delta)}{\ln(1+\varepsilon)}\right\rceil$ iterations or we may conclude that $OPT < \frac{k}{4}$ and restart the algorithm with $u := \frac{k}{8}$. Repeating this at most $O(\ln k)$ times yields the claim, since $\frac{\ln(\xi/\delta)}{\ln(1+\varepsilon)} = O(\varepsilon^{-2}\ln m)$ and because one iteration takes $O(k(m+\beta))$-time. $\qquad\square$

Unfortunately, the above u-scaling process is *not practically efficient*, since the number of iterations in each scaling phase in considerably large (note that we need $\varepsilon \leq \frac{1}{36}$ to guarantee the approximation quality, so even for moderate values of m, say $m := 500$, we may have to await over 8000 iterations before we can decide whether or not $OPT < u$). However, without u-scaling the running-time increases by a factor of $k/\ln k$. Again, we may (practically) improve the running-time by using $\bar{c}_{\max} < (1+6\varepsilon)\sum_{i=1}^{k}l(\tilde{MST}(S_i))/l(G)$ instead of $l(G) \geq \xi$ as the termination criterium. It is an open question whether or not ε-scaling as described in the previous sections can be advantageously applied. (Our experiments strongly indicate that ε-scaling substantially decreases the running-time, but an analysis is still missing.) The following implementation takes care of the fact that due to limited precision it may be necessary to rescale the edge-lengths.

Algorithm APPROXIMATE CONGESTION(3)

$maxdual := 0,\ minprimal := k,\ c_{\max} := 0,\ iteration := 0,\ LG := m$
for $e \in E$: $c(e) := 0,\ l(e) := 1$
do { $LM := 0$
    for $i = 1$ **to** k
      { $T_i = \tilde{MST}(S_i),\ c(T_i) := c(T_i) + 1$
        for $e \in E(T_i)$
          { $LG := LG - l(e),$
            $l(e) := l(e) \cdot \left(1 + \frac{\varepsilon}{k}\right),\ c(e) := c(e) + 1,\ c_{\max} := \max\{c_{\max}, c(e)\},$
            $LG := LG + l(e),\ LM := LM + l(e)$ }
      }
    $iteration := iteration + 1,\ dual := \frac{LM}{LG}$
    $minprimal := \min\{minprimal, \frac{c_{\max}}{iteration}\},\ maxdual := \max\{maxdual, dual\}$
    if $LG > 100000$ { $LG := \frac{LG}{100000}$, **for** $e \in E$: $l(e) := \frac{l(e)}{100000}$ }
} **while** $minprimal \geq (1 + 6\varepsilon)maxdual$
$c_{\max} := \frac{c_{\max}}{iteration}$, **for** $T \in \hat{T}_i$: $c(T) := \frac{c(T)}{|T_i|}$

2.3   Integral Solution

Pure Combinatorial Approach. Klein et al. [9] describe an approximation algorithm for the unit capacity concurrent flow problem that can be modified to approximate an integral solution to the minimum multicast congestion problem without the necessity to round. The algorithm is similar to the one presented in Section 2.2.1. However, instead of updating all trees in one iteration of the while loop, only one "bad" tree per iteration is modified.

Definition 1. *A tree* $T \in \mathcal{T}_i$ *is* r-bad *for* S_i *if* $(1 - \varepsilon')l(T) > rl(MST(S_i)) + \varepsilon' \cdot \frac{c_{max} \cdot l(G)}{k}$.

This allows us to choose σ by a factor of k larger than in the algorithms described previously, and it can be shown that consequently σ never needs to be reduced below 1, if OPT $= \Omega(\log n)$. (We omit the proof due to lack of space.)

Theorem 7. *An integral solution to the minimum multicast congestion problem with congestion* $OPT \cdot O(\sqrt{OPT \cdot \log n})$ *can be found in time* $O(k^2(m+\beta)\log k)$.

Aspes et al. [1] give an online algorithm for approximating our problem within a factor of $O(\log n)$. Since they explicitly consider the multicast congestion problem, we do not restate their algorithm but instead analyze a simpler online algorithm for which the same approximation bound applies, if the value of an optimal solution is known in advance.

Algorithm ONLINE CONGESTION

```
for e ∈ E : { l(e) := 1, c(e) := 0 }
for i = 1 to k { Tᵢ := M̃ST(Sᵢ), for e ∈ E(Tᵢ) {l(e) := l(e) · A, c(e) := c(e) +
1} }
return(T₁,...,Tₖ,cₘₐₓ)
```

Theorem 8. *For* $A := 1 + \frac{1}{2rOPT}$, *the above algorithm terminates with maximal congestion* $O(rOPT \cdot \log(n))$.

Proof. Let $l^{(j)}$ denote the length function at the end of the jth iteration of the for-loop. From $l^{(j)}(G) = l^{(j-1)}(G) + (A-1)l^{(j-1)}(\tilde{M}ST_{l^{(j-1)}}(S_{i-1}))$ we get $l^{(k)}(G) = m + (A-1) \cdot \sum_{i=1}^{k} l^{(i-1)}(\tilde{M}ST_{l^{(i-1)}}(S_i)) \leq m + (A-1) \cdot \sum_{i=1}^{k} l^{(k)}(\tilde{M}ST_{l^{(k)}}(S_i)) \leq m + (A-1) \cdot rOPTl^{(k)}(G)$ by Corollary 1. Since $l^{(k)}(G) \geq A^{c_{max}-1}$, the claim follows by the choice of A. $\quad\square$

Derandomization. A deterministic approximation satisfying the bounds of Theorem 1 can be proved with the tools of Srivastav and Stangier [16]. For a proof we refer to the full paper.

3   A New and Practically Efficient Implementation

In practical experiments the following algorithm was superior to all theoretically efficient approximation algorithms. The algorithm uses different length functions for determining a start set of Steiner trees and for updating trees. The first length function is similar to the one in Aspnes et al.'s online algorithm, but uses a base of n instead of a base $\in (1, 1.5]$. In fact, we found that the quality of the solutions increases substantially as the base grows. To improve near optimal solutions it turned out to be of advantage to replace the exponent $c(e)/c_{max}$ in the length function by $c(e) - c_{max}$, thereby increasing the impact of highly congested edges. On the other hand, the exponential base must be neither too small nor too large to "correctly" take into account edges with lower congestion. Here, the size of the current best tree (experimentally) proved to be a good choice.

Algorithm PRACTICAL CONGESTION

```
λ=1;
for i=1 to k
    { for e ∈ E : l(e) := n^(c(e)/λ - 1)
      T_i := M̃ST(S_i), for e ∈ E(T_i) { c(e) := c(e) + 1, λ = max{λ, c(e)} } }
while (true)
    { for i := 1 to k
        { A := |E(T_i)|, for e ∈ E: l(e) := A^(c(e)-c_max)
          for e ∈ E(T_i) : l(e) := l(e)/A
          T'_i := M̃ST(S_i), for e ∈ E(T_i) : l(e) := l(e) · A
          if l(T_i) > l(T'_i) then T_i := T'_i
          update c(e)
        }
    }
```

4 Experimental Results

We tested our new algorithm against Aspnes et al.'s online algorithm and a version of APPROXIMATE CONGESTION(3), which is theoretically fastest. As discussed in section 2, we did not implement u-scaling for reasons of performance. Instead, we provided APPROXIMATE CONGESTION(3) with the (near) optimal value of u computed by our new algorithm. ε was set to its theoretically maximum value of $1/36$. The number of iterations was restricted to 100. We ran

#requests/ #terminals	ONLINE	APPROXIMATE CONG.(3) (#it.)	New Alg. (# it.)	$\widetilde{\text{Dual}} \geq$
50 / 4	11	12 (1), 2.5 (100)	2 (1)	1.04
100 / 4	17	17 (1), 4.4 (100)	3 (1)	2.01
150 / 4	29	29 (1), 6.1 (100)	5 (1), 4 (2)	2.86
200 / 4	43	44 (1), 8.0 (100)	7 (1), 5 (2)	3.85
300 /4	56	57 (1), 11.5 (100)	9 (1), 7 (3)	5.77
500 / 4	78	79 (1), 19.9 (100)	16 (1), 12 (2)	10.00
1000 / 4	159	161 (1), 36.5 (100)	29 (1), 21 (69)	20.00
2000 / 4	378	383 (1), 76.1 (100)	60 (1), 44 (2)	41.00
500 / 2–100	178	178 (1), 69.1 (100)	41 (1), 32 (9)	31.00
1000 / 2 –100	326	326 (1), 100.5 (100)	79 (1), 65 (3)	64.00

the online algorithm with various values of $A \in (1, 1.5]$ and listed the average outcome. Our test instance was a grid graph on 2079 nodes and 4059 edges with 8 rectangular holes. Such Grid graphs typically arise in the design of VLSI logic chips where the holes represent big arrays on the chip. They are considered as hard instances for path- as well as for tree-packing problems. Multicast requests of size 50 to 2000 were chosen by a random generator. Columns 2–4 show the computed congestion and in brackets the number of iterations. Dual approximates $\max\{\sum_{i=1}^{k} l(\widetilde{MST}(S_i)/l(G) \mid l : E \to \mathbb{R}_{\geq 0}\}$, where $\widetilde{MST}$ is computed by Mehlhorn's algorithm (i.e. $\widetilde{Dual} \leq 2 \cdot OPT$).

References

1. J. Aspnes, Y. Azar, A.Fiat, S. Plotkin, O. Waarts, On-line routing of virtual circuits with applications to load balancing and machine scheduling, *J. of the Association for Computing Machinery* **44** (3), 486–504, 1997.
2. R. Carr, S. Vempala, Randomized Metarounding, *Proc. of the 32nd ACM Symposium on the theory of computing (STOC '00), Portland, USA*, 58–62, 2000
3. N. Garg, J. Könemann, Faster and Simpler Algorithms for Multicommodity Flow and other Fractional Packing Problems. In *Proc. 39th IEEE Annual Symposium on Foundations of Computer Science*, 1998.
4. A.V. Goldberg, A natural randomization strategy for multicommodity flow and related algorithms, *Information Processing Letters* **42**, 249–256, 1992.
5. A.V. Goldberg, A.D. Oldham, S. Plotkin, C. Stein, An Implementation of a Combinatorial Approximation Algorithm for Minimum-Cost Multicommodity Flows. In *Proc. 6th Conf. on Integer Prog. and Combinatorial Optimization*, 1998.
6. M.D. Grigoriadis, L.G. Khachiyan, Fast approximation schemes for convex programs with many blocks and coupling constraints, *SIAM J. on Optimization* **4**, 86–107, 1994.
7. K. Jansen, H. Zhang, An approximation algorithm for the multicast congestion problem via minimum Steiner trees, In *Proc. 3rd Int. Worksh. on Approx. and Random. Alg. in Commun. Netw. (ARANCE'02), Roma, Italy, September 21*, 2002.
8. K. Jansen, H, Zhang, Approximation algorithms for general packing problems with modified logarithmic potential function, In *Proc. 2nd IFIP Int. Conf. on Theoretical Computer Science (TCS'02), Montréal, Québec, Canada, August 25 - 30*, 2002.
9. P. Klein, S. Plotkin, C. Stein, E. Tardos, Faster Approximation Algorithms for the Unit Capacity Concurrent Flow Problem with Applications to Routing and Finding Sparse Cuts, *SIAM J. on Computing* **23** No.3, 466–487, 1994.
10. T. Leighton, F. Makedon, S. Plotkin, C. Stein, E. Tardos, S. Tragoudas, Fast approximation algorithms for multicommodity flow problems, *J. of Comp. and System Sciences* **50**, 228–243, 1995.
11. D.W. Matula, F. Shahrokhi, The maximum concurrent flow problem, *J. of the Association for Computing Machinery* **37**, 318–334, 1990.
12. K. Mehlhorn, A faster approximation algorithm for the Steiner problem in graphs, *Information Processing Letters* **27**, 125–128, 1988.
13. S. Plotkin, D. Shmoys, E. Tardos, Fast approximation algorithms for fractional packing and covering problems, *Math. Oper. Res.* **20**, 257–301, 1995.

14. T. Radzik, Fast deterministic approximation for the multicommodity flow problem, *Math. Prog.* *78*, 43–58, 1997.
15. P. Raghavan, Probabilistic construction of deterministic algorithms: Approximating packing integer programs, *J. of Comp. and System Sciences* **38**, 683–707, 1994.
16. A. Srivastav, P. Stangier, On complexity, representation and approximation of integral multicommodity flows, *Discrete Applied Mathematics* **99**, 183–208, 2000.
17. S. Vempala, B. Vöcking, Approximating Multicast Congestion, *Proc. 10th ISAAC, Chennai, India*, 1999.
18. G. Robins, A. Zelikovsky, Improved Steiner tree approximation in graphs, *Proc. of the 11th Annual ACM-SIAM Symp. on Discrete Algorithms (SODA 2000)*, 770–779, 2000.

Approximation of a Retrieval Problem
for Parallel Disks

Joep Aerts[1], Jan Korst[2], and Frits Spieksma[3]

[1] Centre of Quantitative Methods, P.O. Box 414,
NL-5600 AK Eindhoven, The Netherlands
aerts@cqm.nl
[2] Philips Research Labs, Prof. Holstlaan 4, 5656 AA Eindhoven, The Netherlands
jan.korst@philips.com
[3] Department of Applied Economics, Katholieke Universiteit Leuven,
Naamsestraat 69, B-3000 Leuven, Belgium
frits.spieksma@econ.kuleuven.ac.be

Abstract. We study a number of retrieval problems that relate to effectively using the throughput of parallel disks. These problems can be formulated as assigning a maximum number of jobs to machines of capacity two, where jobs are of size one or two that must satisfy assignment restrictions. We prove that the LP-relaxation of an integer programming formulation is half-integral, and derive an interesting persistency property. In addition, we derive $\frac{2}{3}$-approximation results for two types of retrieval problems.

1 Introduction

When handling large collections of data, the communication between fast internal memory and slow external memory (e.g. disks) can be a major performance bottleneck [18]. Several disk storage strategies have been proposed that use redundancy to balance the load on parallel disks [3,9,16]. Storing multiple copies of each data block on different disks allows one to dynamically determine from which disk to retrieve a requested block. This improves the throughput, i.e., the number of requests served per time unit.

The retrieval of data from parallel disks can be modeled as follows. Given is a set of identical disks on which equal-sized blocks are stored redundantly. Requests for blocks arrive over time and repeatedly a batch of blocks is retrieved. Under the assumption that a disk can only retrieve one block per cycle, the problem of maximizing the number of blocks retrieved in a cycle can be modeled as a maximum flow problem [3,9].

In reality, however, this assumption is not fulfilled. In fact, the disks can be used more efficiently if we exploit the multi-zone character of hard disks [4,13]. Hard disks rotate at a constant angular velocity, while outer tracks contain more data than inner tracks. Hence, one can retrieve data at a higher rate from the outside than from the inside. One can partition a disk's storage space into two halves, an inner half and an outer half. Given this partition, it is reasonable to

R. Petreschi et al. (Eds.): CIAC 2003, LNCS 2653, pp. 178–188, 2003.

assume that the worst-case retrieval rate at the outer half is approximately twice the worst-case retrieval rate achievable at the inner half.

Using the above, we consider the following improvement. Instead of retrieving at most one block per disk per cycle, each disk can now either retrieve two blocks from its outer half or one block from its inner half. In this way, we can considerably increase the average number of blocks read per cycle, with little or no increase in the average cycle duration. Given that blocks are stored redundantly, the problem of maximizing the number of retrieved blocks is no longer polynomially solvable, provided $P \neq NP$.

We consider two variants of the retrieval problem, depending on the way the blocks are stored on the disks. In the first variant, copies of popular blocks, i.e., blocks with a high probability of being requested, are stored on the outer halves, and copies of less popular blocks are stored on the inner halves of the disks. In the second variant, the copies of all blocks can be stored on inner as well as outer halves. If all blocks have approximately the same request probability, the second variant improves upon the first, as each block can be read from a disk's outer half.

If we refer to block requests as jobs and to disks as machines, the retrieval problem can be considered as nonpreemptively assigning jobs to machines, with jobs having a size one or two and machines having capacity two, where each job can only be assigned to a subset of the machines. The objective is to maximize the number of assigned jobs. In the remainder of this paper, we generalize the problem setting as follows. Each job j has a weight w_j, and each machine i has a capacity b_i, where the objective is to maximize the sum of the weights of the assigned jobs. The generalizations of the two variants of the retrieval problem are denoted by RP1 and RP2, respectively.

1.1   Related Work

Storage and retrieval problems for parallel disks have been studied extensively in the area of video on demand and external memory algorithms [18]. Several authors discuss redundant data storage strategies and analyze the corresponding retrieval problems. Korst [9] and Papadopouli and Golubchik [11] discuss duplicate storage strategies and introduce both a max-flow algorithm that minimizes the number of block requests assigned to one disk in a cycle. Santos et al. [17] and Sanders [14] also use redundant storage strategies but assume a nonperiodic retrieval strategy. Santos et al. use shortest queue scheduling to assign the block requests to the disks and Sanders describes alternative online scheduling strategies. All these papers use the number of blocks as optimization criterion and do not exploit the multi-zone character of hard disks. Aerts et al. [3,4] and Sanders [15] describe a more detailed model for the retrieval problem for parallel disks in which the zone location of the stored blocks is taken into account.

Golubchik et al. [7] study a problem related to RP1 where data objects need to be assigned to parallel disks. In their setting each data object can be assigned to any disk, and they are interested in finding an assignment of data objects to disks and an assignment of clients to disks such that the number of served clients is maximized. They derive tight upper and lower bounds on this criterion for different types of disk systems.

Problem RP1 can be seen as a special case of the multiple knapsack problem with assignment restrictions. The special case arises because in our setting the size of each job is 1 or 2. Another special case of this problem has been studied by Dawande et al. [6]. Motivated by applications in inventory problems arising in the steel-industry, they investigate RP1 for the case where the size of a job coincides with its weight. They describe two approaches that each lead to a $\frac{1}{2}$-approximation algorithm.

In Section 2 we show that a formulation of RP1 is so-called *half-integral*, meaning that in an optimal solution of its LP-relaxation each variable equals 0, $\frac{1}{2}$, or 1. This property has been studied by Hochbaum [8]. She describes a technique that, based on first finding a half-integral superoptimal solution, delivers 2-approximations for a large class of minimization problems. In particular, her framework focusses on formulations where there is a unique variable for each constraint in the formulation. Other occurences of half-integrality are described in Chudak and Hochbaum [5] and Ralphs [12].

As far as we are aware, both RP1 and RP2 have not been investigated in the literature.

1.2   Our Results

In this paper we derive the following results. In Section 2 we prove that the LP-relaxation of an integer programming formulation of RP1 is half-integral. We also show that the unweighted version of RP1 satisfies an interesting persistency property. More specifically, it turns out that there exists an optimal solution to RP1 such that jobs of size 1 not assigned in the LP-relaxation are also not assigned in that optimal solution and vice versa.

In addition, we sketch $\frac{2}{3}$-approximation algorithms for RP1 as well as for RP2 in Sections 3 and 4, and we show that these ratios are tight.

2   Problem Description

Consider the following problem setting for RP1. Given is a set J_1 of jobs of size 1 (called the small jobs), a set J_2 of jobs of size 2 (called the large jobs), and a set $M = \{1, \ldots, m\}$ of machines, with machine $i \in M$ having capacity b_i. Let $J = J_1 \cup J_2$. Each job $j \in J$ can be assigned to a job-specific subset of the machines, denoted by $M(j)$. There is a weight w_j given for each job $j \in J$. All data are positive integral numbers. The problem is to find an assignment of jobs to machines that maximizes the weighted number of jobs that are assigned.

Theorem 1 (Aerts et al. [2]). *RP1 is APX-hard, even if each machine has capacity 2, each job is connected to exactly 2 machines, and all weights are 1, i.e., if $b_i = 2$ for all $i \in M$ and $|M(j)| = 2$ and $w_j = 1$ for all $j \in J$.*

2.1   Formulations of RP1

A straightforward formulation of RP1 is as follows. For each $j \in J$ and $i \in M(j)$ we introduce a decision variable

$$x_{ji} = \begin{cases} 1 \text{ if job } j \text{ is assigned to machine } i \\ 0 \text{ otherwise.} \end{cases}$$

The model now is:
(M1) Maximize

$$\sum_{j \in J} \sum_{i \in M(j)} w_j x_{ji}$$

subject to:

$$\forall_{j \in J} \sum_{i \in M(j)} x_{ji} \leq 1, \tag{1}$$

$$\forall_{i \in M} \sum_{j \in J_1(i)} x_{ji} + 2 \sum_{j \in J_2(i)} x_{ji} \leq b_i, \tag{2}$$

$$\forall_{j \in J} \forall_{i \in M(j)} x_{ji} \in \{0, 1\}, \tag{3}$$

with $J_1(i)$ and $J_2(i)$ being the sets of small and large jobs that can be assigned to machine i, respectively. Constraints (1) express that each job can be assigned at most once, (2) state that the capacity of each machine should not be violated, and (3) are the integrality constraints.

To be able to state our approximation results we modify model M1 by 'splitting' each large job j into two jobs of size 1, called $l(j)$ and $r(j)$. The machine sets of the new jobs remain the same, i.e., $M(l(j)) = M(r(j)) = M(j)$ for $j \in J_2$, and the weight of each new job is half of the original weight, i.e., $w_{l(j)} = w_{r(j)} = \frac{1}{2} w_j$ for $j \in J_2$. When we redefine J as the set of all jobs of size 1, we can write down the following, alternative, formulation of RP1.

(M2) Maximize

$$\sum_{j \in J} \sum_{i \in M(j)} w_j x_{ji}$$

subject to:

$$\forall_{j \in J} \sum_{i \in M(j)} x_{ji} \leq 1,$$

$$\forall_{i \in M} \sum_{j \in J(i)} x_{ji} \leq b_i,$$

$$\forall_{j \in J_2} \forall_{i \in M(j)} x_{l(j),i} - x_{r(j),i} = 0, \tag{4}$$

$$\forall_{j \in J} \forall_{i \in M(j)} x_{ji} \in \{0, 1\}, \tag{5}$$

with $J(i)$ being the set of jobs that can be assigned to machine i.

Note that constraints (4) and (5) imply that either both jobs $l(j)$ and $r(j)$ are assigned to machine i or both jobs are not assigned to machine i. Thus, model M2 is a valid formulation of RP1. We use the following terminology to describe our results:

- let LPM1 and LPM2 denote the linear programming relaxation of formulation M1 and M2, respectively, i.e., the model that results when replacing constraints (3) or (5) by $x_{ji} \geq 0$, $\forall_{j,i}$.
- let $\mathcal{I}$ denote an instance of RP1, and let $LP(\mathcal{I})$ and $OPT(\mathcal{I})$ denote the corresponding values of LPM2 and the optimal solution.
- let x^{LP} denote an optimal solution to LPM2, and let x^{OPT} denote an optimal solution to RP1.
- a vertex or a vector x is called *half-integral* iff the value of each component x_i in the vector x is in $\{0, \frac{1}{2}, 1\}$.
- a polyhedron Q is called *half-integral* when each extreme vertex of Q is half-integral.

Theorem 2. *LPM2 is half-integral.*

Proof. We first show that the linear programming relaxation of M2 can be found by solving a min-cost flow problem on a network with $O(n+m)$ nodes and $O(nm)$ arcs where $n = |J|$.

The network is defined as follows. There is a source, a sink, a node for each small job and a node for each machine. There is an arc from the source to each job-node having capacity 1 and cost 0. There is an arc from each machine-node i to the sink having capacity b_i and cost 0. There is an arc from each job-node j to each machine-node $i \in M(j)$ having capacity 1 and cost $-w_j$; we refer to such an arc as a *middle arc*. Finally, there is an arc from the source to the sink with capacity n and cost 0. The supply of each node equals 0, except the supply of the source which equals n and the supply of the sink which equals $-n$.

We now argue that a feasible flow in this network corresponds to a feasible solution to LPM2 with the same value.

Let x be a solution to LPM2. We associate the following flow with this solution. Each element x_{ji} corresponds directly to an arc between a job-node and a machine-node. We set the flow on this arc equal to x_{ji}. In this way we have specified the flows on all middle arcs. The flow on the arcs between the source and the job-nodes as well as the flow on the arcs between machine-nodes and the sink follow immediately. Observe that the resulting flow is feasible since the constraints in M2 ensure that the arc capacities are not exceeded. Also observe that the value of the flow equals the value of the solution x.

Consider now any feasible flow, and concentrate on the flow on the middle arcs. For each job $j \in J_1$ we simply set the corresponding x-elements equal to the flows on the arcs emanating from the corresponding job-node. For each job $j \in J_2$ and a machine i, we have two arcs in the network, namely $(l(j), i)$ and $(r(j), i)$. We distribute the total flow on these two arcs evenly between the variables $x_{l(j),i}$ and $x_{r(j),i}$, thereby ensuring that constraints (4) hold. Note that the constructed x-vector satisfies all constraints in M2 and has the same value.

Observe finally that, due to the integrality of an optimal min-cost flow solution, each extreme vertex of LPM2 is half-integral.

Corollary 1. x^{LP} *can be found by solving a min-cost flow problem on a network with $O(n + m)$ nodes and $O(nm)$ arcs.*

Proof. This follows from Theorem 2.

By establishing a correspondence between a feasible solution to LPM1 and LPM2, one easily deduces the following corollary:

Corollary 2. *LPM1 is half-integral.*

Observe that Theorem 2 implies that each job $j \in J_1$ is integrally assigned in an optimal solution to the LP-relaxation. We can bound the gap that exists between the value of the LP-relaxation and the optimal solution as follows.

Theorem 3. $LP(\mathcal{I}) \leq \frac{3}{2}OPT(\mathcal{I})$ *for all instances $\mathcal{I}$ of RP1.*

Proof. See the proof of Theorem 5.

The following instance of RP1 shows that this bound is asymptotically tight even when we have unit weights, exactly two machines per job and machines with capacity 2.

Instance. We have an instance with k machines, $k+2$ small jobs and $k-2$ large jobs. We number the small jobs from 1 to $k+2$ and the large jobs from $k+3$ to $2k$. We have $M(j) = \{1, 2\}$ for $j = 1, 2, 3, 4$, $M(j) = \{2, j - 2\}$ for $j = 5, \ldots, k+2$, $M(j) = \{j - k, j - k + 1\}$ for $j = k + 3, \ldots, 2k - 1$, and $M(2k) = \{k, 3\}$. An example for $k = 5$ is given in Figure 1. The value of an optimal solution is $k + 2$ (since machines $3, \ldots, k$ each cannot give more than 1), whereas the value of the relaxation equals $\frac{3}{2} \cdot (k - 2) + 4 = \frac{3}{2}k + 1$.

Remark: Note that the results in this subsection go through even when we have machine-dependent profits for a job, i.e., even when assigning job j to machine i yields w_{ji} units of profit.

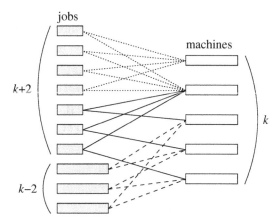

Fig. 1. Example of tightness of Theorem 3

2.2 Persistency

The phenomenon that variables having integral values in a relaxation of the problem attain this same integral value in an optimal solution is called *persistency*. It is a relatively rare phenomenon in integer programming; a well-known example of it is the vertex-packing polytope (Nemhauser and Trotter [10]; see also Adams et al. [1] and the references therein). We show here that the unweighted version of RP1, i.e., with $w_j = 1$ for all $j \in J_1 \cup J_2$, satisfies an interesting form of persistency. We first prove the following two lemmas.

Lemma 1. *There exists an optimal solution to RP1 such that each job $j \in J_1$ that is assigned in an optimal solution to the LP-relaxation, i.e., for which $x_{ji} = 1$ for some $i \in M(j)$, is also assigned in that optimal solution.*

Proof. By contradiction. Suppose that no optimal solution assigns all small jobs that are assigned in x^{LP}. Consider now an optimal solution to RP1 that has the following property: it has a maximum number of small jobs on the same machines as in x^{LP}. By assumption, there exists a small job, say job $j \in J_1$, that is assigned in x^{LP} and not present in this optimal solution. Let i be the machine to which j is assigned in the x^{LP}. Obviously, machine i has no free capacity in the optimal solution (otherwise we could have improved that solution). Also, at least one of the jobs that are assigned to machine i in the optimal solution, say job j', was not present at that machine in x^{LP}. Thus, replacing j' by j increases the number of small jobs that have the same machine both in the optimal solution and in x^{LP}, thereby violating the property. It follows that there exists an optimal solution that assigns each small job that is assigned in x^{LP}.

Lemma 2. *There exists an optimal solution to RP1 such that each job $j \in J_1$ that is not assigned in an optimal solution to the LP-relaxation, i.e., for which $x_{ji} = 0$ for each $i \in M(j)$, is also not assigned in that optimal solution.*

Proof. By contradiction. Suppose that in each optimal solution to RP1 a (nonempty) set of small jobs is assigned that is not assigned in x^{LP}. Let S_{OPT} be an optimal solution to RP1 with a minimal number of such jobs. Let $j \in J_1$ be a small job that is assigned in S_{OPT}, and is not assigned in x^{LP}. It follows that the $|M(j)|$ machines it is connected to, each have b_i small jobs in x^{LP} (otherwise we can improve the relaxation). Moreover, these $\sum_{i \in M(j)} b_i$ small jobs are connected to machines that each must have b_i small jobs in x^{LP} (otherwise we can improve the relaxation). Proceeding along these lines it follows that we can identify in x^{LP} a set of small jobs, say J', that are connected to a set of machines, say M'. The sets J' and M' have the following properties:

- each machine in M' has b_i small jobs from J',
- all jobs from J' are assigned in x^{LP},
- all connections of jobs in J' are to machines in M', i.e., $\cup_{j \in J'} M(j) = M'$.

Now, consider S_{OPT}. By assumption, in S_{OPT} job j is assigned to a machine in M'. But that implies that some small job from J' that was assigned in x^{LP} has not been assigned in this optimal solution. Let us now construct an alternative optimal solution to RP1. This solution is identical to S_{OPT}, except for each machine $i \in M'$, it uses x^{LP}. Note that this is possible since no job from J' is connected to a machine outside M'. As the constructed solution violates the minimality assumption of S_{OPT}, we arrive at a contradiction.

Now we are ready to formulate the persistency property that is valid for the unweighted case of RP1.

Theorem 4. *For each $j \in J_1$:*
$$x_{ji}^{LP} = 0 \ \forall_{i \in M(j)} \Longleftrightarrow x_{ji}^{OPT} = 0 \ \forall_{i \in M(j)}.$$

Proof. This can be shown using the arguments from Lemmas 1 and 2.

Note that Theorem 4 has the potential to reduce the size of the instances that we need to solve. Indeed, any small job that was not assigned in the relaxation can be discarded from consideration.

Finally, we state the following property of problem RP1.

Lemma 3. *There exists an optimal solution to RP1 that assigns a maximum number of small jobs.*

The proof runs along the same lines as the proofs of Lemmas 6 and 7.

3 A $\frac{2}{3}$-Approximation Result for RP1

In this section we derive a $\frac{2}{3}$ approximation result for problem RP1.

Theorem 5. *RP1 admits a $\frac{2}{3}$-approximation algorithm.*

Proof. We first show that the theorem is valid in case $b_i \leq 2$ for all $i \in M$. We assume (wlog) that no large job is connected to a machine i that has $b_i = 1$. The idea is to turn the fractional solution x^{LP} into a feasible solution for RP1 while not losing too much weight. By Theorem 2, we have a fractional solution that is half-integral. It follows that each machine belongs to one of four different types (notice that, if $x_{l(j),i} = x_{r(j),i} = \frac{1}{2}$, we say that machine i *has* a half large job):

- a machine is called *integral* if it has two small jobs, or if it has one small job, or if it has one large job, or if it has no jobs at all.
- a machine is called *half-fractional* if it has one small job and one half large job. There are two types of half-fractional machines, those whose other part of the large job is assigned (half-fractional$^+$) and those whose other part of the large job is not assigned (half-fractional$^-$).
- a machine is called *full-fractional* if it has two half large jobs. There are two types of full-fractional machines, those whose two counterparts are not assigned (full-fractional$^-$), and those for which at least one counterpart is assigned (full-fractional$^+$).
- a machine is called *single* if it has only one half large job.

Let us first modify the current fractional solution to another fractional solution of at least the same value that has the property that there are no single machines and no full-fractional$^-$ machines. If the current solution contains a single machine, we simply schedule its large job completely on that machine. Repeatedly applying this procedure yields a solution without single machines. For full-fractional$^-$ machines both half large jobs have the same weight, otherwise the job with largest weight would have been assigned entirely. So, we can assign an unassigned counterpart of one of the two half large jobs and thereby construct a solution with the mentioned property.

Next, we turn the current fractional solution into a feasible solution for RP1. To start, we keep all integral machines. Then, we construct a graph corresponding to the remaining machines in the solution. This graph has a node for each machine and two machines are connected iff they share a large job. Observe that a node has degree 0, 1 or 2. Consider a node of degree 0. Observe that this must correspond to a half-fractional$^-$ machine. Let j be the small job and j' be the fractional large job present on such a machine. For each of these machines, we simply select the heaviest job. Since

$$\max(w_j, w_{j'}) \geq \frac{2}{3}w_j + \frac{1}{3}w_{j'} = \frac{2}{3}(w_j + \frac{1}{2}w_{j'})$$

it follows that we select at least $\frac{2}{3}$ of the weight of the half-fractional$^-$ machines.

Thus, the remaining graph is a collection of paths and cycles. Consider a path of length k. Such a path involves $k+1$ jobs. Observe that we can select any set of k jobs from such a path. Obviously, we select in our solution the heaviest k jobs from each path. Analogously, from a cycle of length k we can select all k involved jobs. It follows that we can turn the fractional solution into a feasible solution to RP1 while guaranteeing at least $\frac{2}{3}$ of the fractional solution's weight.

Now, assume that the machines have arbitrary integral capacity $b_i, i \in M$. We construct the following instance satisfying $b_i \leq 2$ by using the following trick. We decompose each machine into $\lfloor \frac{b_i}{2} \rfloor$ machines with capacity 2, and, if b_i is odd, one additional machine with capacity 1. We copy the connections of the jobs, except that jobs of size 2 will not be connected to machines with capacity 1. Observe that any feasible solution in the original instance can be transformed to a solution in the new instance and vice versa with the same value. The result follows.

Notice that the proof of Theorem 10 implicitly sketches an algorithm that achieves a 2/3 performance guarantee. The following instance of RP1 shows that this bound is tight even when we have unit weights, exactly two machines per job and machines with capacity 2.

Instance. We have an instance with 2 machines, 2 small jobs and 1 large job, such that each job is connected to both machines. Observe that the optimal solution has value 3, whereas an optimal solution to the relaxation could yield 2 half-fractional$^+$ machines. Then it will schedule 2 out of the 3 jobs.

4   A $\frac{2}{3}$-Approximation Result for RP2

Let us now consider RP2: for each job j, the set of machines to which it is connected is partitioned into two subsets, called $M_1(j)$ and $M_2(j)$ such that the job takes one unit capacity on a machine from $M_1(j)$ and two units capacity on a machine from $M_2(j)$. Notice that, when allowing that $M_1(j)$ or $M_2(j)$ can be empty, RP2 can be considered as a generalization of RP1.

We formulate the problem along the same lines as formulation M2 by introducing for each job j, a *regular* job j and a *dummy* job $d(j)$ of size 1. The regular job j can be assigned to any machine in $M(j)$, whereas the dummy job can only be assigned to a machine in $M_2(j)$. Assigning regular job j to a machine in $M_1(j)$ yields a profit of w_j, whereas assigning a regular job j to a machine in $M_2(j)$ yields a profit of $\frac{1}{2}w_j$. The weight of a dummy job $d(j)$ equals $\frac{1}{2}w_j$.

The model now is:

(M3) Maximize

$$\sum_{j \in J} \sum_{i \in M_1(j)} w_j x_{ji} + \frac{1}{2} \sum_{j \in J} \sum_{i \in M_2(j)} w_j (x_{ji} + x_{d(j),i})$$

subject to:

$$\forall_{j \in J} \sum_{i \in M(j)} x_{ji} \leq 1,$$

$$\forall_{j \in J} \sum_{i \in M_2(j)} x_{d(j),i} \leq 1,$$

$$\forall_{i \in M} \sum_{j \in J_r(i)} x_{ji} + \sum_{j \in J_d(i)} x_{d(j),i} \leq b_i,$$

$$\forall_{j \in J} \forall_{i \in M_2(j)} \quad x_{ji} - x_{d(j),i} = 0, \tag{6}$$

$$\forall_{j \in J} \forall_{i \in M(j)} \quad x_{ji}, x_{d(j),i} \in \{0,1\},$$

with $J_r(i)$ and $J_d(i)$ being the sets of regular and dummy jobs that can be assigned to machine i respectively.

We show that omitting constraints (6) yields a relaxation of M3 that can be solved using min-cost flow. Based on this solution we get a $\frac{2}{3}$-approximation algorithm for RP2.

The network is very similar to the one described in Section 2: a node corresponding to a regular job j is connected via an arc to machine-nodes corresponding to machines in $M_1(j)$. Such an arc has capacity 1 and cost $-w_j$. Also, this node is connected to machines in $M_2(j)$ via arcs that have capacity 1 and cost $-\frac{1}{2}w_j$. Each dummy job is connected to machines in $M_2(i)$ with weight $-\frac{1}{2}w_j$. Since the proof of the following result is similar in spirit as the proof of Theorem 5, we omit the proof.

Theorem 6. *RP2 admits a $\frac{2}{3}$-approximation algorithm.*

References

1. Adams, W.P., J. Bowers Lassiter, and H.D. Sherali (1998), *Persistency in 0-1 polynomial programming*, Mathematics of Operations Research **23**, 359–389.
2. Aerts, J., J. Korst, F. Spieksma, W. Verhaegh, and G. Woeginger (2002), *Load balancing in disk arrays: complexity of retrieval problems*, to appear in IEEE Transactions on Computers.
3. Aerts, J., J. Korst, and W. Verhaegh (2001), *Load balancing for redundant storage strategies: Multiprocessor scheduling with machine eligibility*, Journal of Scheduling **4**, 245–257.
4. Aerts, J., J. Korst, and W. Verhaegh (2002), *Improving disk efficiency in video servers by random redundant storage*, To appear in the Proceedings of the 6th IASTED International Conference on Internet and Multimedia Systems and Applications (August 2002).
5. Chudak, F. and D.S. Hochbaum, (1999), *A half-integral linear programming relaxation for scheduling precedence-constrained jobs on a single machine*, Operations Research Letters **25**, 199–204.
6. Dawande, M., J. Kalagnanam, P. Keskinocak, F.S. Salman, and R. Ravi (2000), *Approximation algorithms for the multiple knapsack problem with assignment restrictions*, Journal of Combinatorial Optimization **4**, 171–186.
7. Golubchik, L., S. Khanna, S. Khuller, R. Thurimella, and A. Zhu (2000), *Approximation algorithms for data placement on parallel disks*, Proceedings of the 11th ACM-SIAM Symposium on Discrete Algorithms, pp. 223–232, New York.
8. Hochbaum, D.S. (2002), *Solving integer programs over monotone inequalities in three variables: A framework for half integrality and good approximations*, European Journal of Operational Research **140**, 291–321.
9. Korst, J. (1997), *Random duplicated Assignment: An alternative to striping in video servers*, Proceedings of the ACM Multimedia Conference, pp. 219–226, Seattle.
10. Nemhauser, G.L. and L.E. Trotter, Jr. (1975), *Vertex packings: structural properties and algorithms*, Mathematical Programming **8**, 232–248.
11. Papadopouli, M. and L. Golubchik (1998), *A scalable video-on-demand server for a dynamic heterogeneous environment*, Proceedings of the 4th International Workshop on Advances in Multimedia Information Systems, 4–17.
12. Ralphs, T.K. (1993), *On the mixed chinese postman problem*, Operations Research Letters **14**, 123–127.
13. Ruemmler, C. and J. Wilkes (1994), *An introduction to disk drive modeling*, IEEE Computer **27**, 17–28.
14. Sanders, P. (2000), *Asynchronous scheduling for redundant disk arrays*, Proceedings of the 12th ACM Symposium on Parallel Algorithms and Architectures, 98–108.
15. Sanders, P. (2001), *Reconciling simplicity and realism in parallel disk models*, Proceedings of the 12th ACM-SIAM Symposium on Discrete Algorithms, 67–76.
16. Sanders, P., S. Egner and J. Korst (2000), *Fast concurrent access to parallel disks* Proceedings of the 11th ACM-SIAM Symposium on Discrete Algorithms, 849–858.
17. Santos, J., R. Muntz, and B. Ribeiro-Neto (2000), *Comparing random data allocation and data striping in multimedia servers*, Proceedings of the ACM Sigmetrics conference on measurements and modelling of computer systems, 44–55.
18. Vitter, J.S. (2001), *External memory algorithms and data structures: Dealing with massive data*, ACM Computing Surveys **33**, 1–75.

On k-Edge-Connectivity Problems with Sharpened Triangle Inequality[*]

Extended Abstract

Hans-Joachim Böckenhauer[1], Dirk Bongartz[1], Juraj Hromkovič[1], Ralf Klasing[2], Guido Proietti[3], Sebastian Seibert[1], and Walter Unger[1]

[1] Lehrstuhl für Informatik I, RWTH Aachen, D-52056 Aachen, Germany
{hjb,bongartz,jh,seibert,quax}@cs.rwth-aachen.de
[2] Project MASCOTTE, CNRS/INRIA, BP 93, F-06902 Sophia-Antipolis, France
klasing@sophia.inria.fr
[3] Dipartimento di Informatica, Università di L'Aquila, I-67010 L'Aquila, Italy,
and Istituto di Analisi dei Sistemi ed Informatica "Antonio Ruberti", CNR, Roma, Italy
proietti@di.univaq.it

Abstract. The edge-connectivity problem is to find a minimum-cost k-edge-connected spanning subgraph of an edge-weighted, undirected graph G for any given G and k. Here we consider its APX-hard subproblems with respect to the parameter β, with $\frac{1}{2} \leqslant \beta < 1$, where $G = (V, E)$ is a complete graph with a cost function c satisfying the sharpened triangle inequality

$$c(\{u, v\}) \leqslant \beta \cdot (c(\{u, w\}) + c(\{w, v\}))$$

for all $u, v, w \in V$.

First, we give a linear-time approximation algorithm for these optimization problems with approximation ratio $\frac{\beta}{1-\beta}$ for any $\frac{1}{2} \leqslant \beta < 1$, which does not depend on k.

The result above is based on a rough combinatorial argumentation. We sophisticate our combinatorial consideration in order to design a $\left(1 + \frac{5(2\beta-1)}{9(1-\beta)}\right)$-approximation algorithm for the 3-edge-connectivity subgraph problem for graphs satisfying the sharpened triangle inequality for $\frac{1}{2} \leqslant \beta \leqslant \frac{2}{3}$.

1 Introduction

In order to attack hard optimization problems that do not admit any polynomial-time approximation scheme (PTAS) or α-approximation algorithm for a reasonable constant α (or with an even worse approximability) one can consider the concept of *stability of approximation* [6,16,17]. The idea behind this concept is to find a parameter (characteristic) of the input instances that captures the hardness of particular inputs. An approximation algorithm is called *stable* with respect to this parameter, if its approximation ratio grows with this parameter but not with the size of the input instances. This approach is similar to the concept of parameterized complexity introduced by Downey and Fellows [12,13]. (The difference is in that we relate the parameter to the approximation ratio while Downey and Fellows relate the parameter to the time complexity.) A nice example is the Traveling Salesman Problem

---
[*] This work was partially supported by DFG-grant Hr 14/5-1, the CNR-Agenzia 2000 Program, under Grants No. CNRC00CAB8 and CNRG003EF8, and the Research Project REAL-WINE, partially funded by the Italian Ministry of Education, University and Re-

(TSP) that does not admit any polynomial-time approximation algorithm with an approximation ratio bounded by a polynomial in the size of the input instance, but is $\frac{3}{2}$-approximable for metric input instances. Here, one can characterize the input instances by their "distance" to metric instances. This can be expressed by the so-called β-triangle inequality for any $\beta \geqslant \frac{1}{2}$. For any complete graph $G = (V, E)$ with a cost function $c : E \to \mathbb{Q}^{>0}$ we say that (G, c) satisfies the β-triangle inequality, if

$$c(\{u, v\}) \leqslant \beta \cdot (c\{u, w\}) + c(\{w, v\}))$$

for all vertices $u, v, w \in V$. In the case of $\beta < 1$ we speak about the *sharpened triangle inequality*, and if $\beta > 1$ we speak about the *relaxed triangle inequality*. Note that in the case of $\beta = 1$ we have the well-known metric TSP, and if $\beta = \frac{1}{2}$, the problem becomes trivial since all edges must have the same cost. Experimental investigations [21] show that the relaxed triangle inequality is of practical interest because many TSP instances in the TSP libraries satisfy the β-triangle inequality for β close to 1. For a detailed motivation of the study of TSP instances satisfying sharpened triangle inequalities, see [5].

In a sequence of papers [1,2,3,5,6,7,8] it was shown that

1. there are stable approximation algorithms for the TSP whose approximation ratio grows with β, but is independent of the size of the input, and
2. for every $\beta > \frac{1}{2}$ one can prove explicit lower bounds on the polynomial-time approximability growing with β.

Thus, one can partition the set of all input instances of the TSP into infinitely many classes with respect to their hardness, and one gains the knowledge that hard TSP input instances have to have small edge costs as well as edge costs of exponential size in the size of G.

A natural question is whether there are other problems for which the triangle inequality can serve as a measure of hardness of the input instances. In [4] it is shown that this is the case for the problem of constructing 2-connected spanning subgraphs of a given complete edge-weighted graph. For both the 2-edge-connectivity and the 2-vertex-connectivity, the best known approximation ratio in the general case is 2 [19,20].

In [4] it is proved (with an explicit lower bound on the approximability) that these minimization problems are APX-hard even for graphs satisfying a sharpened triangle inequality for any $\beta > \frac{1}{2}$, i.e., even if the edge costs are from an interval $[1, 1 + \varepsilon]$ for an arbitrarily small $\varepsilon > 0$. On the other hand, an upper bound of $\frac{2}{3} + \frac{1}{3} \cdot \frac{\beta}{1-\beta}$ on the approximability is achieved in [4] for all inputs satisfying the sharpened triangle inequality, which is an improvement over the previously known results for $\beta < \frac{5}{7}$.

Here, we consider a more general problem: For a given positive integer $k \geqslant 2$ and a complete edge-weighted graph G one has to find a minimum k-edge-connected spanning subgraph. This problem is well-known to be NP-hard [15]. Concerning approximability results, the problem is approximable within 2 [19].[1]

We design a linear-time $\frac{\beta}{1-\beta}$-approximation algorithm for this general problem for graphs satisfying the sharpened β-triangle inequality. The achieved approximation ratio is smaller than 2 for $\beta < \frac{2}{3}$, and it does not depend on k. This result is based on the simple observation that the costs of two edges adjacent to the same vertex may not differ too much, if the input satisfies the sharpened triangle inequality. Some rough combinatorial calculations show that the cost of an optimal k-edge-connected subgraph does not differ too much from the cost of any k-regular subgraph.

---

[1] Some better results can be obtained for some restricted cases like unweighted graphs

The main contribution of this paper is to develop this approach to a more sophisticated technique providing a polynomial-time $(1+\frac{5}{9}\cdot\frac{2\beta-1}{1-\beta})$-approximation algorithm for the 3-edge-connectivity problem on graphs satisfying the sharpened triangle inequality for some $\frac{1}{2}\leqslant\beta<\frac{2}{3}$. Note that this approximation ratio equals 1 for $\beta=\frac{1}{2}$, and $\frac{14}{9}$ for $\beta=\frac{2}{3}$.

The paper is organized as follows. In Section 2 we will formally define the edge-connectivity problem and provide some useful facts about graphs satisfying a sharpened triangle inequality. Section 3 is devoted to a linear-time approximation algorithm for the edge-connectivity problem, and in Section 4 we will present our main result, namely an improved approximation algorithm for the 3-edge-connectivity problem. We note that the proof techniques used in this paper essentially differ from the approaches used in the previous papers devoted to the approximability of input instances satisfying sharpened or relaxed triangle inequalities.

2 Preliminaries

Recall that a graph is said to be k-edge-connected if the removal of any $k-1$ edges leaves the graph connected. Furthermore, note that all graphs throughout this paper are considered to be undirected.

Now we give a formal definition of the Edge-Connectivity-Problem.

Definition 1. Edge-Connectivity-Problem (ECP)

Input: A weighted complete graph (G,c), where $G=(V,E)$, and $c:E\to\mathbb{Q}^{>0}$ is a cost function, and a positive integer k.
Output constraints: For all inputs $x=((G,c),k)$, the set of feasible solutions for x is $\mathcal{M}(x)=\{G'\mid G'$ is a spanning k-edge-connected subgraph of $G\}$.
Costs: For every feasible solution $G'\in\mathcal{M}(x)$, the costs are defined as

$$\mathrm{cost}(G',x)=\sum_{e'\in E'}c(e'),\ \ where\ G'=(V,E').$$

Goal: Minimization.

Furthermore, we will denote by k-ECP the subproblem of the ECP where the input is restricted to instances with a fixed value of k.

In this paper, we will focus on the ECP and the 3-ECP for graphs obeying the sharpened triangle inequality as defined in the introduction, and therefore we give some basic properties of weighted graphs satisfying the sharpened triangle inequality.

Lemma 1. [5] Let (G,c) be a weighted complete graph with $G=(V,E)$, $c:E\to\mathbb{Q}^{>0}$, and $\frac{1}{2}<\beta<1$. Let $c_{\min}$ and $c_{\max}$ denote the minimum and the maximum edge cost occurring in G, respectively. If the cost function c obeys the sharpened triangle inequality, then:

(a) For all adjacent edges $e_1,e_2\in E$, the inequality $c(e_1)\leqslant\frac{\beta}{1-\beta}c(e_2)$ holds.

(b) $c_{\max}\leqslant\frac{2\beta^2}{1-\beta}c_{\min}$. $\qquad\qquad\square$

From this we can derive the following:

Corollary 1. [4] Let (G,c) be a weighted complete graph with $G=(V,E)$, $c:E\to\mathbb{Q}^{>0}$, where c obeys the sharpened triangle inequality for $\frac{1}{2}\leqslant\beta\leqslant\frac{2}{3}$. Then $c(e_3)\leqslant$

This result implies that in the case where $\beta \leqslant \frac{2}{3}$ holds, we can replace two edges by one adjacent edge without increasing the cost.

It has been shown in [4] that the k-ECP for graphs obeying the sharpened triangle inequality for $\beta < \frac{2}{3}$ and $k = 2$ corresponds to the problem of finding a minimum cost Hamiltonian cycle, i.e., to the TSP in that graph. In [5] a $(\frac{2}{3} + \frac{1}{3} \cdot \frac{\beta}{1-\beta})$-approximation algorithm for the TSP for graphs obeying the sharpened triangle inequality, i.e. for $\frac{1}{2} \leqslant \beta < 1$, has been proposed, which we will shortly recall here, since we will apply it in Section 4 of this paper.

Algorithm Cycle Cover

Input: A weighted complete graph (G, c), where $c : E \to \mathbb{Q}^{>0}$ obeys the sharpened triangle inequality.
Step 1: Construct a minimum cost cycle cover $C = \{C_1, \ldots, C_k\}$ of G.
Step 2: For $1 \leqslant i \leqslant k$, find the cheapest edge $\{a_i, b_i\}$ in every cycle C_i of C.
Step 3: Obtain a Hamiltonian cycle H of G from C by replacing the edges $\{\{a_i, b_i\} \mid 1 \leqslant i \leqslant k\}$ by the edges $\{\{b_i, a_{i+1}\} \mid 1 \leqslant i \leqslant k-1\} \cup \{\{b_k, a_1\}\}$.
Output: H.

Furthermore, let us recall that a graph is called k-regular, iff each of its vertices has degree exactly k. Additionally note that for any graph G that is k-regular, it must hold that $k \cdot n$ is even, where n is the number of vertices in G. This is due to the following argument. If $G = (V, E)$ is k-regular then the number of edges in G is $|E| = \frac{k \cdot n}{2}$. Thus either n or k has to be even.

3   An Approximation Algorithm for the Edge-Connectivity-Problem

In this section we will investigate an algorithm for the ECP for graphs obeying the sharpened triangle inequality for $\frac{1}{2} \leqslant \beta < 1$. We will proceed as follows.

Let (G, c) be a weighted complete graph such that c satisfies the sharpened triangle inequality.

(i) We will prove that the cost of an arbitrary k-regular graph (G', c) differs from the cost of any k-edge-connected graph (G'', c) at most by a factor $\frac{\beta}{1-\beta}$, where G' and G'' are both spanning subgraphs of the same complete weighted graph (G, c).
(ii) We will describe a strategy to construct a spanning subgraph of (G, c) that is both k-edge-connected and k-regular.

Thus, the algorithm proposed in (ii) will compute a k-edge-connected spanning subgraph that $\frac{\beta}{1-\beta}$-approximates a minimum k-edge-connected spanning subgraph.

Theorem 1. *Let (G, c) be a weighted complete graph obeying the sharpened β-triangle inequality $(\frac{1}{2} \leqslant \beta < 1)$. Let $G' = (V, E')$ be an arbitrary k-regular spanning subgraph of G, and let $G'' = (V, E'')$ be an arbitrary k-edge-connected spanning subgraph of G. Then*

$$\text{cost}(G') \leqslant \frac{\beta}{1 - \beta} \cdot \text{cost}(G'').$$

Proof. The idea is to pairwise compare adjacent edges in G' and G'', since, if an edge $e' \in E'$ is adjacent to an edge $e'' \in E''$, according to Lemma 1, their costs cannot differ by more than a factor $\frac{\beta}{1-\beta}$.

Hence, for each vertex v in V, denote by I'_v all edges in E' that are incident with exactly k due to the k-regularity of G', and the number of edges in I_v'' is at least k due to the k-edge-connectivity of G''. Since the edges in I_v' and I_v'' are pairwise adjacent we can apply Lemma 1 and obtain

$$\text{cost}(I_v') \leqslant \frac{\beta}{1-\beta} \cdot \text{cost}(I_v'') . \tag{1}$$

Thus, we can estimate the costs of G' as follows:

$$\text{cost}(G') = \sum_{e' \in E'} \text{cost}(e') = \frac{1}{2} \sum_{v \in V} \sum_{e' \in I_v'} \text{cost}(e')$$

$$\leqslant \frac{1}{2} \sum_{v \in V} \sum_{e'' \in I_v''} \frac{\beta}{1-\beta} \text{cost}(e'') = \frac{\beta}{1-\beta} \sum_{e'' \in E''} \text{cost}(e'') = \frac{\beta}{1-\beta} \text{cost}(G'').$$

$\square$

Next, we present an algorithm that, for all integers k and n, where $n \geqslant k+1$ and $k \cdot n$ even, constructs a graph that has n vertices and is k-regular as well as k-edge-connected.

The idea is to start with a cycle of n vertices and to iteratively add all edges that connect vertices at distance i for $2 \leqslant i \leqslant \lfloor \frac{k}{2} \rfloor$. In the case of an odd value of k we additionally connect the vertices in a spoke-like way.

Algorithm RC-Graph

Input: Integers k and n, $k \geqslant 2$ and $n \geqslant k+1$, where $k \cdot n$ is even.
Step 1: Let $V = \{v_0, \ldots, v_{n-1}\}$.
Step 2: Let $E' = \{\{v_i, v_{(i+j) \bmod n}\} \mid 1 \leqslant j \leqslant \lfloor \frac{k}{2} \rfloor, 0 \leqslant i < n\}$
Step 3: If k is odd, then let $E = E' \cup \{\{v_i, v_{i+\frac{n}{2}}\} \mid 0 \leqslant i < \frac{n}{2}\}$,
      else let $E = E'$.
Output: Graph $G = (V, E)$.

Thus, each vertex v_i in the graph G produced by the above algorithm is directly connected to all vertices that are within distance $\lfloor \frac{k}{2} \rfloor$ according to the basic cycle $v_0, v_1, \ldots, v_{n-1}, v_0$. For an example see Figure 1.

It is immediately clear from the construction that the graph constructed by algorithm RC-Graph contains exactly n vertices and is k-regular. The formal proof that the graph is also k-edge-connected will be given in the full paper.

We obtain the following result. Note that linear time means linear in the number of edges.

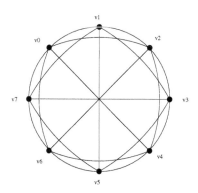

Fig. 1. Graph constructed by Algorithm RC-Graph for $k = 5$ and $n = 8$.

Theorem 2. *For all inputs $((G, c), k)$ for the ECP, where c obeys the sharpened triangle inequality, there exists a linear-time $\frac{\beta}{1-\beta}$-approximation algorithm for the ECP.*

Proof. If $k \cdot n$ is even, where n is the number of vertices in G, we can take the output of Algorithm RC-Graph and interpret it as spanning subgraph of G. Then, the claim is a direct consequence from Theorem 1 and the correctness of the Algorithm RC-Graph, which runs in linear-time.

In the case where $k \cdot n$ is odd, one can consider an algorithm that is similar to Algorithm RC-Graph. This algorithm first finds an edge $e_{\min}$ of minimum cost in the graph, puts it into the solution, and proceeds by determining a graph by a similar construction as in Algorithm RC-Graph in such a way that $e_{\min}$ is not considered there and the vertex not incident to a spoke edge is incident to $e_{\min}$.[2] Now we can estimate the costs of the resulting k-edge-connected graph against the optimal solution in a similar way as in Theorem 1 by additionally taking into account that the optimal solution has at least $\lfloor \frac{k \cdot n}{2} \rfloor + 1$ edges while the constructed solution has exactly $\lfloor \frac{k \cdot n}{2} \rfloor + 1$ edges. Thus, the only edge which could not be estimated like in Theorem 1 is $e_{\min}$ having minimal cost, and hence can be effectively estimated against the additional edge occurring in the optimal solution.   □

The original idea behind the presented Algorithm RC-Graph (now hidden behind the combinatorial argument in the proof of Theorem 1) was based on the following two-step consideration.

1. For any optimal k-edge-connected subgraph G_{opt} of G satisfying the sharpened triangle inequality for $\frac{1}{2} \leqslant \beta \leqslant \frac{2}{3}$, one can construct a k-regular subgraph G_k (not necessarily k-edge-connected) with $\mathrm{cost}(G_k) \leqslant \mathrm{cost}(G_{\mathrm{opt}})$.
2. For all k-regular subgraphs G' and G'' of G,

$$\mathrm{cost}(G') \leqslant \frac{\beta}{1 - \beta} \mathrm{cost}(G'').$$

Thus, any 3-regular 3-edge-connected spanning subgraph is a good feasible solution for 3-ECP. In the next section we improve this concept for $k = 3$ in the sense, that we apply the rough combinatorial argument (providing the approximation ratio $\frac{\beta}{1-\beta}$) only for a subpart of the graph G_k. More precisely, we exchange the steps 1 and 2 for

1' For any optimal 3-edge-connected subgraph G_{opt} of G satisfying the sharpened triangle inequality for $\frac{1}{2} \leqslant \beta \leqslant \frac{2}{3}$, one can construct a 3-regular 2-edge-connected subgraph G_3 with

$$\mathrm{cost}(G_3) \leqslant \mathrm{cost}(G_{\mathrm{opt}}).$$

2' G_3 can be partitioned into a 1-factor M and a 2-factor C. The cost of the 1-factor can be bounded from above in the same way as in 2, but the cost of C can be better approximated.

The core of this approach is that the cost of the 2-factor of G_3 (for which we have a better approximation) cannot be dominated by the cost of the 1-factor of G_3. The main technical difficulty is in proving 1'.

---

[2] It is not difficult to verify that it is always possible to construct a graph with these properties. Note that by naming the vertices appropriately, we may choose $e_{\min} = \{v_1, v_{\frac{n+1}{2}}\}$, thus avoid needing it in edge set E'. A proof that the result is k-connected will be given

4   An Approximation Algorithm for the 3-Edge-Connectivity-Problem

In this section, we will present an approximation algorithm for 3-ECP with an approximation ratio of $1 + \frac{5(2\beta-1)}{9(1-\beta)}$ in case $\beta \leqslant \frac{2}{3}$.

We know from Lemma 1 that, for each two adjacent edges e, f of a graph obeying the sharpened triangle inequality with $\frac{1}{2} \leqslant \beta < 1$, $c(e) \leqslant \frac{\beta}{1-\beta}c(f)$, and vice versa. Since $\frac{\beta}{1-\beta} = 1 + \frac{2\beta-1}{1-\beta}$, we define

$$\gamma := \frac{2\beta-1}{1-\beta}$$

and obtain the reformulation

$$c(e) \leqslant (1+\gamma)c(f) \tag{2}$$

for each two adjacent edges e, f of a graph obeying the sharpened triangle inequality with $\frac{1}{2} \leqslant \beta < 1$.

The idea of our algorithm is as follows. It first constructs a Hamiltonian cycle (using Algorithm Cycle Cover given in Section 2) and then connects each pair of opposite vertices of this cycle by an edge.

Algorithm 3C

Input: A complete weighted graph (G, c), with $G = (V, E)$, $|V| = n$ even, and
$c : E \to \mathbb{Q}^{>0}$.
Step 1: $H :=$ Hamiltonian cycle in G; (using Algorithm Cycle Cover)
Step 2: Let $v_1, v_2, \ldots, v_n$ be the order of vertices in H;
Step 3: $E_m := \{\{v_i, v_{i+\frac{n}{2}}\} \mid i = 1, \ldots, \frac{n}{2}\}$;
Output: $H \cup E_m$.

It is easy to see that Algorithm 3C outputs a 3-edge-connected subgraph. A formal proof will be given in the full paper.

To show that Algorithm 3C has the approximation ratio stated above, we will show that any optimal solution of 3-ECP can be transformed into a 3-regular 2-edge-connected subgraph without increasing the cost. The main reason for doing this is that such a subgraph can be split into a 1-factor and a 2-factor.[3] By basically comparing these two parts separately to the corresponding parts of the constructed solution (the 1-factor vs E_m, the 2-factor vs H), we get the claimed result.

Theorem 3. *Let (G, c) be an input for 3-ECP, where $G = (V, E)$, $|V| = n$ is even, and $c : E \to \mathbb{Q}^{>0}$ obeys the sharpened triangle inequality for $\beta \leqslant \frac{2}{3}$. Then Algorithm 3C obtains an approximation ratio of $1 + \frac{5}{9}\gamma = 1 + \frac{5(2\beta-1)}{9(1-\beta)}$.*

Note that the approximation ratio is 1 for $\beta = \frac{1}{2}$, and it is $1 + \frac{5}{9}$ for $\beta = \frac{2}{3}$.

Furthermore, one can obtain an asymptotically equivalent approximation ratio also for the case where n is odd, but due to additional technical details we omit the formal proof in this extended abstract and present it in the full paper.

Proof. As we will show in Theorem 4, there exists a 3-regular and 2-edge-connected subgraph $G' = (V, E')$ of G whose cost is less or equal the cost of any 3-edge-connected subgraph of G.

---

[3] Note that for splitting off a 2-factor (or even a 1-factor) it is not sufficient, that the considered graph is 3-edge-connected. A counterexample will be given in the full paper.

By Petersen's Theorem (a consequence of Tutte's Theorem, see e.g. [11]), G' contains a 1-factor (or perfect matching) M, and consequently the remainder of G' is a 2-factor (or cycle cover) C. In the following, we obtain the claimed bound on the approximation ratio by essentially comparing E_m to M and H to C. But we have to distinguish two cases depending on which has the higher cost per edge, C or M.

Case 1. Let

$$\text{cost}(C) \geqslant 2 \cdot \text{cost}(M). \tag{3}$$

By (2), each edge $\{v, w\}$ of E_m costs at most $\frac{1}{2}(1+\gamma)(c(e_v)+c(e_w))$, where e_v and e_w are the edges of M incident to v and w, respectively. (Note that this includes the case $e_v = e_w = \{v, w\}$.) When we sum this up over all edges of E_m, each edge of M occurs exactly twice, which yields

$$\text{cost}(E_m) \leqslant (1+\gamma)\text{cost}(M). \tag{4}$$

Using the approximation ratio of Algorithm Cycle Cover, we have

$$\text{cost}(H) \leqslant (1 + \frac{1}{3}\gamma)\text{cost}(C), \tag{5}$$

since exactly this inequality was established in [5] to obtain the approximation ratio of Algorithm Cycle Cover by using the cost of an optimal cycle cover as a lower bound on the cost of an optimal Hamiltonian cycle.
Put together this gives the claimed bound

$$
\begin{aligned}
\text{cost}(H) + \text{cost}(E_m) &\underset{(4),(5)}{\leqslant} (1 + \frac{1}{3}\gamma)(\text{cost}(C) + \text{cost}(M)) + \frac{2}{3}\gamma \cdot \text{cost}(M) \\
&\underset{(3)}{\leqslant} (1 + \frac{1}{3}\gamma)(\text{cost}(C) + \text{cost}(M)) + \\
&\quad \frac{2}{3}\gamma \cdot \frac{1}{3}(\text{cost}(C) + \text{cost}(M)) \\
&= (1 + \frac{5}{9}\gamma)(\text{cost}(C) + \text{cost}(M)).
\end{aligned}
$$

Case 2. Let

$$\text{cost}(C) < 2 \cdot \text{cost}(M). \tag{6}$$

Here, we compare each edge $\{v, w\}$ of E_m to all edges $e_{v,1}, e_{v,2}, e_{v,3}, e_{w,1}, e_{w,2}, e_{w,3}$ of $C \cup M$ incident to v and w, respectively. (Again, two of these may be identical to $\{v, w\}$.) By (2)

$$c(\{v, w\}) \leqslant \frac{1}{6}(1 + \gamma)\left(\sum_{i=1}^{3} c(e_{v,i}) + \sum_{i=1}^{3} c(e_{w,i})\right).$$

When we sum this up over all edges of E_m, each edge of $C \cup M$ occurs exactly twice, which yields

$$\text{cost}(E_m) \leqslant \frac{1}{3}(1 + \gamma)(\text{cost}(C) + \text{cost}(M)). \tag{7}$$

Using the approximation ratio of Algorithm Cycle Cover in the same way as described in (5), we have

$$\text{cost}(H) \leqslant (1 + \frac{1}{3}\gamma)\text{cost}(C) \underset{(6)}{\leqslant} (1 + \frac{1}{3}\gamma)\frac{2}{3}(\text{cost}(C) + \text{cost}(M)). \tag{8}$$

Put together this gives as before the claimed bound

$$\mathrm{cost}(H) + \mathrm{cost}(E_m) \underset{(7),(8)}{\leqslant} \left(\frac{1}{3}(1+\gamma) + \frac{2}{3}(1+\frac{1}{3}\gamma)\right)(\mathrm{cost}(C) + \mathrm{cost}(M))$$

$$= (1+\frac{5}{9}\gamma)(\mathrm{cost}(C) + \mathrm{cost}(M)).$$

$\square$

Instrumental to the proof of the approximation ratio of Algorithm 3C is mainly the following theorem. It states that there exists a subgraph G'', that is not more expensive than any 3-edge-connected one, such that the structure of G'' is close to the solution constructed by our algorithm.

Theorem 4. *Let (G, c) be a complete weighted graph, where $G = (V, E)$, $|V| = n$ is even, and $c : E \to \mathbb{Q}^{>0}$ obeys the sharpened triangle inequality for $\beta \leqslant \frac{2}{3}$. Let $G' = (V, E')$ be a minimal 3-edge-connected subgraph of G.*
Then, there exists a subgraph $G'' = (V, E'')$ of G that is 3-regular and 2-edge-connected such that

$$\mathrm{cost}(G'') \leqslant \mathrm{cost}(G').$$

We will construct G'' from G' by successively deleting and adding edges in order to obtain degree exactly 3 for all vertices. This will be done in a way as to always maintain 2-edge-connectivity by making use of the following observation. It tells us which connections only need to be checked after modifying the subgraph.

Remark 1. Let G_1 be 2-edge-connected, and let G_2 result from deleting a few edges of G_1 (as well as potentially adding others).
Then G_2 is 2-edge-connected iff for every pair of vertices v and w such that $\{v, w\}$ was deleted, there exist two edge-disjoint paths from v to w in G_2.

It is not trivial but easy to see that for any other pair of vertices x, y that needed the edge $\{v, w\}$ in one of its two connecting paths in G_1, two paths can be reestablished in G_2 by using the connections between v and w.
There is another easy observation which will be used frequently in the proof of Theorem 4.

Remark 2. Let v be a vertex in a subgraph G' of G such that $G' \backslash v$ is 2-edge-connected. If there are two edges in G' between v and $G' \setminus v$, then G'' is 2-edge-connected, too.

Proof of Theorem 4. As outlined above we want to construct G'' from G' by successively deleting and adding edges. This will be done in such a way that the cost is never increased, since whenever we introduce a new edge, it replaces two edges adjacent to its endpoints. By Corollary 1, this does not increase the cost in case $\beta \leqslant \frac{2}{3}$, even if the deleted edges do not have their other endpoint in common.
Since G' is 3-edge-connected, it is also a 2-edge-connected graph where every vertex has degree at least 3. Thus, all we have to do is to reduce vertex degrees higher than 3 while maintaining the 2-edge-connectivity. The procedure obviously terminates since any modification decreases the number of remaining edges.
First, we deal with each vertex of degree 5 or more individually, and later, we will see how to handle vertices of degree 4.

A. Vertices of degree $\geqslant 5$. Let v be a vertex in a 2-edge-connected subgraph G_1 of G having at least five neighbors $w_1, \ldots, w_l, l \geqslant 5$ in G_1. We distinguish three cases.

A.1 Assume $G_1 \setminus v$ has a 2-edge-connected subgraph that contains at least four vertices out of $w_1, \ldots, w_l$, say w_1, w_2, w_3, w_4.

Then, we may delete two edges from v to these four vertices, without affecting the 2-edge-connectivity (cf. Remark 2). If w_1, w_2, w_3, w_4 form a clique, their degree still is at least 3. Otherwise, we chose two non-adjacent vertices, say w_1 and w_2. The edges $\{v, w_1\}$ and $\{v, w_2\}$ are deleted, and the edge $\{w_1, w_2\}$ is added to keep vertex degree at least 3 for w_1 and w_2 (see Figure 2 (a)).

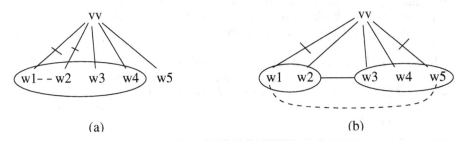

(a)   (b)

Fig. 2. Vertices of degree at least five: The ticked edges are removed, and the dashed edges are added.

A.2 If $G_1 \setminus v$ is connected, but the previous case does not apply, it contains at least one bridge, such that the deletion of that bridge would separate $G_1 \setminus v$ into two components each of which contains at least two vertices from $\{w_1, \ldots, w_l\}$. Without loss of generality, let those components contain $\{w_1, w_2\}$ and $\{w_3, w_4, w_5\}$, respectively. (It doesn't matter where $w_6, \ldots, w_l$ are located if $l > 5$.)

Then we pick one vertex from each set such that the two are not adjacent. Without loss of generality, let those vertices be w_1 and w_5. Now we replace $\{v, w_1\}$ and $\{v, w_5\}$ by $\{w_1, w_5\}$, obtaining G_2, see Figure 2 (b).

By symmetry and Remark 1, we only need to show that there are two edge-disjoint paths from w_1 to v in G_2 for to prove that G_2 is 2-edge-connected.

By construction, there is a path from w_1 to w_2 that stays on one side of the bridge. Similarly a path $\langle w_5, \ldots, w_4 \rangle$ exists on the other side. Consequently, the paths $\langle w_1, \ldots, w_2, v \rangle$ and $\langle w_1, w_5, \ldots, w_4, v \rangle$ are edge-disjoint (even internally vertex-disjoint).

A.3 If $G_1 \setminus v$ is not connected, each vertex from $\{w_1, \ldots, w_l\}$ needs to be in the same component as at least one other from this set. The reason for this is the 2-edge-connectivity of G_1 which implies that from each component of $G_1 \setminus v$ there need to be two edges connecting that component with v.

Now, we may proceed exactly as in the previous case (where in fact we have handled the bridge as if it would divide $G_1 \setminus v$ into two components).

B. Adjacent vertices of degree 4. Let u, v be two adjacent vertices of degree 4 in a 2-edge-connected subgraph G_1. We will distinguish four cases.

B.1 If u and v have at least two adjacent vertices in common, we simply can delete the edge $\{u, v\}$ (see Figure 3 (a)).

In the following, let x_1, x_2, x_3 be adjacent to u, and let y_1, y_2, y_3 be adjacent to v. Only one x_i may be identical to one y_j.

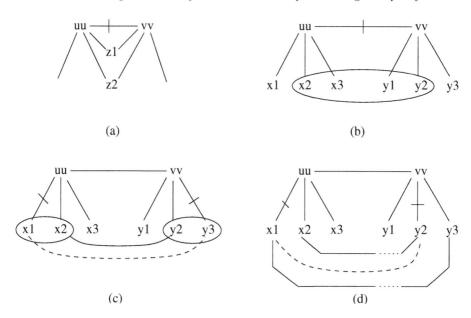

Fig. 3. Adjacent vertices of degree four.

B.2 Assume that there is a 2-edge-connected component in $G_1 \setminus u, v$ that contains at least two vertices each from the sets $\{x_1, x_2, x_3\}$ and $\{y_1, y_2, y_3\}$ respectively. Again, we simply can delete $\{u, v\}$ (see Figure 3 (b)).

B.3 Otherwise, if $G_1 \setminus u, v$ is connected, there is a bridge whose deletion would leave two vertices from $\{x_1, x_2, x_3\}$, say x_1, x_2, in one component, and two vertices from $\{y_1, y_2, y_3\}$, say y_2, y_3, in another. Only one edge from x_1 or x_2 to y_2 or y_3 may exist in G_1 (namely the bridge). Thus, let x_1 and y_3 be non-adjacent. Now, we delete $\{u, x_1\}$ and $\{v, y_3\}$ from G, and we insert $\{x_1, y_3\}$, obtaining G_2 (see Figure 3 (c)). By symmetry and Remark 1, we only need to show that there are two edge-disjoint paths from u to x_1 for to prove G_2 to be 2-edge-connected. These paths are $\langle u, x_2, \ldots, x_1 \rangle$ and $\langle u, v, y_2, \ldots, y_3, x_1 \rangle$.

B.4 Finally, let $G_1 \setminus u, v$ be not connected. By the 2-edge-connectivity of G_1, each of $\{x_1, x_2, x_3, y_1, y_2, y_3\}$ must be in the same component with another vertex from that same set. Either there are at least two "mixed" pairs, say x_1, y_3 and x_2, y_2, or two pairs like x_1, x_2 and y_2, y_3. In the latter case, we proceed exactly like when having a bridge, replacing $\{u, x_1\}$ and $\{v, y_3\}$ by $\{x_1, y_3\}$.

In the former case, we replace $\{u,x_1\}$ and $\{v,y_2\}$ by $\{x_1,y_2\}$ (see Figure 3 (d)). Again, by symmetry and Remark 1, we only need to show that there are two edge-disjoint paths from u to x_1. These paths are $\langle u,x_2, \ldots ,y_2,x_1 \rangle$ and $\langle u, v, y_3, \ldots , x_1 \rangle$.

C. Independent vertices of degree 4. Let u, v be two independent vertices in a 2-edge-connected subgraph G_1 having four neighbors each.

We have to distinguish similar cases as for adjacent vertices of degree 4 with some additional technical details. Due to space limitations we move this part of the proof to the full paper. $\qquad\square$

References

1. T. Andreae: On the traveling salesman problem restricted to inputs satisfying a relaxed triangle inequality. *Networks* 38(2) (2001), pp. 59-67.
2. T. Andreae, H.-J. Bandelt: Performance guarantees for approximation algorithms depending on parameterized triangle inequalities. *SIAM Journal on Discrete Mathematics* 8 (1995), pp. 1–16.
3. M.A. Bender, C. Chekuri: Performance guarantees for the TSP with a parameterized triangle inequality. *Information Processing Letters* 73 (2000), pp. 17–21.
4. H.-J. Böckenhauer, D. Bongartz, J. Hromkovič, R. Klasing, G. Proietti, S. Seibert, W. Unger: On the hardness of constructing minimal 2-connected spanning subgraphs in complete graphs with sharpened triangle inequality. Proc. *FSTTCS 2002*, LNCS 2556, Springer 2002, pp. 59–70.
5. H.-J. Böckenhauer, J. Hromkovič, R. Klasing, S. Seibert, W. Unger: Approximation algorithms for the TSP with sharpened triangle inequality. *Information Processing Letters* 75, 2000, pp. 133–138.
6. H.-J. Böckenhauer, J. Hromkovič, R. Klasing, S. Seibert, W. Unger: Towards the Notion of Stability of Approximation for Hard Optimization Tasks and the Traveling Salesman Problem (Extended Abstract). Proc. *CIAC 2000*, LNCS 1767, Springer 2000, pp. 72–86. Full version in *Theoretical Computer Science* 285(1) (2002), pp. 3–24.
7. H.-J. Böckenhauer, J. Hromkovič, R. Klasing, S. Seibert, W. Unger: An Improved Lower Bound on the Approximability of Metric TSP and Approximation Algorithms for the TSP with Sharpened Triangle Inequality (Extended Abstract). Proc. *STACS 2000*, LNCS 1770, Springer 2000, pp. 382–394.
8. H.-J. Böckenhauer, S. Seibert: Improved lower bounds on the approximability of the traveling salesman problem. *RAIRO - Theoretical Informatics and Applications* 34, 2000, pp. 213–255.
9. A. Czumaj, A. Lingas: On approximability of the minimum-cost k-connected spanning subgraph problem. *SODA'99*, 1999, pp. 281–290.
10. J. Cheriyan, R. Thurimella: Approximating minimum-size k-connected spanning subgraphs via matching. *SIAM Journal on Computing* 30(2): 528-560 (2000), pp. 528–560.
11. R. Diestel: *Graph Theory*. Second Edition, Springer 2000.
12. R.G. Downey, M.R. Fellows: Fixed-parameter tractability and completeness. *Congressus Numerantium* 87 (1992), pp. 161–187.
13. R.G. Downey, M.R. Fellows: *Parameterized Complexity*. Springer 1999.
14. C.G. Fernandes: A better approximation ratio for the minimum size k-edge-connected spanning subgraph problem. *J. Algorithms*, 28(1) (1998), pp. 105–124.
15. M.R. Garey, D.S. Johnson: *Computers and Intractability: A guide to the theory of NP-completeness*. W. H. Freeman and Company, San Francisco, 1979.
16. J. Hromkovič: Stability of approximation algorithms and the knapsack problem. In: J. Karhumäki, H. Maurer, G. Paun, G. Rozenberg (Eds.) *Jewels are Forever*, Springer 1999, pp. 238–249.
17. J. Hromkovič: *Algorithmics for Hard Problems - Introduction to Combinatorial Optimization, Randomization, Approximation, and Heuristics*. Springer 2001.
18. S. Khuller, B. Raghavachari: Improved approximation algorithms for uniform connectivity problems. *J. Algorithms* 21(2): (1996), pp. 434-450.
19. S. Khuller, U. Vishkin: Biconnectivity approximations and graph carvings. *Journal of the ACM* 41 (1994), pp. 214–235.
20. M. Penn, H. Shasha-Krupnik: Improved approximation algorithms for weighted 2- and 3-vertex connectivity augmentation. *Journal of Algorithms* 22 (1997), pp. 187–196.
21. D. Weckauf: Experimental Analysis of Approximation Algorithms for the Traveling Salesperson Problem with Relaxed Triangle Inequality. Diploma thesis, RWTH Aachen 2002 (in German).

The Complexity
of Detecting Fixed-Density Clusters

Klaus Holzapfel, Sven Kosub, Moritz G. Maaß*, and Hanjo Täubig**

Fakultät für Informatik, Technische Universität München
Boltzmannstraße 3, D-85748 Garching b. München, Germany
{holzapfe,kosub,maass,taeubig}@in.tum.de

Abstract. We study the complexity of finding a subgraph of a certain size and a certain density, where density is measured by the average degree. Let $\gamma : \mathbb{N} \to \mathbb{Q}_+$ be any density function, i.e., γ is computable in polynomial time and satisfies $\gamma(k) \leq k - 1$ for all $k \in \mathbb{N}$. Then γ-CLUSTER is the problem of deciding, given an undirected graph G and a natural number k, whether there is a subgraph of G on k vertices which has average degree at least $\gamma(k)$. For $\gamma(k) = k - 1$, this problem is the same as the well-known clique problem, and thus **NP**-complete. In contrast to this, the problem is known to be solvable in polynomial time for $\gamma(k) = 2$. We ask for the possible functions γ such that γ-CLUSTER remains **NP**-complete or becomes solvable in polynomial time. We show a rather sharp boundary: γ-CLUSTER is **NP**-complete if $\gamma = 2 + \Omega(\frac{1}{k^{1-\varepsilon}})$ for some $\varepsilon > 0$ and has a polynomial-time algorithm for $\gamma = 2 + O(\frac{1}{k})$.
Keywords: Density-based clustering, computational complexity, graph algorithms, fixed-parameter problems.

1   Introduction

Density-based approaches are highly natural to network-clustering issues. Web communities, for instance, both well-established and emerging have in common that they show a significantly high ratio of linkage among their members (see, e.g. [17,18,19]). And, in VLSI design, collapsing subgraphs of density beyond a certain threshold into one node provides the basis for hierarchical graph-representation of large circuits to be decomposed (see, e.g., [7,16,3]).

The fundamental task in density-based clustering is finding a dense subgraph (cluster) of a certain size. Density of a graph might be defined in several different ways. One can define the density of an undirected graph on n vertices to be the ratio of the number of edges in the graph and the maximum edge-number of an n-vertex graph. Thus, an n-vertex-clique has density one and n isolated vertices have density zero. The definition is very intuitive and, in particular, it allows to compare graphs of different sizes directly, regarding their densities. On the other

---

*   Research of the third and of the fourth author supported by DFG, grant Ma 870/5-1 (Leibnizpreis Ernst W. Mayr).
** Research supported by DFG (Deutsche Forschungsgemeinschaft), grant Ma 870/6-1 (SPP 1126 Algorithmik großer und komplexer Netzwerke).

R. Petreschi et al. (Eds.): CIAC 2003, LNCS 2653, pp. 201–212, 2003.

hand, since a number of edges quadratic in the number of vertices is required for a graph to be dense, small graphs are biased. Therefore, density of undirected graphs is usually measured by the average degree. A clique of size n clearly has average degree $n - 1$.

The problem of deciding, given a graph G and natural numbers k and γ, if there exists a k-vertex subgraph of G having average degree at least γ, is easily seen to be **NP**-complete. In contrast to this *variable* cluster-detection problem, we focus in this paper on the *fixed-parameter* cluster-detection problem, which we call γ-CLUSTER. That is, we fix one parameter, namely the function γ (i.e. the required average degree depending on the size of the subgraph) and want to decide, given graph G and natural number k, if there exists a k-vertex subgraph of G with average degree at least $\gamma(k)$. We are interested in what choices of γ still admit polynomial-time algorithms, and for which γ the problem becomes **NP**-complete.

Studying the complexity of the fixed-parameter problem is motivated by at least two reasons. First, knowing the precise boundary between polynomial-time and **NP**-complete cases is essential to obtain efficient methods for the above-mentioned clustering issues in several settings, e.g., web graphs, where good choices of γ describe reality sufficiently. Second, if the polynomial-time cases can be realized by a uniform algorithm (i.e., parameters $t \leq \gamma$ may be given to the input), then we can approximate the maximum average degree reachable on n vertices in a graph within factor $\frac{n}{\gamma}$. The best known algorithm guarantees approximation within a factor $n^{\frac{1}{3}-\varepsilon}$ for some $\varepsilon > 0$ [10]. Thus, we would outperform this algorithm if we find a (uniform) polynomial-time algorithm for γ-CLUSTER up to little over $n^{\frac{2}{3}+\varepsilon}$. Unfortunately, the boundary with the **NP**-complete cases turns out to be much lower.

Previous Work

The problem of finding dense subgraphs has attracted a lot of attention in the context of combinatorial optimization.

Gallo, Grigoriadis, and Tarjan [12] showed, by using flow techniques, that there is a polynomial-time algorithm for the densest subgraph problem, in which we are supposed to find a subgraph of arbitrary size with highest average degree. Feige, Kortsarz, and Peleg [10] studied a restricted version, which they called the dense k-subgraph problem, where we have to find a subgraph with highest average degree among all subgraphs on k vertices. They provide a polynomial-time algorithm that approximates the maximum average degree of such k-vertex subgraphs within factor $n^{\frac{1}{3}-\varepsilon}$ for some $\varepsilon > 0$. Several authors proved approximation results for the dense k-subgraph problem using different techniques, mainly greedy algorithms [2,6] and semidefinite programming [11,21]. For special graph classes, they obtained partly better approximation results. Arora, Karger, and Karpinski [4] showed that the dense k-subgraph problem on dense graphs (i.e., with quadratic number of edges) admits a polynomial-time approximation scheme, which has been improved by Czygrinow to a fully polynomial-time approximation scheme [8]. In contrast to this, it is not known whether the dense

k-subgraph problem on general graphs is not approximable within factor $(1+\varepsilon)$ for all $\varepsilon > 0$ (unless $\mathbf{P} = \mathbf{NP}$ or similar complexity-theoretic collapses), although it is even conjectured that the problem is hard to approximate within factor n^ε for some $\varepsilon > 0$ [10].

Fixed-parameter problems were also considered in our setting. Nehme and Yu [20] investigated the complexity of the constrained maximum value sub-hypergraph problem, which contains the dense k-subgraph problem as a special case. They obtained bounds on the number of (hyper)edges a (hyper)graph may have, such that the problem is still solvable in polynomial time, namely, $n - s + \alpha \log n$, where n is the number of vertices, s the number of connected components, and α any constant. Similar fixed parameter-restrictions to the input graphs were also considered in [2,1]. Note that this scenario has no consequences for our problem since these restrictions affect the graph outside of possible dense subgraphs, and we are interested in the existence of dense subgraphs of fixed quality inside an arbitrary graph.

Most recently, Asahiro, Hassin, and Iwama [1] studied the k-$f(k)$ dense subgraph problem, $(k, f(k))$-DSP for short, which asks whether there is a k-vertex subgraph of a given graph G with at least $f(k)$ edges. This problem is almost the same problem as our γ-cluster problem, since a k-$f(k)$ subgraph has average degree at least $\frac{2f(k)}{k}$. The authors proved that the problem remains $\mathbf{NP}$-complete for $f(k) = \Theta(k^{1+\varepsilon})$ for all $0 < \varepsilon < 1$ and is polynomial-time solvable for $f(k) = k$. From these results we can conclude that γ-CLUSTER is $\mathbf{NP}$-complete for $\gamma = \Theta(k^\varepsilon)$ for any $0 < \varepsilon < 1$, and is decidable in polynomial time for $\gamma = 2$. Feige and Seltser [11] even proved that $(k, f(k))$-DSP is $\mathbf{NP}$-complete if $f(k) = k + k^\varepsilon$ (which, in our notation, is $\gamma = 2 + 2k^{\varepsilon-1}$) for any $0 < \varepsilon < 2$. We will enhance those bounds to more general settings.

This Work

In this paper we prove that γ-CLUSTER is solvable in polynomial time for $\gamma = 2 + O(\frac{1}{k})$ and that γ-CLUSTER is $\mathbf{NP}$-complete for $\gamma = 2 + \Omega(\frac{1}{k^{1-\varepsilon}})$ for $0 < \varepsilon < 2$. We thus establish a rather sharp boundary between polynomial time solvable and $\mathbf{NP}$-complete cases. As a corollary we obtain, for the more intuitive case of γ constant, that detecting a k-vertex subgraph of average degree at least two (which is nearly the case of any connected graph) can be done in polynomial time whereas finding a k-vertex subgraph of slightly-higher average degree at least $2 + \varepsilon$ is already $\mathbf{NP}$-complete. Thus, density-based clustering is inherently hard as a general methodology.

In terms of the $(k, f(k))$-DSP our results mean that $(k, f(k))$-DSP remains $\mathbf{NP}$-complete if $f(k) = k + \Omega(k^\varepsilon)$ for any $0 < \varepsilon < 2$, which, for $\varepsilon \leq 1$, is more precise than in [1,11], and is polynomial-time decidable for $f(k) = k + c$ for all (constant) integers c.

The proof of the polynomial-time cases is mainly based on dynamic programming over collections of minimal subgraphs having certain properties. For instance, for the above-mentioned polynomial-time result for $(k, f(k))$-DSP with $f(k) = k$ [1], we simply need to find shortest cycles in a graph, which is easy.

For functions $f(k) = k + c$ with $c > 0$, the search for similar minimal subgraphs is not obvious to solve and is the main difficulty to overcome in order to obtain polynomial-time algorithms. In the **NP**-hardness proofs we adapt techniques used by [1,11], that are well suited for Θ-behavior of functions but lead to different reductions according to the different growth classes. Thus the main issue for getting results for Ω-behavior is to unify reductions by a non-trivial choice of the parameters involved.[1]

2   Definitions and Main Results

Throughout this paper we consider undirected graphs without loops. Let G be any graph. $V(G)$ denotes the set of vertices of G and $E(G)$ denotes the set of edges of G. The size of a graph is $|V(G)|$, i.e., the cardinality of $V(G)$. For any function $\gamma : \mathbb{N} \to \mathbb{Q}_+$, graph G is said to be a γ-*cluster* if and only if $d(G) \geq \gamma(|V(G)|)$ where $d(G)$ denotes the average degree of G, i.e., $d(G) = 2|E(G)|/|V(G)|$.

We study the complexity of the following problem. Let $\gamma : \mathbb{N} \to \mathbb{Q}_+$ be any function computable in polynomial time.

> *Problem:* γ-CLUSTER
> *Input:*   A graph G and a natural number k
> *Question:* Does G contain a γ-cluster of size k?

In this context "contains" means, that there exists a subgraph on k vertices with average degree at least $\gamma(k)$. Obviously, the existence of such a subgraph implies the existence of an induced subgraph with the same property and vice versa. Note that 0-CLUSTER is a trivial problem and that $(k-1)$-CLUSTER = CLIQUE. Moreover, it is easily seen that γ-CLUSTER is in **NP** whenever γ is computable in polynomial time. The following theorem expresses our main results.

Theorem 1. *Let $\gamma : \mathbb{N} \to \mathbb{Q}_+$ be computable in polynomial time, $\gamma(k) \leq k - 1$.*

1. *If $\gamma = 2 + O(\frac{1}{k})$, then γ-CLUSTER is solvable in polynomial time.*
2. *If $\gamma = 2 + \Omega(\frac{1}{k^{1-\varepsilon}})$ for some $\varepsilon > 0$, then γ-CLUSTER is **NP**-complete.*

In the remainder of the paper we prove Theorem 1. Section 3 contains the polynomial-time cases. Section 4 establishes the **NP**-completeness statements

---

[1] Basically, having Turán's theorem [22] in mind, one could ask whether it is possible, at least in the case of dense graphs, to deduce intractability results using inapproximability of MAXIMUM CLIQUE due to Håstad [14]: there is no polynomial-time algorithm finding cliques of size at least $n^{\frac{1}{2}+\varepsilon}$ (where n is the size of the maximum clique) unless $\mathbf{P} = \mathbf{NP}$. Assume we would have a polynomial-time algorithm for γ-CLUSTER with, e.g., $\gamma(k) = \beta\binom{k}{2}$ and $0 < \beta < 1$, are we now able to decide whether there is a clique of size $k^{\frac{1}{2}+\varepsilon}$? Turán's theorem says that there is a clique of size k in a graph with n vertices and m edges, if $m > \frac{1}{2}n^2\frac{k-1}{k-2}$. Unfortunately this implies that we can only assure that in a graph with n vertices and at least $\beta\binom{n}{2}$ edges, there is a clique of size at most $\frac{3-2\beta}{1-\beta}$, which is constant and makes the argument fail.

of Theorem 1. Due to space restrictions most of the proofs are omitted (for detailed proofs see [15]).

3 Computing $(2 + O(\frac{1}{k}))$-Dense Subgraphs in Polynomial Time

In this section we show how to solve γ-CLUSTER for $\gamma = 2 + O(\frac{1}{k})$ in polynomial time. In other words, we prove that searching a k-vertex subgraph with at least $k + c$ edges with c constant is a polynomial-time problem. We will formalize this issue in the problem EXCESS-c SUBGRAPH, which is more intuitive in this setting.

For a graph G, let the *excess of G*, denoted by $\nu(G)$, be defined as $\nu(G) = |E(G)| - |V(G)|$. A (sub)graph G with $\nu(G) \geq c$ is said to be an *excess-c (sub)graph*.

Problem: EXCESS-c SUBGRAPH
Input: A graph G and natural number k
Question: Does G contain an excess-c subgraph of size k?

We will show how to find excess-c subgraphs in polynomial time. The general solution is based on the case of a connected graph which is handled by the following lemmas:

Lemma 1. *Let $c \geq 0$ be any integer. Given a connected graph G on n vertices, an excess-c subgraph of minimum size can be computed in time $O(n^{2c+2})$.*

Proof. Let G be any connected graph with $\nu(G) \geq c$. Then there exists a subgraph G_c of minimum size with excess equal to c. Note, that the induced subgraph on the same vertices might have higher excess. For the degree-sum of G_c we obtain

$$\sum_{v \in V(G_c)} \deg_{G_c}(v) = 2|E(G_c)| = 2(|V(G_c)| + c).$$

Since G_c is minimal subject to the number of vertices, there exists no vertex with degree less than two. Therefore the number of vertices with degree greater than two in G_c is at most

$$\sum_{v \in V(G_c)} (\deg_{G_c}(v) - 2) = 2(|V(G_c)| + c) - 2|V(G_c)| = 2c.$$

Let S be the set of all vertices with degree greater than two in G_c. If there is a path connecting vertices $u, v \in S$ using only vertices from $V(G_c) \setminus S$ (u and v are not necessarily distinct), then there can be no shorter path connecting u and v containing vertices from $V(G) \setminus V(G_c)$. Otherwise G_c would not be minimal subject to the number of vertices. In the following we will describe how to find such a subgraph G_c if it exists.

We examine all sets $S' \subseteq V(G)$ of size at most $2c$ such that S' contains only vertices with degree greater than two in G, i.e., the elements of S' are those vertices where paths can cross. For such a set we can iteratively construct a candidate $H(S')$ for G_c. In each step we include a path which has minimum length among all paths that connect any two vertices in S'. We may restrict ourselves to those paths that do not intersect or join common edges, since otherwise $H(S')$ can be also obtained by one of the other possible choices of a set S'. This process is done until either excess c is reached or no further connecting path exists. In the latter case the set S' does not constitute a valid candidate for G_c. Otherwise $H(S')$ is kept as a possible choice for G_c. After considering all possibilities for S', the graph G_c can be chosen as a vertex-minimal subgraph among all remaining candidates. Note that G_c is not unique with respect to exchanging paths of the same length.

Since, $|S'| \leq 2c$, there are

$$\sum_{i=1}^{2c} \binom{n}{i} = O(n^{2c})$$

possible choices for S'. For the verification of a chosen set S' consisting of i vertices we have to find iteratively $i + c$ shortest non-crossing paths, e.g., by using $i + c \leq 3c$ parallel executions of a breadth-first-search, which takes time $O(3c|E(G)|) = O(n^2)$.

Finally, this implies that determining an excess-c subgraph of minimum size by testing all possible choices of S' can be done in total time $O(n^{2c+2})$. Note that for $c = 0$ we only have to find a shortest cycle (e.g., by breadth-first search) which can be done in time $O(n^2)$.   □

Unfortunately the algorithm of Lemma 1 cannot directly be used for the general case of possibly non-connected graphs. For those graphs vertices from different connected components may be chosen. Therefore our algorithm is based on solving the subproblem of maximizing the excess for a given number of vertices within a connected graph.

Lemma 2. *Let $c \geq 0$ be any integer. Given a connected graph G with n vertices. Let ν_i be the maximum excess of an i-vertex subgraph of G. Calculating $\min\{\nu_i, c\}$ for all values of $i \in \{0, 1, \ldots, n\}$ can be done in time $O(n^{2c+2})$.*

Before we proceed to the main theorem we have to discuss a further property which allows to restrict to a suitable subset of the connected components when deciding about the existence of EXCESS-c SUBGRAPH for an unconnected graph.

Let (G, k) be the instance of the EXCESS-c SUBGRAPH problem, i.e., we have to find a subgraph of G on k vertices with at least $k + c$ edges. In linear time we can (as a preprocessing step) partition G into its connected components and calculate their excess. Let $C_1, \ldots, C_r$ be the list of the components, sorted non-increasingly by their excess. Note that $\nu(C_j) \geq -1$ since all components are connected. Let j_0 denote the maximum index of the components with non-negative excess and k_0 the total number of all vertices of those components.

Lemma 3. *1. If $k > k_0$ then there is a maximum excess subgraph of size k comprising all vertices from the non-negative excess components $C_1, \ldots, C_{j_0}$.*

2. *If $k \leq k_0$ then there always exists a subgraph of size k having maximum excess within G and consisting only of vertices from components with non-negative excess.*

With these results we are able to state the main theorem of this section. The omitted proof is based on a combination of the above stated lemmas combined with dynamic programming which allows to locate subgraphs of sufficient size and excess to decide the existence of a suitable excess-c subgraph.

Theorem 2. *Let c be any integer. EXCESS-c SUBGRAPH can be decided in time $O(n^{2|c|+4})$.*

So far we only considered the EXCESS-c SUBGRAPH problem for constant values c. If we are interested in a k-vertex subgraph with excess $f(k) = O(1)$, the same method can be applied. From $f = O(1)$ we know that $f(k)$ is bounded from above by a constant c'. Obviously the time complexity for our algorithm is $O(n^{2c'+4})$, if $f(k)$ can be computed in the same time. This problem corresponds to finding a $(k+O(\frac{1}{k}))$-dense subgraph. Applying some modifications the method can be used to find such a subgraph instead of only deciding its existence.

Corollary 1. *For polynomial-time computable $\gamma = 2 + O(\frac{1}{k})$, γ-CLUSTER is is solvable in polynomial time and, moreover, finding a γ-cluster is solvable in polynomial time.*

Similarly, the problem can be examined for $f = O(\log k)$ which leads to a quasi-polynomial time algorithm.

Corollary 2. 1. *For polynomial-time computable $\gamma = 2 + O(\frac{\log k}{k})$, finding γ-clusters can be done in time $n^{O(\log n)}$.*
2. *Let $\gamma = 2 + \Theta(\frac{\log k}{k})$ be polynomial-time computable. If γ-CLUSTER is **NP**-complete, then $\mathbf{NP} \subseteq \mathbf{DTIME}(n^{O(\log n)})$.*

4 Finding $(2 + \Omega(\frac{1}{k^{1-\varepsilon}}))$-Dense Subgraphs Is NP-Complete

In this section we prove that all γ-CLUSTER problems are complete for **NP** if $\gamma = 2 + \Omega(\frac{1}{k^{1-\varepsilon}})$ for some $\varepsilon > 0$. In doing so, we focus on the $(k, f(k))$-DSP, namely, we show that $(k, f(k))$-DSP is **NP**-complete whenever $f = k + \Omega(k^{\varepsilon})$.

For this, we need the concept of a quasi-regular graph. A graph G is said to be *quasi-regular* if and only if the difference between the maximal and the minimal degree of the vertices in G is at most one.

Proposition 1. *For every $n \geq 0$ and $0 \leq m \leq \binom{n}{2}$ both given in unary (i.e., as inputs 1^n and 1^m), a quasi-regular graph having exactly n vertices and m edges can be computed in time polynomial in the input length.*

Theorem 3. *Let $f : \mathbb{N} \to \mathbb{N}$ be a polynomial-time computable function such that $f = k + \Omega(k^\varepsilon)$ for some $\varepsilon > 0$ and $f(k) \leq \binom{k}{2}$. Then, $(k, f(k))$-DSP is* **NP**-*complete.*

Proof. Due to space limitations, we only describe the construction part of the proof without correctness part.

Let f be a polynomial-time computable function with $f = k + \Omega(k^\varepsilon)$ for some rational $\varepsilon > 0$ and $f(k) \leq \binom{k}{2}$. Containment of $(k, f(k))$-DSP in **NP** is obvious. We prove the **NP**-hardness of $(k, f(k))$-DSP by reduction from a special version of CLIQUE which will be explained below. Since there are several cases to be handled we need different constructions. However, in each of this constructions the following three operations (with parameters from $\mathbb{N}$) on graphs are involved (in exactly the order they are listed): R_s, S_t, and $T^\alpha_{r,N(r)}$.

- R_s is defined as follows. Let G be any undirected graph. Define the following sequence of graphs: $G_0 =_{\text{def}} G$ and, for $j > 0$, $G_j =_{\text{def}} h(G_{j-1})$ where h transforms a graph I by adding to I a new vertex which has an edge to each vertex in I. Define $R_s(G) =_{\text{def}} G_s$. Obviously, the following property holds:

$$G \text{ has a clique of size } k \Longleftrightarrow R_s(G) \text{ has a clique of size } k + s.$$

The operator R_s can be used to define a special **NP**-complete version of CLIQUE (see, e.g., [1]). Define $\text{CLIQUE}_{\frac{1}{2}}$ to be the set of all instances (G, k) such that G has a clique of size k and it holds $|V(G)| \leq (1 + \frac{1}{2})k$. It is easily seen that CLIQUE can be reduced to $\text{CLIQUE}_{\frac{1}{2}}$, namely by applying R_s to a graph G with parameter $s = 2|V(G)| - 3k$ for each instance (G, k) with $|V(G)| > (1 + \frac{1}{2})k$. The transformed graph G_s now has $|V(G)| + s = \frac{1}{2}s + (1 + \frac{1}{2})k + s = (1 + \frac{1}{2})(k + s)$ vertices, and the new clique-size G_s is asked for, is $k + s$.
- S_t is defined as follows. Let G be any undirected graph. S_t transforms G into a graph G_t by replacing each edge in G by a path of length $t + 1$ involving t new vertices. The new vertices are referred to as *inner vertices* and the old vertices are referred to as *outer vertices*. Note that inner vertices always have degree two and that an outer vertex has equal degrees in G_t and in G. It is easily seen that cliques in G of size $k \geq 3$ are related to subgraphs of G_t as follows (for formal proofs, see, e.g., [9,13,11]):

$$G \text{ has a clique of size } k \Longleftrightarrow S_t(G) \text{ has a subgraph with } k + t\binom{k}{2} \text{ vertices}$$
$$\text{and } (t + 1)\binom{k}{2} \text{ edges.}$$

- $T^\alpha_{r,N(r)}$ (with $\alpha \in \{0, 1\}$) is defined as follows. Let G be any undirected graph. $T^0_{r,N(r)}$ transforms G by the disjoint union with a quasi-regular graph $A(r, N(r))$ with r vertices and $N(r)$ edges. In the case of $\alpha = 1$, i.e. transformations by $T^1_{r,N(r)}$, we additionally have edges between each vertex in $A(r, N(r))$ and each vertex in G.

Now let (G, k) be any instance to $\text{CLIQUE}_{\frac{1}{2}}$ with $|V(G)| \leq \frac{3}{2}k$. Define $G' =_{\text{def}}$ $T^{\alpha}_{r,N(r)}(S_t(R_s(G)))$ to be the transformed graph and let the parameter $N(r)$ be defined as

$$N(r) =_{\text{def}} f\left(k + s + t\binom{k+s}{2} + r\right) - (t+1)\binom{k+s}{2} - \alpha r\left(k + s + t\binom{k+s}{2}\right).$$

By an appropriate choice of the parameters r, s, t, and α (which is a non-trivial problem) we can prove that G has a clique of size k if and only if G' has a subgraph on $k + s + t\binom{k+s}{2} + r$ vertices and at least $f(k + s + t\binom{k+s}{2} + r)$ edges, thus establishing the correctness of the reduction. This can be achieved by choosing the parameters in such a way that a densest subgraph of G' of the desired size can be derived from the densest subgraph of G with size k.

Since $f = k + \Omega(k^{\varepsilon})$ for some rational $\varepsilon > 0$, there exists a natural number $D > 1$ such that for some $k_0 \in \mathbb{N}$, $k + D^{-1}k^{\varepsilon} \leq f(k)$ for all $k \geq k_0$. We may suppose that $\varepsilon < \frac{1}{8}$ and $D \geq 3$. Since we will have to respect several finer growth-classes the function f might belong to, we choose one argument k' to distinguish between these different classes. Define

$$k' =_{\text{def}} \left\lceil (D^6 k^2)^{\frac{1}{\varepsilon}} \right\rceil.$$

Clearly, k' is computable in time polynomial in the length of k. Depending on the function value $f(k')$ we choose different parameters to obtain the graph G', in such a way that $k' = k + s + t\binom{k+s}{2} + r$. We distinguish between five cases that represent a partitioning of the interval between $k' + D^{-1}k'^{\varepsilon}$ and $\binom{k'}{2}$.

Case I. Let $k' + D^{-1}k'^{\varepsilon} \leq f(k') < k' + Dk'$. We split this case in several subcases. We consider, depending on j with $0 \leq j < \log_{\frac{7}{6}} \frac{1}{\varepsilon}$, the ranges $k' + D^{-1}k'^{(\frac{7}{6})^j \varepsilon} \leq$ $f(k') \leq k' + Dk'^{(\frac{7}{6})^{j+1}\varepsilon}$. We can combine those subcases to cover the complete range from $k' + D^{-1}k'^{\varepsilon}$ to $k + Dk'$ as required for Case I. For each value of j we apply R_s, S_t, and $T^{\alpha}_{r,N(r)}$ with the following parameters:

$$s = 0, \qquad t = (k' - r - k)/\binom{k}{2}, \qquad \alpha = 0,$$

$$r = \left\lceil (4D^4 k^2)^{(\frac{7}{6})^j} \right\rceil + \left[\left(k' - \left\lceil (4D^4 k^2)^{(\frac{7}{6})^j} \right\rceil - k \right) \mod \binom{k}{2} \right]$$

Note that $t \in \mathbb{N}$ because of the modular term in the definition of r.

Case II. Let $k + Dk' = (1 + D)k' \leq f(k') < (1 + D)k'^{\frac{3}{2}}$. Apply R_s, S_t, and $T^{\alpha}_{r,N(r)}$ with the following parameters:

$$s = 0, \qquad t = 1, \qquad \alpha = 0, \qquad r = k' - \binom{k}{2} - k$$

Case III. Let $(1 + D)k'^{\frac{3}{2}} \leq f(k') < \binom{k'}{2} - k'^{\frac{9}{8}}$. Apply R_s, S_t, and $T^{\alpha}_{r,N(r)}$ with the following parameters:

$$s = 0, \qquad t = 0, \qquad \alpha = 0, \qquad r = k' - k$$

Case IV. Let $\binom{k'}{2} - k'^{\frac{9}{8}} \leq f(k') < \binom{k'}{2} - \frac{k'}{3}$. Apply R_s, S_t, and $T^\alpha_{r,N(r)}$ with parameters:

$$s = \left\lceil \frac{1}{3}k'^{\frac{1}{4}} \right\rceil - k, \qquad t = 1, \qquad \alpha = 1, \qquad r = k' - k - s - \binom{k+s}{2}.$$

Case V. Let $\binom{k'}{2} - \frac{k'}{3} \leq f(k') \leq \binom{k'}{2}$. Apply R_s, S_t, and $T^\alpha_{r,N(r)}$ with parameters:

$$s = 0, \qquad t = 0, \qquad \alpha = 1, \qquad r = k' - k.$$

The omitted part of the proof (see [15]) shows that using these parameters the graph G' can always be constructed and that the above stated property (G' contains a subgraph on k' vertices with at least $f(k')$ edges if and only if G contains a clique of size k) is correct.   □

Now we are able to formulate the result of Theorem 3 in terms of γ-CLUSTER.

Corollary 3. Let $\gamma = 2 + \Omega(\frac{1}{k^{1-\varepsilon}})$ for some $\varepsilon > 0$ be polynomial-time computable, $\gamma(k) \leq k - 1$. Then γ-CLUSTER is **NP**-complete.

5   Conclusion

In this paper we proved that density-based clustering in graphs is inherently hard. The main result states that finding a k-vertex subgraph with average degree at least $\gamma(k)$ is **NP**-complete if $\gamma = 2 + \Omega(\frac{1}{k^{1-\varepsilon}})$ and solvable in polynomial time if $\gamma = 2 + O(\frac{1}{k})$. In particular, for constant average-degree that means that detecting whether there is a k-vertex subgraph with average degree at least two is easy but with average degree at least $2 + \varepsilon$ it is intractable. Since the **NP**-threshold is so tremendously low, it seems inevitable to explore how the problem behaves in relevant graph classes, e.g., in sparse graphs or graphs with small diameter. Sparsity, however, is not be expected to lift the **NP**-threshold.

Though detecting a subgraph of a certain size and a certain density is an important algorithmic issue, the original problem intended to be solved is MAXIMUM γ-CLUSTER: compute the largest subgraph with average degree over some γ-value. Of course, this problem is intimately related to γ-CLUSTER, and in fact, we have the same tractable-intractable threshold as for the decision problem. The main open question is: how good is MAXIMUM γ-CLUSTER approximable depending on γ? For instance, for $\gamma(k) = k - 1$ (i.e., MAXIMUM CLIQUE), it is known to be approximable within $O(\frac{n}{(\log n)^2})$ [5] but not approximable within $n^{\frac{1}{2}-\varepsilon}$ unless **P** = **NP** [14]. How do these results translate to intermediate densities?

Acknowledgment

For helpful hints and discussions we are grateful to Christian Glaßer, Ernst W. Mayr, and Alexander Offtermatt-Souza. We also thank Yuichi Asahiro, Refael Hassin, and Kazuo Iwama for providing us with an early draft of [1].

References

1. Y. Asahiro, R. Hassin, and K. Iwama. Complexity of finding dense subgraphs. *Discrete Applied Mathematics*, 121(1-3):15–26, 2002.
2. Y. Asahiro, K. Iwama, H. Tamaki, and T. Tokuyama. Greedily finding a dense subgraph. *Journal of Algorithms*, 34(2):203–221, 2000.
3. C. J. Alpert and A. B. Kahng. Recent developments in netlist partitioning: A survey. *Integration: The VLSI Journal*, 19(1-2):1–81, 1995.
4. S. Arora, D. Karger, and M. Karpinski. Polynomial time approximation schemes for dense instances of NP-hard problems. *Journal of Computer and System Sciences*, 58(1):193–210, 1999.
5. R. B. Boppana and M. M. Halldórsson. Approximating maximum independent sets by excluding subgraphs. *BIT*, 32(2):180–196, 1992.
6. M. S. Charikar. Greedy approximation algorithms for finding dense components in a graph. In *Proceedings 3rd International Workshop on Approximation Algorithms for Combinatorial Optimization*, volume 1913 of *Lecture Notes in Computer Science*, pages 84–95. Springer-Verlag, Berlin, 2000.
7. J. Cong and M. Smith. A parallel bottom-up clustering algorithm with applications to circuit partitioning in VLSI design. In *Proceedings 30th ACM/IEEE Design Automation Conference*, pages 755–760. ACM Press, New York, 1993.
8. A. Czygrinow. Maximum dispersion problem in dense graphs. *Operations Research Letters*, 27(5):223–227, 2000.
9. U. Faigle and W. Kern. Computational complexity of some maximum average weight problems with precedence constraints. *Operations Research*, 42(4):1268–1272, 1994.
10. U. Feige, G. Kortsarz, and D. Peleg. The dense k-subgraph problem. *Algorithmica*, 29(3):410–421, 2001.
11. U. Feige and M. Seltser. On the densest k-subgraph problem. Technical Report CS97-16, Department of Applied Mathematics and Computer Science, The Weizmann Institute of Science, Rehovot, Israel, 1997.
12. G. Gallo, M. D. Grigoriadis, and R. E. Tarjan. A fast parametric maximum flow algorithm and applications. *SIAM Journal on Computing*, 18(1):30–55, 1989.
13. O. Goldschmidt, D. Nehme, and G. Yu. On the set union knapsack problem. *Naval Research Logistics*, 41(6):833–842, 1994.
14. J. Håstad. Clique is hard to approximate within $n^{1-\varepsilon}$. *Acta Mathematica*, 182(1):105–142, 1999.
15. K. Holzapfel, S. Kosub, M. G. Maaß, and H. Täubig. The complexity of detecting fixed-density clusters. Technical Report TUM-I0212, Fakultät für Informatik, Technische Universität München, 2002.
16. D. J.-H. Huang and A. B. Kahng. When cluster meet partitions: New density-based methods for circuit decomposition. In *Proceedings European Design and Test Conference*, pages 60–64. IEEE Computer Society Press, Los Alamitos, 1995.
17. J. M. Kleinberg, S. R. Kumar, P. Raghavan, S. Rajagopalan, and A. S. Tomkins. The Web as a graph: measurements, models, and methods. In *Proceedings 5th International Conference on Computing and Combinatorics*, volume 1627 of *Lecture Notes in Computer Science*, pages 1–17. Springer-Verlag, Berlin, 1999.
18. S. R. Kumar, P. Raghavan, S. Rajagopalan, and A. S. Tomkins. Trawling the Web for emerging cyber-communities. *Computer Networks*, 31(3):1481–1493, 1999.
19. T. Murata. Discovery of Web communities based on the co-occurrence of references. In *Proceedings 9th International Conference on Discovery Science*, volume 1967 of *Lecture Notes in Artificial Intelligence*, pages 65–75. Springer-Verlag, Berlin, 2000.

20. D. Nehme and G. Yu. The cardinality and precedence constrained maximum value sub-hypergraph problem and its applications. *Discrete Applied Mathematics*, 74(1):57–68, 1997.
21. A. Srivastav and K. Wolf. Finding dense subgraphs with semidefinite programming. In *Proceedings International Workshop on Approximation Algorithms for Combinatorial Optimization*, volume 1444 of *Lecture Notes in Computer Science*, pages 181–191. Springer-Verlag, Berlin, 1998.
22. P. Turán. On an extremal problem in graph theory. *Matematikai és Fizikai Lapok*, 48:436–452, 1941. In Hungarian.

Nearly Bounded Error Probabilistic Sets

Extended Abstract

Tomoyuki Yamakami

School of Information Technology and Engineering
University of Ottawa, Ottawa, Ontario, Canada K1N 6N5

Abstract. We study polynomial-time randomized algorithms that solve problems on "most" inputs with "small" error probability. The sets that have such algorithms are called nearly BPP sets, which naturally expand BPP sets. Notably, sparse sets and average BPP sets are typical examples of nearly BPP sets. It is, however, open whether all NP sets are nearly BPP. The nearly BPP sets can be captured by Nisan-Wigderson's approximation scheme as well as viewed as a special case of promise BPP problems. Moreover, nearly BPP sets are precisely described in terms of Sipser's distinguishing complexity. These sets have a connection to average-case complexity and cryptography. Nevertheless, unlike BPP, the class of nearly BPP sets is not closed even under honest polynomial-time one-one reductions. In this paper, we study a more general notion of nearly BP[$\mathcal{C}$] sets, analogous to Schöning's probabilistic class BP[$\mathcal{C}$] for any complexity class $\mathcal{C}$. The "infinitely-often" version of nearly BPP sets shows a direct connection to cryptographic one-way partial functions.

1 Errors of Randomized Algorithms

Randomized algorithms have drawn wide attention due to their simplicity and speed as well as their direct implication to approximation algorithms and average-case analysis. This paper focuses on two different types of *errors* that are of major importance in the designing of randomized algorithms. The first one is an error associated with each individual computation on a single input. As seen in a BPP algorithm, its computation is allowed to produce on each input a false outcome with probability at most $1/3$. The second one is related to cumulative errors over all inputs of each fixed length. Such errors are seen in a P-close set A, which has a polynomial-time deterministic algorithm that can decide the membership of A only on all but polynomially-many inputs of each length.

The notion of *nearly BPP sets* stems from the aforementioned two types of errors and it embodies these errors that may occur in a run of polynomial-time randomized algorithms. A nearly BPP set is the set recognized by a certain polynomial-time randomized Turing machine that errs with *small probability* only on *most inputs*. To describe these errors, we use two parameters $\epsilon(n)$ and $\delta(n)$. The first parameter $\epsilon(n)$ is used to indicate an error bound of a randomized

R. Petreschi et al. (Eds.): CIAC 2003, LNCS 2653, pp. 213–226, 2003.
© Springer-Verlag Berlin Heidelberg 2003

computation on a single input of length n whereas $\delta(n)$ designates an error rate accumulated over all randomized computations on any inputs of length n.

Fix our alphabet Σ to be $\{0,1\}$ throughout this paper. Note that *polynomials* used in this paper are all assumed to have only positive integer coefficients. For any set A, the notation $A(x)$ denotes the value of the *characteristic function* of A on input x; namely, $A(x) = 1$ if $x \in A$ and $A(x) = 0$ otherwise.

Definition 1. [21,22] *Let A be any set and M be any multi-tape randomized Turing machine. Let ϵ and δ be any functions from $\mathbb{N}$ to the real interval $[0,1]$.*

1. The machine M recognizes A with two-sided $(\epsilon(n), \delta(n))$-error if the two-sided $\epsilon(n)$-discrepancy set $D = \{x \mid \mathrm{Prob}_M[M(x) \neq A(x)] > \epsilon(|x|)\}$ for (M,A) has density at most $\delta(n) \cdot 2^n$ almost everywhere.[1]

2. The machine M recognizes A with one-sided $(\epsilon(n), \delta(n))$-error if the one-sided $\epsilon(n)$-discrepancy set $D = \{x \mid \mathrm{Prob}_M[M(x) \neq A(x) = 1] > \epsilon(|x|)\} \cup \{x \mid \mathrm{Prob}_M[M(x) \neq A(x) = 0] > 0\}$ for (M,A) has density at most $\delta(n) \cdot 2^n$ almost everywhere.

3. A set S is called nearly BPP if, for every polynomial p, there exists a polynomial-time randomized Turing machine that recognizes S with two-sided $(1/3, 1/p(n))$-error. A nearly RP set is defined similarly using the phrase "one-sided $(1/2, 1/p(n))$-error" instead. Let Nearly-BPP and Nearly-RP denote respectively the collection of all nearly BPP sets and that of all nearly RP sets.

In retrospect, the notion of nearly BPP sets was first discussed in [21,22] in connection to average-case complexity theory developed by Levin [13] and Gurevich [10] in the 1980s. This notion was later spotlighted by Schindelhauer and Jakoby [17] under the name of error complexity. Recent work of Aida and Tsukiji [1] showed a tight connection between nearly BPP sets and polynomial-time samplable distributions. This paper aims for enriching the knowledge of nearly BPP sets.

For later use, it is useful to introduce the functional version of Nearly-BPP by expanding set A in Definition 1 into polynomially-bounded[2] total function f from Σ^* to Σ^*. We use the notation Nearly-FBPP to denote this function class.

Classic but principal examples of nearly BPP sets are *polynomially sparse sets* and *average BPP sets*. We use a capital letter together with subscript, such as X_n, to denote a random variable distributed uniformly over $\{0,1\}^n$.

1. Let g be any function from $\mathbb{N}$ to $\mathbb{N}$ such that (i) $g(n)$ is computable in time polynomial in n and (ii) $g(n) \in \omega(n^k)$ for all positive integers k. Every set of density at most $2^n/g(n)$ almost everywhere is clearly nearly RP. In particular, every polynomially sparse set is nearly RP.

---

[1] For a function d from $\mathbb{N}$ to $\mathbb{N}$ and a set S, we say that a set S has *density at most (at least, resp.) $d(n)$ almost everywhere* if $|S^{=n}| \leq d(n)$ ($|S^{=n}| \geq d(n)$, resp.) for all but finitely-many natural numbers n, where $S^{=n} = S \cap \Sigma^n$ and $|\cdot|$ means the cardinality.

[2] A (partial) function f from Σ^* to Σ^* is *polynomially bounded* if there exists a polynomial p such that $|f(x)| \leq p(|x|)$ for all $x \in \Sigma^*$.

2. A set S is called *average BPP*[3] if there exist a polynomial p and a randomized Turing machine M such that (i) $\text{Prob}_M[M(x) = S(x)] \geq 2/3$ for all $x \in \Sigma^*$ and (ii) $\text{Exp}_{X_n, R}[\text{Time}_M(X_n; R)] \leq p(n)$ for each natural number n, where $\text{Time}_M(x; r)$ denotes the running time of M on input x following r, a series of coin tosses made by M on x. All average BPP sets are nearly BPP.

We further generalize the definition of Nearly-BPP into a class operator. In late 1980s, Schöning [18] introduced the BP[·]-operator, which schematically generates a new complexity class BP[$\mathcal{C}$] from any given class $\mathcal{C}$. The BP[·]-operator greatly expands the class BPP. Moreover, the BP[·]-operator can be viewed as a special form of "BPP-m-reducibility." Similarly, we expand Nearly-BPP by extracting its probabilistic nature into the Nearly-BP[·]-operator. Its precise definition is given as follows.

Definition 2. *Let A and B be any sets,[4] q be any polynomial, and $\mathcal{C}$ be any complexity class of languages.*

1. We say that the pair (B, q) defines A with two-sided $(\epsilon(n), \delta(n))$-error if the two-sided $\epsilon(n)$-discrepancy set $D = \{z \mid \text{Prob}_{X_{q(n)}}[B(z, X_{q(n)}) \neq A(z)] > \epsilon(|z|)\}$ for (B, q, A) has density at most $\delta(n) \cdot 2^n$ almost everywhere.

2. We say that the pair (B, q) defines A with one-sided $(\epsilon(n), \delta(n))$-error if the one-sided $\epsilon(n)$-discrepancy set $D = \{z \mid \text{Prob}_{X_{q(n)}}[B(z, X_{q(n)}) \neq A(z) = 1] > \epsilon(|z|)\} \cup \{x \mid \text{Prob}_{X_{q(n)}}[B(z, X_{q(n)}) \neq A(z) = 0] > 0\}$ for (B, q, A) has density at most $\delta(n) \cdot 2^n$ almost everywhere.

3. A set A is in Nearly-BP[$\mathcal{C}$] if, for every polynomial p, there exist a set $B \in \mathcal{C}$ and a polynomial q such that (B, q) defines A with two-sided $(1/3, 1/p(n))$-error. Nearly-R[$\mathcal{C}$] is defined similarly using one-sided $(1/2, 1/p(n))$-errors.

It is straightforward to prove that Nearly-BP[BPP] equals Nearly-BPP by simulating B (together with $X_{q(n)}$) in Definition 2 on a certain randomized Turing machine M. The following lemma is also immediate from the definition of the Nearly-BP[·]-operator.

Lemma 1. *Let $\mathcal{C}$ and $\mathcal{D}$ be any complexity classes.*
1. BP[$\mathcal{C}$] $\subseteq$ Nearly-BP[$\mathcal{C}$].
2. Nearly-BP[co-$\mathcal{C}$] = co-Nearly-BP[$\mathcal{C}$].
3. If $\mathcal{C} \subseteq \mathcal{D}$, then Nearly-BP[$\mathcal{C}$] $\subseteq$ Nearly-BP[$\mathcal{D}$].

The Nearly-BP[·]-operator enables us to introduce a variety of new complexity classes. Rooted in the polynomial hierarchy [15], for instance, we can define Nearly-BPΣ_k^P to be Nearly-BP[Σ_k^P] for each integer $k \geq 0$. Likewise, Nearly-BPΔ_k^P and Nearly-RΔ_k^P are defined. For practicality, we introduce the functional version of Nearly-BPΔ_k^P, denoted Nearly-FBPΔ_k^P, which expands the function class FΔ_k^P $(= \text{FP}^{\Sigma_{k-1}^P})$ for each positive integer k.

---

[3] The class of all average BPP sets is known to coincide with the "quintessential" complexity class BPP$_{\{\nu\}}$, where ν is the standard distribution. See [19,22].

[4] By use of an appropriate pairing function, we often identify Σ^* with $\Sigma^* \times \Sigma^*$.

2   Fundamental Features of Nearly-BPP

We first show close associations between Nearly-BPP and the concepts of approximation schemes [2,16], promise problems [6,7], and distinguishing complexity [20].

In their study of pseudorandom generators, Nisan and Wigderson [16] considered the notion of "approximation by randomized algorithms." Following [2,16], we define the notion of $\epsilon(n)$-approximation as follows. Let ϵ be any function from $\mathbb{N}$ to $[0,1]$ and A be any set. A randomized Turing machine M $\epsilon(n)$-*approximates* A if $\mathrm{Prob}_{X_n,M}[A(X_n) \neq M(X_n)] \leq \epsilon(n)$ for all but finitely-many natural numbers n. We thus say that A is $\epsilon(n)$-*approximated by polynomial-time randomized algorithms* if there exists a polynomial-time randomized Turing machine that $\epsilon(n)$-approximates A.

Even and Yacobi [7] studied resource-bounded partial decision problems, known as *promise problems*. This paper follows [6,7] and define the promise complexity class Promise-BPP. A promise problem in Promise-BPP is a pair $(Q,A) \subseteq \Sigma^* \times \Sigma^*$ such that there exists a polynomial-time randomized Turing machine M satisfying $\mathrm{Prob}_M[M(x) = A(x)] \geq 2/3$ only for each x in Q.

Recently, Buhrman and Torenvliet [5] demonstrated a new characterization of probabilistic classes in terms of distinguishing complexity introduced by Sipser [20]. We fix a universal deterministic Turing machine M_U with three input tapes.[5] For any function t from $\mathbb{N}$ to $\mathbb{N}$, any oracle A, and any strings x and y, the *t-time bounded conditional Kolmogorov complexity of x conditional to y relative to A*, denoted $\mathrm{C}^{t,A}(x|y)$, is the length of a shortest binary string w (conventionally called a *program*) such that $M_U^A(w,y) = x$ within $t(|x|+|y|)$ steps. We now say that w *t-distinguishes x conditional to y relative to A* if $M_U^A(w,x,y) = 1$ within $t(|x| + |y|)$ steps and also $M_U^A(w,z,y) = 0$ within $t(|z| + |y|)$ steps for any $z \in \Sigma^* - \{x\}$. The *t-time bounded conditional distinguishing complexity of x conditional to y relative to A*, $\mathrm{CD}^{t,A}(x|y)$, is the length of a shortest program w that t-distinguishes x conditional to y relative to A. The *unconditional* version of the above two complexity measures are defined as follows: $\mathrm{C}^{t,A}(x) = \mathrm{C}^{t,A}(x|\lambda)$ and $\mathrm{CD}^{t,A}(x) = \mathrm{CD}^{t,A}(x|\lambda)$, where λ is the empty string.

The following proposition provides three different characterizations of Nearly-BPP in terms of $\frac{1}{p(n)}$-approximation, Promise-BPP, and $\mathrm{CD}^t(x)$. For any set A, let $\overline{A}$ denote $\Sigma^* - A$.

Proposition 1. *Let A be any subset of Σ^*. The following statements are all logically equivalent.*

1. A is in Nearly-BPP.

2. For every polynomial p, A is $\frac{1}{p(n)}$-approximated by polynomial-time randomized algorithms.

3. For every polynomial p, there exists a set Q such that $(Q,A) \in$ Promise-BPP and $\overline{Q}$ has density at most $2^n/p(n)$ almost everywhere.

---

[5] When we write $M_U(x,y,z)$, we understand that x is given on the first input tape, y is on the second input tape, and z is on the third input tape. The notation $M_U(x,y)$, however, indicates that the third input tape is empty.

4. For every polynomial p and every constant $\epsilon \in (0,1]$, there exist a polynomial q and a set B in P such that the discrepancy set $D = \{x \mid \exists y \in \Sigma^{q(|x|)}[\mathrm{CD}^q(y) \geq \epsilon|y| \wedge A(x) \neq B(x,y)]\}$ has density at most $2^n/p(n)$ almost everywhere.

The proof of Proposition 1(4) especially follows an argument in [5] with help of Lemma 2.

We next address a link between Nearly-BPP and nonadaptive random-self-reducibility discussed by Feigenbaum, Kannan, and Nisan [9] and by Feigenbaum and Fortnow [8]. For any functions h, k from Σ^* to Σ^*, we say that h is *BPP-tt-reducible* to k if there exist polynomials p, q and total FP-functions f, g such that, for every x, the probability, over all strings $r \in \Sigma^{q(|x|)}$, that $f(x, r, k(g(1, x, r)), k(g(1^2, x, r)), \ldots, k(g(1^{p(|x|)}, x, r))) = h(x)$ is at least 2/3. The quadruple (f, g, q, p) is called a *BPP-tt-reduction*. When h and k are respectively the characteristic functions of A and of B, we also say that A is BPP-tt-reducible to B. A set A is *uniformly nonadaptively random-self-reducible* if A is BPP-tt-reducible to A via (f, g, q, p) with the condition that, for every x and every $i \in [1, p(|x|)]_{\mathbb{Z}}$,[6] the value $g(1^i, x, r)$ is uniformly distributed over $\{0,1\}^{q(|x|)}$ when r is chosen uniformly at random from $\Sigma^{q(|x|)}$. The notation uniformly-RSR$_{tt}$ denotes the collection of all uniformly nonadaptively random-self-reducible sets.

The following proposition implicates that any nearly BPP set not in BPP cannot be uniformly nonadaptively random-self-reducible.

Proposition 2.   uniform-RSR$_{tt}$ ∩ Nearly-BPP = BPP.

Earlier, Karp and Lipton [11] introduced the advice class P/poly, which is also known as the collection of all sets that have non-uniform polynomial-size Boolean circuits. Similar to Nearly-BPP, this class P/poly contains all polynomially sparse sets and all BPP sets. Nonetheless, Nearly-BPP and P/poly are essentially different. In the following proposition, we show that P/poly and Nearly-BPP are incomparable.

Proposition 3.   Nearly-BPP $\not\subseteq$ P/poly *and* P/poly $\not\subseteq$ Nearly-BPP.

Relativizations are useful tools, imported from recursion theory, in studying the behaviors of underlying computations and the roles of oracles by way of (oracle) queries. A *relativized nearly BPP set* as well as a *relativized RP set* is naturally introduced by using a randomized oracle Turing machine in place of a randomized Turing machine in Definition 1. We use the notations Nearly-BPP^A and Nearly-RP^A to denote respectively the relativizations of Nearly-BPP and of Nearly-RP relative to oracle A. For any collection $\mathcal{C}$ of oracles, Nearly-BPP$^{\mathcal{C}}$ and Nearly-RP$^{\mathcal{C}}$ denote the classes $\bigcup_{A \in \mathcal{C}}$ Nearly-BPP^A and $\bigcup_{A \in \mathcal{C}}$ Nearly-RP^A, respectively. It is not difficult to show that Nearly-BPP^{BPP} = Nearly-BPP and Nearly-RP^{ZPP} = Nearly-RP. The first equality asserts that the oracle class BPP

---

[6] The notation $[n,m]_{\mathbb{Z}}$ denotes the set $\{n, n+1, n+2, \ldots, m\}$ for any integers n, m with $n \leq m$.

does not enhance the computational power of Nearly-BPP. However, BPP cannot be expanded to Nearly-BPP because Nearly-BPP$^{\text{Nearly-BPP}} \neq$ Nearly-BPP (see Lemma 6).

Before closing this section, we exhibit a relativized world where BPP and Nearly-BPP are different within NP.

Proposition 4. *There exists an oracle A such that $\text{NP}^A \not\subseteq \text{BPP}^A$ and $\text{NP}^A \subseteq$ Nearly-BPP^A.*

3   Amplification Properties

Given a randomized algorithm, it is important to know whether we can amplify its success probabilities. As was shown in [21,22], all nearly BPP sets enjoy this amplification property. This section shows the amplification property for a more general class Nearly-BP[$\mathcal{C}$]. We first give a restricted form of majority reductions.

Definition 3. *A set A is* nicely majority reducible to B *(denoted $A \leq^p_{maj,nice} B$) if there exist a total function f and a polynomial k such that, for every string x of length $k(n)n$ for some n, (i) $f(x) = \langle y_1, y_2, \ldots, y_{k(n)} \rangle$, where $|y_i| = n$ for any $i \in [1, k(n)]_{\mathbb{Z}}$, (ii) f is a bijection from $\Sigma^{k(n)n}$ to $\{\langle y_1, \ldots, y_{k(n)} \rangle \mid y_i, \ldots, y_{k(n)} \in \Sigma^n\}$, and (iii) (ii) $A(x) = 1$ if $|\{i \in [1, k(n)]_{\mathbb{Z}} \mid y_i \in B\}| > \lfloor k(n)/2 \rfloor$ and $A(x) = 0$ otherwise.*

Although we can further relax the above constraint, this definition serves well to prove the desired amplification lemma for Nearly-BP[$\mathcal{C}$]. Note that many complexity classes are indeed closed downward under nice majority reductions.

In order to enhance the success probabilities of randomized computations with a relatively small number of random bits, we utilize Zuckerman's universal oblivious samplers [23]. The following slightly simpler form suffices for our purpose. A *universal $(r, d, m, \epsilon, \gamma)$-oblivious sampler* is a deterministic Turing machine M that takes a binary string x of length r and outputs a code $\langle z_1^{(y)}, z_2^{(y)}, \ldots, z_d^{(y)} \rangle$ of sample points $z_1^{(y)}, z_2^{(y)}, \ldots, z_d^{(y)}$ with $z_i \in \{0,1\}^m$ such that, for any function f from $\{0,1\}^m$ to $[0,1]$, $\text{Prob}_{Y_r}[|\frac{1}{d}\sum_{i=1}^d f(z_i^{(Y_r)}) - \text{Exp}[f]| \leq \epsilon] \geq 1 - \gamma$, where $\text{Exp}[f] = 2^{-m}\sum_{z:|z|=m} f(z)$.

Lemma 2. (Amplification Lemma) *Let $\mathcal{C}$ be any complexity class closed downward under $\leq^p_{maj,nice}$. Assume that $A \in$ Nearly-BP[$\mathcal{C}$]. Let p be any polynomial and let k and α be any positive numbers satisfying $0 < \alpha \leq 1/2$. There exist a polynomial q and a set $B \in \mathcal{C}$ such that (B, s) defines A with two-sided $(2^{-kq(n)}, 1/p(n))$-error, where $s(n) = (1 + \alpha)(1 + k)q(n)$.*

The proof of Lemma 2 is an adaptation of the argument given in [5] with universal oblivious samplers. Lemma 2 implies the closure property of Nearly-BP[$\mathcal{C}$] under nice majority reductions.

Corollary 1. *Let $\mathcal{C}$ be any complexity class of languages. If $\mathcal{C}$ is closed downward under $\leq^p_{maj,nice}$ then so is Nearly-BP[$\mathcal{C}$].*

The following proposition also follows from Lemma 2.

Proposition 5. *Let C be any complexity class of languages. If C is closed under $\leq^p_{maj,nice}$, then* Nearly-BP[Nearly-BP[C]] = BP[Nearly-BP[C]] = Nearly-BP[C].

Applying Proposition 5 to the class Δ^P_k, we obtain the following consequence: Nearly-BP[Nearly-BPΔ^P_k] = BP[Nearly-BPΔ^P_k] = Nearly-BPΔ^P_k for each positive integer k. This exemplifies the robustness of the class Nearly-BPΔ^P_k.

4   Nice Truth Table Reductions

We have shown the closure property of Nearly-BP[C] under nice majority reductions. In this section, we seek more general and helpful reductions.

Most of the well-known complexity classes (sitting above P) are closed downward under P-m-reductions. Such a closure property is often regarded as a structural asset; for example, since SAT is P-m-complete for NP, the assumption SAT $\in$ P makes the whole NP collapse to P. Disappointingly, Nearly-BPP is not even closed under length-regular[7] h-P-1-reductions, where a set A is said to be *h-P-1-reducible to* another set B if there exists a function f from Σ^* to Σ^* such that (i) f is one-one and honest[8] and (ii) $A = \{x \mid f(x) \in B\}$.

Proposition 6. Nearly-BPP *is not closed downward under length-regular h-P-1-reductions.*

The above proposition indicates that a much weaker form of polynomial-time reductions is required for Nearly-BP[C]. We thus bring in the concepts of *nice BPΔ^P_k-tt-reductions* and *nice RP-m-reductions*. Conventionally, a polynomial-time randomized Turing machine M is said to *RP-m-reduce* A to B if there exists a polynomial p such that, for every x, if $x \in A$ then $\text{Prob}_M[M(x) \in B] \geq 1/p(n)$ and otherwise $\text{Prob}_M[M(x) \in B] = 0$. Additionally, a randomized Turing machine M is called $t(n)$-*to-one* if, for every y, the set $C_y = \{x \mid \text{Prob}_M[M(x) = y] > 0\}$ has cardinality at most $t(|y|)$.

Definition 4. *1. For any set $A, B \subseteq \Sigma^*$, A is said to be* nicely RP-m-reducible *to B if there exist a function t and a length-regular[9] randomized Turing machine M with length function ℓ such that (i) M RP-m-reduces A to B, (ii) M is $t(n)$-to-one, and (iii) for a certain constant $c > 0$, $\ell(n) + \log_2 t(\ell(n)) \leq s(n) + c\log_2 n$ for all but finitely-many natural numbers n, where $s(n)$ is the minimal number of coin tosses (along each computation path) of M on any input of length n.*

---

[7]   A (partial) function f from Σ^* to Σ^* is *length-regular* if, for every pair $x, y \in \text{dom}(f)$, $|x| = |y|$ implies $|f(x)| = |f(y)|$.

[8]   For a (partial) function f from Σ^* to Σ^*, f is *polynomially honest* (honest, for short) if there exists a polynomial p satisfying that $|x| \leq p(|f(x)|)$ for all $x \in \text{dom}(f)$.

[9]   A randomized Turing machine M is *length-regular* if there exists a function ℓ from $\mathbb{N}$ to $\mathbb{N}$ such that $\text{Prob}_M[|M(x)| = \ell(|x|)] = 1$ for all x. This ℓ is called the *length function* of M.

2. *Let d be any positive integer. For any functions h, k from Σ^* to Σ^*, we say that h is nicely $BP\Delta_d^P$-tt-reducible to k if there exist polynomials p, q, s and total $F\Delta_d^P$-functions f, g such that (i) h is BPP-tt-reducible to k via (f, g, q, p), (ii) $s(|g(1^i, x, r)|) \geq |x|$ for all x, all $i \in [1, p(|x|)]_\mathbb{Z}$, and all $r \in \Sigma^{q(|x|)}$ (we call g honest), and (iii) for every n, every $y \in \Sigma^*$, and every $i \in [1, p(n)]_\mathbb{Z}$, $\mathrm{Prob}_{X_n, R_{q(n)}}[g(1^i, X_n, R_{q(n)}) = y] \leq s(n)2^{-|y|}$. In this case, write $h \leq_{tt,nice}^{BP\Delta_d^P} k$. In particular, if h and k are respectively the characteristic functions of A and of B, we write $A \leq_{tt,nice}^{BP\Delta_d^P} B$.*

Unfortunately, nearly-PR is not closed downward under nice RP-m-reductions although RP is closed under RP-m-reductions. This is because, for each set A in Nearly-RP, its associated one-sided $1/2$-discrepancy set D may contain elements in $\overline{A}$ and thus, no reduction correctly reduces these elements. To avoid this inconvenience, we introduce a slightly restricted class, named Nearly-RP$_*$, as follows.

Definition 5. *Nearly-RP$_*$ is the collection of all sets A such that, for every polynomial p, there exists a polynomial-time randomized Turing machine M such that the one-sided $1/2$-discrepancy set D for (M, A) has density at most $\frac{1}{p(n)} \cdot 2^n$ almost everywhere and also satisfies $D \subseteq A$.*

Note that Nearly-RP$_*$ is *properly* contained in Nearly-RP. In the end, we can prove that Nearly-RP$_*$ and Nearly-BPΔ_k^P are indeed closed downward under nicely RP-m-reductions and BPΔ_k^P-tt-reductions, respectively.

Lemma 3. *Let A and B be any sets. Let k be any positive integer and let $\mathcal{C}$ be any complexity class that is closed downward under $\leq_{maj,nice}^{BPP}$ and $\leq_{tt,nice}^{BPP}$.*

1. *If $A \leq_{m,nice}^{RP} B$ and $B \in$ Nearly-RP$_*$, then $A \in$ Nearly-RP$_*$.*
2. *If $A \leq_{tt,nice}^{BPP} B$ and $B \in$ Nearly-BP$[\mathcal{C}]$, then $A \in$ Nearly-BP$[\mathcal{C}]$.*
3. *If $A \leq_{tt,nice}^{BP\Delta_k^P} B$ and $B \in$ Nearly-BPΔ_k^P, then $A \in$ Nearly-BPΔ_k^P.*

To show the usefulness of nice BPP-tt-reducibility, we consider the existence of complete sets for NP under this reducibility. The BOUNDED HALTING PROBLEM (BHP) is the collection of all strings of the form $\langle M, x, 1^t \rangle$, where M is a nondeterministic Turing machine, x is a string, and t is a positive integer, such that M accepts x within t steps. It is known that BHP is P-m-complete for NP. We can straightforwardly prove that BHP is also nicely BPP-tt-complete for NP.

In late 1990s, Buhrman and Fortnow [3] proved that BPP $\subseteq$ RP$^{\text{Promise-RP}}$, where Promise-RP is the one-sided version of Promise-BPP. By refining their proof (which is based on the proof of Lautemann [12]), we can show that BPP $\subseteq$ RP$^{\text{Nearly-RP}}$. Since Nearly-RP $\subseteq$ Promise-RP, our result gives a slight improvement to the upper bound of BPP. More strongly, we have:

Proposition 7. *BPP $\subseteq$ RP$^{\text{Nearly-RP}_*[1]}$, where $[\cdot]$ stands for the number of queries made along each computation path.*

Subsequently, we show a direct application of nice $\mathrm{BP}\Delta_k^{\mathrm{P}}$-tt-reducibility to Σ_k^{P}-search problems. Since search problems are in general partial functions, we need to expand the scope of $\leq_{tt,nice}^{\mathrm{BP}\Delta_k^{\mathrm{P}}}$ from sets to *partial multi-valued functions* in the following fashion. For convenience, we use the notation $\mathrm{F}\Delta_k^{\mathrm{P}}$ for *partial single-valued functions*; namely, f is in $\mathrm{F}\Delta_k^{\mathrm{P}}$ if there exist a polynomial-time deterministic Turing machine M and a set $A \in \Sigma_{k-1}^{\mathrm{P}}$ such that, for every x, (i) if $x \in \mathrm{dom}(f)$ then M^A on input x outputs $f(x)$ and (ii) if $x \notin \mathrm{dom}(f)$ then M^A on input x outputs $\perp$, where $\perp$ is a distinguished tape symbol not in Σ. For any partial multi-valued function h and any set B, we say that h is *nicely $\mathrm{BP}\Delta_k^{\mathrm{P}}$-tt-reducible* to B (denoted $h \leq_{tt,nice}^{\mathrm{BP}\Delta_k^{\mathrm{P}}} B$) if there exist polynomials p, q, s, a partial $\mathrm{F}\Delta_k^{\mathrm{P}}$-function f, and a total $\mathrm{F}\Delta_k^{\mathrm{P}}$-function g such that (i) for every $x \in \mathrm{dom}(h)$, the probability, over all strings $r \in \Sigma^{q(|x|)}$, of $f(x, r, B(g(1, x, r)), B(g(1^2, x, r)), \ldots, B(g(1^{p(|x|)}, x, r))) \in h(x)$ is at least $2/3$, (ii) $s(|g(1^i, x, r)|) \geq |x|$ for all x, $i \in [1, p(|x|)]_{\mathbb{Z}}$, and $r \in \Sigma^{q(|x|)}$, (iii) for every n, every $y \in \Sigma^*$, and every $i \in [1, p(n)]_{\mathbb{Z}}$, $\mathrm{Prob}_{X_n, R_{q(n)}}[g(1^i, X_n, R_{q(n)}) = y] \leq s(n)2^{-|y|}$, and (iv) for every $x \notin \mathrm{dom}(h)$, the probability, over all strings $r \in \Sigma^{q(|x|)}$, that $f(x, r, B(g(1, x, r)), B(g(1^2, x, r)), \ldots, B(g(1^{p(|x|)}, x, r)))$ is undefined is at least $2/3$.

For any set $A \subseteq \Sigma^*$ and any polynomial p, let $Search_{A,p}$ be the partial multi-valued function defined by $Search_{A,p}(x) = \{y \in \Sigma^{p(|x|)} \mid (x, y) \in A\}$ for every x. In the case where A belongs to Δ_k^{P}, $Search_{A,p}$ represents a Σ_k^{P}-search problem. The following lemma shows that Σ_k^{P}-search problems can be nicely $\mathrm{BP}\Delta_k^{\mathrm{P}}$-tt-reducible to certain sets in Σ_k^{P}.

Lemma 4. *Let k be any positive integer. For every set $A \in \Delta_k^{\mathrm{P}}$ and every polynomial p, there exists a set B in Σ_k^{P} such that $Search_{A,p} \leq_{tt,nice}^{\mathrm{BP}\Delta_k^{\mathrm{P}}} B$.*

5   Relationships between Σ_k^{P} and Nearly-$\mathrm{BP}\Delta_k^{\mathrm{P}}$

A crucial open problem is whether Nearly-BPP includes NP, or more generally, Nearly-$\mathrm{BP}\Delta_k^{\mathrm{P}}$ includes Σ_k^{P} for each $k \geq 1$. The inclusion NP $\subseteq$ Nearly-BPP, for instance, results in the non-existence of cryptographic one-way functions [21,22]. This section exhibits five new statements, which are all logically equivalent to the $\Sigma_k^{\mathrm{P}} \subseteq$?Nearly-$\mathrm{BP}\Delta_k^{\mathrm{P}}$ question.

For our purpose, Nearly-$\mathrm{FBP}\Delta_k^{\mathrm{P}}$ is expanded into a class of *partial multi-valued functions*. Henceforth, Nearly-$\mathrm{FBP}\Delta_k^{\mathrm{P}}$ denotes the collection of all partial multi-valued functions f from Σ^* to Σ^* satisfying the following condition:[10] for every polynomial p, there exist a polynomial-time randomized Turing machine M and a set D such that, for every $x \in \Sigma^*$, (i) if $x \in \mathrm{dom}(f) \cap \overline{D}$, then $f(x) = \{y \mid \mathrm{Prob}_M[M^{\mathrm{QBF}_{k-1}}(x) = y] > 0\}$ and $\mathrm{Prob}_M[M^{\mathrm{QBF}_{k-1}}(x) \in f(x)] \geq 2/3$, (ii) if $x \in \overline{D} - \mathrm{dom}(f)$, then $\mathrm{Prob}_M[M^{\mathrm{QBF}_{k-1}}(x) = \perp] \geq 2/3$, and (iii)

---

[10] For each positive integer k, let QBF_k be any P-m-complete set for Σ_k^{P}, for example, the set of satisfiable quantified Boolean formulas with at most k alternations of quantifiers starting with $\exists$. For convenience, let QBF_0 be the empty set.

$|D^{=n}| \leq 2^n/p(n)$ for all but finitely-many natural numbers n. The classes $\Sigma_k^P SV$ and $\Sigma_k^P MV$ are respectively the natural extensions of NPSV and NPMV; namely, $\Sigma_k^P MV$ is the collection of all partial multi-valued functions f such that $f(x) = \{g(x, y) \mid y \in \Sigma^{p(|x|)} \wedge (x, y) \in \text{dom}(g)\}$ for certain partial functions g in $F\Delta_k^P$, and $\Sigma_k^P SV$ is the collection of all single-valued $\Sigma_k^P MV$-functions. For any two partial multi-valued functions f and g, we say that f is a *refinement of* g if, for every x, (i) $f(x) \subseteq g(x)$ and (ii) $f(x) = \emptyset$ implies $g(x) = \emptyset$. Finally, for any class $\mathcal{F}$ of partial multi-valued functions, g *has an* $\mathcal{F}$ *refinement* if there exists an element $f \in \mathcal{F}$ such that f is a refinement of g.

Theorem 1. *Let k be any positive integers. The following statements are all logically equivalent.*

1. $\Sigma_k^P \subseteq \text{Nearly-BP}\Delta_k^P$.
2. $\Sigma_k^P \subseteq \text{Nearly-R}\Delta_k^P$.
3. $\text{BP}\Sigma_k^P \subseteq \text{Nearly-BP}\Delta_k^P$.
4. $\text{Nearly-BP}\Sigma_k^P = \text{Nearly-BP}\Delta_k^P$.
5. $\Sigma_k^P SV \subseteq \text{Nearly-FBP}\Delta_k^P$.
6. *Every* $\Sigma_k^P MV$-*function has a* $\text{Nearly-FBP}\Delta_k^P$ *refinement.*

Theorem 1 is proven by Propositions 1 and 5 and Lemmas 2, 3, and 4.

In addition to the six statements of Theorem 1, it is possible to give a Kolmogorov-complexity characterization of the $\Sigma_k^P \subseteq?\text{Nearly-BP}\Delta_k^P$ question in a manner similar to Theorem 2(5). With a less complex argument, we also prove in the next proposition a new characterization of the $\Sigma_k^P \subseteq?\text{BP}\Delta_k^P$ question. This compares with the result in [4].

Proposition 8. *For each positive integer k, $\Sigma_k^P \subseteq \text{BP}\Delta_k^P$ iff, for every polynomial q and every number $\epsilon \in (0, \frac{1}{2}]$, there exist a polynomial s and a positive real number c such that $\forall x \forall y \in \Sigma^{\leq q(|x|)} \forall r \in \Sigma^{s(|x|)} [\text{CD}^{s, \text{QBF}_{k-1}}(r) \geq \epsilon |r| \to C^{s, \text{QBF}_{k-1}}(y|x, r) \leq \text{CD}^{q, \text{QBF}_{k-1}}(y|x) + c \log_2(|x| + |y|)]$.*

The proof of Proposition 8 uses a special case of Proposition 1 and Theorem 1 as well as basic lemmas from [14].

Analogous to Theorem 1, we can prove the following.

Proposition 9. *Let k be any positive integer.*

1. $\Pi_k^P \subseteq \text{Nearly-BP}\Sigma_k^P$ *iff* $\text{Nearly-BP}\Pi_k^P = \text{Nearly-BP}\Sigma_k^P$.
2. $\Sigma_{k+1}^P \subseteq \text{Nearly-BP}\Sigma_k^P$ *iff* $\text{Nearly-BP}\Sigma_{k+1}^P = \text{Nearly-BP}\Sigma_k^P$.

Regarding the function class #P, it is open whether $NP \subseteq \text{Nearly-BPP}$ is equivalent to #P $\subseteq \text{Nearly-FBPP}$. Here, we give a partial answer to this question by considering a subset of #P, denoted #FewP, which is the collection of all total functions f from Σ^* to $\mathbb{N}$ satisfying the following: there exist a polynomial p and a polynomial-time nondeterministic Turing machine M such that, for every x, (i) $f(x)$ equals the number of accepting computation paths of M on input x and (ii) $f(x) \leq p(|x|)$. Let h be any #FewP-function. Parallel to Lemma 4, we can construct a set B in FewP satisfying $h \leq_{tt, nice}^{\text{BPP}} B$. This fact implies the following result.

Proposition 10. *FewP $\subseteq \text{Nearly-BPP}$ iff #FewP $\subseteq \text{Nearly-FBPP}$.*

6   Infinitely Often Nearly BPP Sets

Complexity classes usually impose the "almost all" clause on various parts of their definitions. Typically, an underlying machine for a given language is required to satisfy a certain acceptance criteria on *almost all* inputs. For example, any BPP set must be witnessed by a certain polynomial-time randomized Turing machine whose error probability is bounded below by $1/3$ for *almost all* inputs. Recent research has also investigated the complexity classes defined by substituting the "infinitely often" clause for the "almost all" clause. In this line of research, we consider the "infinitely-often" version of Nearly-BPP and present its close connection to the existence of cryptographic one-way functions.

We begin with formulating the class io-Nearly-BPP.

Definition 6. *Let ϵ, δ be any function from $\mathbb{N}$ to $[0, 1]$, M be any randomized Turing machine, and S be any subset of Σ^*.*

1. The machine M recognizes *S with infinitely-often two-sided $(\epsilon(n), \delta(n))$-error if the two-sided ϵ-discrepancy set for (M, S) has density at most $\delta(n) \cdot 2^n$ infinitely often.*[11]

2. The set S is infinitely-often nearly BPP *if, for every polynomial p, there exists a polynomial-time randomized Turing machine that recognizes S with infinitely-often two-sided $(1/3, 1/p(n))$-error. Let* io-Nearly-BPP *denote the collection of all infinitely-often nearly BPP sets.*

Apparently, Nearly-BPP is a *proper* subclass of io-Nearly-BPP. It also follows that Nearly-BP[io-Nearly-BPP] = BP[io-Nearly-BPP] = io-Nearly-BPP.

Hereafter, we focus on cryptographic one-way functions. Although it is not proven to exist, a one-way function has been an important component in building a secure cryptosystem. Our goal is to connect io-Nearly-BPP to such functions. In particular, we cast our interest onto "partial" one-way functions for which we require the one-wayness only on the domain of the functions.

Definition 7. *1. A partial single-valued function f from Σ^* to Σ^* is called (uniformly)* weakly one-way *if (i) $dom(f)$ is dense,[12] (ii) f is in FP (as a partial function), and (iii) there exists a polynomial p such that, for every polynomial-time randomized Turing machine M, $\mathrm{Prob}_{X_n, M}[f(M(1^n, f(X_n))) = f(X_n) \mid X_n \in dom(f)] \leq 1 - 1/p(n)$ for all but finitely-many natural numbers n. Note that if $x \notin dom(f)$ then $f(x) = \bot$ and thus, we may assume that $M(1^n, \bot)$ always outputs $\bot$.*

2. Likewise, f is called (uniformly) strongly one-way *if (i) $dom(f)$ is dense, (ii) f is in FP (as a partial function), and (iii) for every polynomial-time randomized Turing machine M and for any positive polynomial p, it holds that $\mathrm{Prob}_{X_n, M}[f(M(1^n, f(X_n))) = f(X_n) \mid X_n \in dom(f)] < 1/p(n)$ for all but finitely-many natural numbers n.*

---

[11] A set S is said to have *density at most $d(n)$ infinitely often* if there exists an infinite set N of natural numbers such that $|S^{=n}| \leq d(n)$ holds for all $n \in N$.

[12] A set S is *dense* if there exists a polynomial p such that S has density at least $2^n/p(n)$ almost everywhere.

For a *total* one-way function, it is proven in [21,22] that the existence of such a function leads to the consequence NP $\not\subseteq$ Nearly-BPP. However, it is not yet known that the converse holds as well. Theorem 2 in contrast establishes a link between partial one-way functions and io-Nearly-BPP.

To describe our result, we introduce a subclass of NP, called ωNP ("omega" NP). First, notice that any NP set A can be written as $A = \{x \mid \exists y \in \mathrm{dom}(f)[f(y) = x]\}$ for a certain partial FP-function f. In this case, f is said to *witness* A. This witness function f is in general not necessarily effectively honest.[13] The class ωNP is the collection of all sets A that have effectively-honest witness functions f in FP. Similarly, ωNPMV is the collection of all partial multi-valued functions g such that there exist a polynomial p and a partial effectively-honest function $f \in$ FP satisfying $g(x) = \{y \in \mathrm{dom}(f) \cap \Sigma^{\leq p(|x|)} \mid f(y) = x\}$ for every x.

Theorem 2. *The following statements are all logically equivalent.*
  1. *A partial weakly one-way function exists.*
  2. *A partial strongly one-way function exists.*
  3. *There exists an ωNPMV-function that has no io-Nearly-FBPP refinement.*
  4. ωNP $\not\subseteq$ io-Nearly-BPP.
  5. *There exist constants d, ϵ and polynomials p, q such that, for every polynomial s and every number c, the discrepancy set $D=\{x \mid \exists y \in \Sigma^{\leq |x|+d\,\mathrm{ilog}(|x|)} \exists r \in \Sigma^{s(|x|)} [\mathrm{CD}^s(r) \geq \epsilon|r| \wedge \mathrm{CD}^q(y|x) \leq c\log_2(|x|+|y|) \wedge \mathrm{C}^q(x|y) \leq c\log_2(|x|+|y|) \wedge \mathrm{C}^s(y|x,r) > \mathrm{CD}^q(y|x) + \mathrm{C}^q(x|y) + c\log_2(|x|+|y|)]\}$ has density at least $2^n/p(n)$ almost everywhere.*

To discuss total one-way functions, we further draw our attention to a subclass of ωNP by requiring corresponding witness functions to be *total*. To be precise, a set S is in ωNP$_t$ (the subscript "t" stands for "total") if there exists a total effectively-honest function $f \in$ FP such that $A = \{x \mid \exists y[f(y) = x]\}$. Notice that ωNP$_t \subseteq$ NP but P $\not\subseteq \omega$NP$_t$ because $\emptyset \notin \omega$NP$_t$ and $\Sigma^* \in \omega$NP$_t$. The following significantly improves the result in [21,22].

Corollary 2. *A total one-way function exists iff ωNP$_t \not\subseteq$ io-Nearly-BPP.*

Note that it is not known whether we can replace the assumption ωNP$_t \not\subseteq$ io-Nearly-BPP in Corollary 2 by a much weaker assumption, such as NP $\not\subseteq$ Nearly-BPP.

---

[13] A (partial) function f from Σ^* to Σ^* is *effectively honest* if there exists a real number $c \geq 0$ such that $|x| \leq |f(x)| + c\,\mathrm{ilog}(|f(x)|)$ for all but finitely-many x in $\mathrm{dom}(f)$, where $\mathrm{ilog}(n) = \lceil \log_2 n \rceil$.

References

1. S. Aida and T. Tsukiji, P-comp versus P-samp questions on average polynomial domination. *IEICE Trans. Inf. & Syst.*, **E84-D** (2001), 1402–1410.
2. L. Babai, L. Fortnow, N. Nisan, and A. Wigderson, BPP has subexponential time simulation unless EXPTIME has publishable proofs, *Comput. Complexity*, **3** (1993), 307–318.
3. H. Buhrman and L. Fortnow, One-sided versus two-sided error in probabilistic computation, *Proc. 16th Symposium on Theoretical Aspects of Computer Science*, Lecture Notes in Computer Science, Vol.1563, pp.100–109, 1999.
4. H. Buhrman, L. Fortnow, and S. Laplante, Resource-bounded Kolmogorov complexity revisited, *SIAM J. Comput.*, **31** (2002), 887–905.
5. H. Buhrman and L. Torenvliet, Randomness is hard, *SIAM J. Comput.*, **30** (2000), 1485–1501.
6. S. Even, A. Selman, and Y. Yacobi, The complexity of promise problems with applications to public-key cryptography, *Inform. and Control*, **61** (1984), 159–173.
7. S. Even and Y. Yacobi, Cryptocomplexity and NP-completeness, in *Proc. 7th Colloquium on Automata, Languages and Programming*, Lecture Notes in Computer Science, Vol.85, pp.195–207, 1980.
8. J. Feigenbaum and L. Fortnow, Random-self-reducibility of complete sets, *SIAM J. Comput.*, **22** (1993), 994–1005.
9. J. Feigenbaum, S. Kannan, and N. Nisan, Lower bounds on random-self-reducibility, in *Proc. 5th Structure in Complexity Theory Conference*, pp.100–109, 1990.
10. Y. Gurevich, Average case complexity, *J. Comput. System Sci.*, **42** (1991), 346–398.
11. R. M. Karp and R. Lipton, Turing machines that take advice, *L'enseigment Mathematique*, **28** (1982), 191–209.
12. C. Lautemann, BPP and the polynomial hierarchy, *Inform. Process. Lett.*, **17** (1983), 215–217.
13. L. A. Levin, Average case complete problems, *SIAM J. Comput.*, **15** (1986), 285–286.
14. M. Li and P. Vitányi, *An Introduction to Kolmogorov Complexity and Its Applications* (second edition), Springer, 1997.
15. A. Meyer and L. Stockmeyer, The equivalence problem for regular expressions with squaring requires exponential space, in *Proc. 13th Symposium on Switching and Automata*, pp.125–129, 1972.
16. N. Nisan and A. Wigderson, Hardness vs. randomness, *J. Comput. System Sci.*, **49** (1994), 149–167.
17. C. Schindelhauer and A. Jakoby, The non-recursive power of erroneous computation, in *Proc. 19th Conference on Foundations of Software Technology and Theoretical Computer Science*, pp.394–406, 1999.
18. U. Schöning, Probabilistic complexity classes and lowness, *J. Comput. System Sci.*, **39** (1989), 84–100.
19. R. Schuler and T. Yamakami, Sets computable in polynomial time on average, in *Proc. 1st International Computing and Combinatorics Conference*, Lecture Notes in Computer Science, Vol.959, pp.400–409, 1995.
20. M. Sipser, A complexity theoretic approach to randomness, in *Proc. 15th ACM Symposium on Theory of Computing*, pp.330–335, 1983.

21. T. Yamakami, Polynomial time samplable distributions, *J. Complexity*, **15** (1999), 557–574. A preliminary version appeared in *Proc. 21th International Symposium on Mathematical Foundations of Computer Science*, Lecture Notes in Computer Science, Vol.1113, pp.566–578, 1996.

22. T. Yamakami, *Average Case Complexity Theory*, Ph.D. dissertation, Department of Computer Science, University of Toronto, 1997. Available as Technical Report 307/97, University of Toronto. Also available at ECCC Thesis Listings.

23. D. Zuckerman, Randomness-optimal sampling, extractors, and constructive leader election, in *Proc. 28th ACM Symposium on Theory of Computing*, pp.286–295, 1996.

Some Properties
of MOD_m Circuits Computing Simple Functions

Kazuyuki Amano and Akira Maruoka

Graduate School of Information Sciences, Tohoku University
Aoba 05, Aramaki, Aoba-ku, Sendai 980–8579, Japan
{ama|maruoka}@ecei.tohoku.ac.jp

Abstract. We investigate the complexity of circuits consisting solely of modulo gates and obtain results which might be helpful to derive lower bounds on circuit complexity: (i) We describe a procedure that converts a circuit with only modulo $2p$ gates, where p is a prime number, into a depth two circuit with modulo 2 gates at the input level and a modulo p gate at the output. (ii) We show some properties of such depth two circuits computing symmetric functions. As a consequence we might think of the strategy for deriving lower bounds on modular circuits: Suppose that a polynomial size constant depth modulo $2p$ circuit C computes a symmetric function. If we can show that the circuit obtained by applying the procedure given in (i) to the circuit C cannot satisfy the properties described in (ii), then we have a super-polynomial lower bound on the size of a constant depth modulo $2p$ circuit computing a certain symmetric function.

Keywords: modular circuits, lower bounds, symmetric functions, composite modulus, Fourier analysis

1   Introduction

To derive a strong lower bound on the size complexity of Boolean functions is a big challenge in theoretical computer science. Exponential lower bounds have been obtained so far on the size of Boolean circuits to compute a certain Boolean functions when we place restriction on the Boolean circuits such as constant depth or monotone. In this paper we consider Boolean circuits consisting of modulo gates, which will be referred to as modular circuits, and derive statements concerning finite depth modular circuits which might be helpful to derive lower bounds on that type of circuits computing certain functions.

A MOD_m gate is a Boolean gate with unbounded fan-in whose output is 0 if and only if the sum of its inputs is divisible by m, and a MOD_m circuit is an acyclic circuit with only MOD_m gates. It is known so far that if modulus m is a prime number or a power of a prime, then the AND function of n variables $x_1, \ldots, x_n$, i.e., $\wedge_{i=1}^{n} x_i$, can not be computed by any constant depth MOD_m circuit even if we are allowed to use an arbitrarily large number of such gates[8]. On the other hand we have little knowledge about the complexity on MOD_m

R. Petreschi et al. (Eds.): CIAC 2003, LNCS 2653, pp. 227–237, 2003.

circuits in the case that m is a composite number such as $m = 6$. Krause and Waack[7] provide an exponential lower bound for the size of depth two MOD_6 circuits computing the AND function. A simpler proof for the statement was given by Caussinus[3]. To derive a super-polynomial lower bound on the size complexity of depth three MOD_m circuits with m being a composite number is unsolved for a long time. The best lower bound obtained so far is $\Omega(n)$ for the n fan-in AND function for any m by Thérien[9]. See [1] for a survey on the complexity of constant depth circuits.

In this paper we explore new approaches to obtain a good lower bound on the size of modular circuits. Most of the techniques, such as the random restrictions or the random clustering, for deriving strong lower bounds of modular circuits rely on the probabilistic arguments(see e.g., [1],[2], [5]). In contrast, our approaches are not relying on the probabilistic arguments but relying fully on the constructible arguments.

What is established in the paper consists mainly of two parts: (i) For a prime number p, a $(\text{MOD}_p - \text{MOD}_2)$ circuit is a depth two circuit with MOD_2 gates at the input level and a MOD_p gate at the output level. We give a procedure that converts a MOD_{2p} circuit with an arbitrary finite depth to a $(\text{MOD}_p - \text{MOD}_2)$ circuit without changing the function that the original MOD_{2p} circuit computes. (ii) We derive properties of $(\text{MOD}_p - \text{MOD}_2)$ circuits computing any symmetric function by employing Fourier analysis. So we may think of the following strategy to derive lower bounds on MOD_{2p} circuits: Suppose that a polynomial size constant depth MOD_{2p} circuit C computes a symmetric function. If we can show that the circuit obtained by applying the procedure given in (i) to the circuit C cannot satisfy the properties described in (ii), then we have a super-polynomial lower bound on the size of MOD_{2p} circuit computing a certain symmetric function. But unfortunately we have not yet succeeded in obtaining lower bound along the lines.

The paper is organized as follows. In Section 2 we give some basic notations and definitions. In Section 3 we describe a procedure that converts a MOD_{2p} circuit with $p \geq 3$ being a prime number, into a depth two $(\text{MOD}_p - \text{MOD}_2)$ circuit that computes the function computed by the MOD_{2p} circuit. In Section 4 we investigate the properties of $(\text{MOD}_p - \text{MOD}_2)$ circuits computing symmetric functions must have. Finally, in Section 5, we show some lower bounds obtained by using the techniques developed in this paper.

2   Preliminaries

A MOD_m gate is a Boolean gate with unbounded fan-in whose output is 0 if and only if the sum of its inputs is divisible by m, i.e.,

$$\text{MOD}_m(x_1, x_2, \ldots, x_n) = \begin{cases} 0 \text{ if } \sum_{i=1}^n x_i \equiv 0 \pmod{m}, \\ 1 \text{ otherwise.} \end{cases}$$

Remark that there are several possible definitions for "modulo" gates. Let S be a subset of $\{0, \ldots, m-1\}$. A *general* MOD$_m$ gate is defined as follows:

$$\text{MOD}_m^S(x_1, x_2, \ldots, x_n) = \begin{cases} 1 \text{ if } \sum_{i=1}^n x_i \bmod m \in S \\ 0 \text{ otherwise.} \end{cases}$$

In what follows, we restrict ourselves to the case $S = \{1, 2, \ldots, m-1\}$. A MOD$_m$ circuit is an acyclic circuit with inputs $\{1, x_1, x_2, \ldots, x_n\}$ which uses only MOD$_m$ gates. We allow to connect inputs to gates or gates to gates through multiple wires. Note that several proofs for the exponential lower bounds on the size of depth two MOD$_m^{\{1,2,\ldots,m-1\}}$ circuits were known (e.g., [7,3]), but no superlinear lower bounds for depth two general MOD$_m$ circuits are known[3]. Following the standard notations, the class of constant depth circuits with MOD$_m$ gates is denoted by $CC^0(m)$.

A (MOD$_q$ – MOD$_p$) circuit is a depth two circuit with MOD$_p$ gates at the input level and a MOD$_q$ gate at the output level. More generally, we denote a depth k circuit with L_i gates at the ith level, for $i = 1, 2, \ldots, k$, by $(L_k - \cdots - L_2 - L_1)$. The *size* of a circuit is defined to be the number of gates in it.

For $x \in \{0,1\}^n$, $\sharp_1(x)$ denotes the number of 1's in x. For $x, y \in \{0,1\}^n$, $x \cdot y$ denotes the inner product mod 2, i.e., $x \cdot y = \oplus_{i=1}^n x_i y_i$, where x_i (respectively, y_i) is the ith bit of x (respectively, y), and $x \oplus y$ denotes the bitwise exclusive OR's of x and y. For a set S, $|S|$ denotes the number of elements in S.

3   Conversion from CC(2p) to (MOD$_p$ – MOD$_2$)

In this section, we describe a procedure that converts a MOD$_{2p}$ circuit into a (MOD$_p$ – MOD$_2$) circuit that computes the function computed by the MOD$_{2p}$ circuit. First, we describe three procedures that will be used in our conversion and in the last subsection of this section, we will give the entire procedure by putting all them together.

3.1   From MOD$_{2p}$ to (MOD$_p$ – MOD$_2$)

In this subsection, we describe a procedure that converts a MOD$_{2p}$ gate into a depth two (MOD$_p$ – MOD$_2$) circuit, where $p \geq 3$ is a prime number. First, we deal with the simplest case with $p = 3$.

Lemma 1. *Let C be a circuit consisting of a single MOD$_6$ gate with inputs $I = \{y_1, y_2, \ldots, y_k\}$, where $y_i \in \{1, x_1, x_2, \ldots, x_n\}$. Note that I is supposed to be a multiset : if x_i is connected to the MOD$_6$ gate with j wires, then x_i appears j times in I. Then the function computed by C can be computed by a depth two (MOD$_3$ – MOD$_2$) circuit given by*

$$MOD_3\left(MOD_2(I), (k \bmod 3)\sum_{i=1}^k y_i + 2\sum_{1 \leq i < j \leq k}(MOD_2(y_i, y_j))\right)$$

Proof. The lemma follows from the series of the equations :

$$\mathrm{MOD}_6(I) = \mathrm{OR}(\mathrm{MOD}_2(I), \mathrm{MOD}_3(I)) = \mathrm{MOD}_3(\mathrm{MOD}_2(I), \mathrm{MOD}_3(I))$$

$$= \mathrm{MOD}_3\left(\mathrm{MOD}_2(I), \left(\sum_{i=1}^{k} y_i\right)^2 (\mathrm{mod}\ 3)\right)$$

$$= \mathrm{MOD}_3\left(\mathrm{MOD}_2(I), \left(\sum_{i=1}^{k} y_i\right)^2\right) \tag{1}$$

$$= \mathrm{MOD}_3\left(\mathrm{MOD}_2(I), \sum_{i=1}^{k} y_i^2 + 2\sum_{1\le i<j\le k} y_i y_j\right)$$

$$= \mathrm{MOD}_3\left(\mathrm{MOD}_2(I), \sum_{i=1}^{k} y_i + 2\sum_{1\le i<j\le k} (2y_i + 2y_j + (y_i \oplus y_j))\right)$$

$$= \mathrm{MOD}_3\left(\mathrm{MOD}_2(I), (k\ \mathrm{mod}\ 3)\sum_{i=1}^{k} y_i + 2zw\sum_{1\le i<j\le k} (\mathrm{MOD}_2(y_i, y_j))\right).$$

The second to last equality holds since $y^2 = y$ for $y = 0, 1$, and $y_i y_j \equiv 2y_i + 2y_j + (y_i \oplus y_j) \pmod 3$ for any $y_i, y_j \in \{0, 1\}$. ☐

For the general case, to convert a MOD_{2p} gate, where $p \ge 3$ is a prime number, to an equivalent (MOD_p – MOD_2) circuit, we use the equation:

$$\mathrm{MOD}_{2p}(I) = \mathrm{MOD}_p\left(\mathrm{MOD}_2(I), \left(\sum_{i=1}^{k} y_i\right)^{p-1}\right),$$

in place of the equation (1). By this equation and the lemma described below, it is easy to construct a desired depth two (MOD_p – MOD_2) circuit.

Lemma 2. *Let $p \ge 3$ be a prime number, and let J be a subset of $\{1, 2, \ldots, n\}$ such that $2 \le |J| \le p - 1$. Then for any $y_j \in \{0, 1\}$ ($j \in J$),*

$$\prod_{j \in J} y_j \equiv e\left(\sum_{\substack{S \subseteq J \\ |S|:even}} \mathrm{MOD}_2(\sum_{j \in S} y_j)\right) + o\left(\sum_{\substack{S \subseteq J \\ |S|:odd}} \mathrm{MOD}_2(\sum_{j \in S} y_j)\right) \pmod p, \tag{2}$$

where o and e are positive integers satisfying $2^{|J|-1}o \equiv 1 \pmod p$ and $e+o \equiv 0 \pmod p$.

Proof. Since

$$\mathrm{RHS\ of\ (2)} = \begin{cases} 0 & \text{if } |y| = 0, \\ 2^{|J|-2}(e+o) & \text{if } 1 \le |y| \le |J| - 1, \\ 2^{|J|-1}o & \text{if } |y| = |J|, \end{cases}$$

where $|y|$ denotes $\sum_{j \in J} y_j$, the lemma easily follows. ☐

3.2   From (MOD$_p$ – MOD$_p$ – MOD$_2$) to (MOD$_p$ – MOD$_2$)

In this subsection, we describe a procedure that converts a depth three (MOD$_p$ – MOD$_p$ – MOD$_2$) circuit into a depth two (MOD$_p$ – MOD$_2$) circuit, where $p \geq 3$ is a prime number. As in the last subsection, to start with, we start with describing the simplest case with $p = 3$.

Lemma 3. *Let C be a depth three (MOD$_3$ – MOD$_3$ – MOD$_2$) circuit. Let $g_1, g_2, \ldots, g_l$ be all the gates in the second level, and for $i = 1, 2, \ldots, l$, let $g_{i,1}, g_{i,2}, \ldots, g_{i,k_i}$ be all the gates connecting to the gate g_i. Then the function computed by C can be computed by the depth two (MOD$_3$ – MOD$_2$) circuit described as*

$$MOD_3\left(\sum_{i=1}^{l}\left((k_i \bmod 3)\sum_{j=1}^{k_i} g_{i,j} + 2\sum_{j_1=1}^{k_i-1}\sum_{j_2=j_1+1}^{k_i} MOD_2(g_{i,j_1}, g_{i,j_2})\right)\right).$$

Proof. The lemma follows from

$$\text{Output of } C \tag{3}$$
$$= MOD_3(MOD_3(g_{1,1}, \ldots, g_{1,k_1}), \cdots, MOD_3(g_{l,1}, \ldots, g_{l,k_l}))$$
$$= MOD_3\left(\left(\sum_{j=1}^{k_1} g_{1,j}\right)^2 (\bmod\ 3), \ldots, \left(\sum_{j=1}^{k_l} g_{l,j}\right)^2 (\bmod\ 3)\right)$$
$$= MOD_3\left(\left(\sum_{j=1}^{k_1} g_{1,j}\right)^2, \ldots, \left(\sum_{j=1}^{k_l} g_{l,j}\right)^2\right) \tag{4}$$
$$= MOD_3\left(\sum_{i=1}^{l}\left(\sum_{j=1}^{k_i} g_{i,j}^2 + 2\sum_{j_1=1}^{k_i-1}\sum_{j_2=j_1+1}^{k_i} g_{i,j_1} g_{i,j_2}\right)\right)$$
$$= MOD_3\left(\sum_{i=1}^{l}\left(\sum_{j=1}^{k_i} g_{i,j} + 2\sum_{j_1=1}^{k_i-1}\sum_{j_2=j_1+1}^{k_i} (2g_{i,j_1} + 2g_{i,j_2} + (g_{i,j_1} \oplus g_{i,j_2}))\right)\right)$$
$$= MOD_3\left(\sum_{i=1}^{l}\left((k_i \bmod 3)\sum_{j=1}^{k_i} g_{i,j} + 2\sum_{j_1=1}^{k_i-1}\sum_{j_2=j_1+1}^{k_i} MOD_2(g_{i,j_1}, g_{i,j_2})\right)\right) \tag{5}$$

The second to last equality holds since $g^2 = g$ for $g = 0, 1$, and $g_i g_j \equiv 2g_i + 2g_j + (g_i \oplus g_j) (\bmod\ 3)$ for any $g_i, g_j \in \{0, 1\}$. Recall that every $g_{i,j}$ is the output of a MOD$_2$ gate at the first level in C, Hence each MOD$_2(g_{i,j_1}, g_{i,j_2})$ function in the formula (5) can be computed by a single MOD$_2$ gate. This completes the proof of the lemma.   □

For the general case, to convert a (MOD$_p$ – MOD$_p$ – MOD$_2$) circuit C to an equivalent (MOD$_p$ – MOD$_2$) circuit, we use the equation:

$$\text{Output of } C = MOD_p\left(\left(\sum_{j=1}^{k_1} g_{1,j}\right)^{p-1}, \ldots, \left(\sum_{j=1}^{k_l} g_{l,j}\right)^{p-1}\right),$$

in place of (4). By this equation and Lemma 2, it is easy to construct a desired depth two $(\mathrm{MOD}_p - \mathrm{MOD}_2)$ circuit.

3.3   From $(\mathrm{MOD}_2 - \mathrm{MOD}_p - \mathrm{MOD}_2)$ to $(\mathrm{MOD}_p - \mathrm{MOD}_2)$

A procedure that converts a $(\mathrm{MOD}_2 - \mathrm{MOD}_p - \mathrm{MOD}_2)$ circuit into a $(\mathrm{MOD}_p - \mathrm{MOD}_2)$ circuit is as follows :

Let C be a depth three $(\mathrm{MOD}_2 - \mathrm{MOD}_p - \mathrm{MOD}_2)$ circuit. We assume that the number of inputs of the top gate of C is 2^k for some positive integer k. First, we replace the top gate of C by a depth k circuit with only MOD_2 gates, each having fan-in two. It is easy to check that $\mathrm{MOD}_2(g_1, g_2) = \mathrm{MOD}_p(g_1, (p-1)g_2)$ for any Boolean function g_1, g_2. Thus we can replace each MOD_2 gate by the MOD_p gate and apply the procedure described in subsection 3.2 recursively to obtain a depth two $(\mathrm{MOD}_p - \mathrm{MOD}_2)$ circuit computing the same function as the one computed by C.

We note that the resulting circuit may be exponentially large when the fan-in of the top MOD_2 gate is $\Omega(n)$. As long as we follow the arguments mentioned above this fact is the largest obstacle to get a good lower bound on the size of MOD_{2p} circuits at this moment.

3.4   Putting All Together

In this subsection, we describe a procedure that converts a MOD_{2p} circuit into a $(\mathrm{MOD}_p - \mathrm{MOD}_2)$ circuit by using three procedures described in the previous three subsections. It is worthwhile to note here that every Boolean function can be computed by a depth two $(\mathrm{MOD}_p - \mathrm{MOD}_2)$ circuit for any prime $p \geq 3$. First, we focus on the lowest two levels of a MOD_{2p} circuit. By the procedure described in subsection 3.1, we get a depth four $(\mathrm{MOD}_p - \mathrm{MOD}_2 - \mathrm{MOD}_p - \mathrm{MOD}_2)$ circuit from the lowest two levels. This circuit can be converted into a depth three $(\mathrm{MOD}_p - \mathrm{MOD}_p - \mathrm{MOD}_2)$ circuit by using the procedure described in subsection 3.3 which in turn can be converted into a depth two $(\mathrm{MOD}_p - \mathrm{MOD}_2)$ circuit by using the procedure described in subsection 3.2. By recursively continuing above process, we can finally convert a MOD_{2p} circuit with arbitrarily finite depth into a $(\mathrm{MOD}_p - \mathrm{MOD}_2)$ circuit that computes the same function computed by the original MOD_{2p} circuit.

In the remainder of this subsection, we show that every Boolean function can be computed by a $(\mathrm{MOD}_3 - \mathrm{MOD}_2)$ circuit. This fact may be known, but we were unable to find a published paper that states the fact, so we show the statement together with its proof.

Fact 1. *Every Boolean function can be computed by a depth two $(\mathrm{MOD}_3 - \mathrm{MOD}_2)$ circuit.*

Proof. It was known that the OR function of n variables, i.e., $\vee_{i=1}^{n}x_i$, can be computed by a depth two (MOD$_3$ – MOD$_2$) circuit. See [5] for the simple construction of size 2^n and see [6] for the slightly effective construction of size $2^{n/2+1}$. This implies that every disjunction of n literals, i.e., $\vee_{i=1}^{n}l_i$ where $l_i \in \{x_i, \bar{x}_i\}$, also can be computed by a depth two (MOD$_3$ – MOD$_2$) circuit. Let f be an arbitrary Boolean function on n variables. Consider a CNF formula representing f, where each clause contains exactly n literals. Without loss of generality, we can assume that the formula has $m \equiv 1 \pmod 3$ clauses. (by adding some "always true" clauses if necessary). If the value of f is 0, then $m - 1 \equiv 0 \pmod 3$ clauses of f takes value 1, and if the value of f is 1, then all of the $m \not\equiv 0 \pmod 3$ clauses takes value 1. Thus the MOD$_3$ gate, whose inputs are all (MOD$_3$ – MOD$_2$) circuit computing each clause in the CNF formula, evaluates the function f. Finally, using the procedure described in subsection 3.2, we obtain a (MOD$_3$ – MOD$_2$) circuit computing f. □

4   Properties of Depth Two Circuits for Symmetric Functions

In this section, we investigate the properties of (MOD$_p$ – MOD$_2$) circuits computing symmetric functions where p is a prime number. The technique we use is the Fourier analysis that is an extension of these by Kahn and Meshulam[6] and Yan and Parberry[10].

Definition 1. *A Boolean function f on n variables is symmetric if $f(x) = f(y)$ for any $x, y \in \{0,1\}^n$ such that $\sharp_1(x) = \sharp_1(y)$. In other words, a symmetric function is a function whose value depends only on the number of 1's in its input.* □

For a prime number p, Z_p denotes the residue-class field modulo p. The Fourier transform of a function $f : Z_2^n \to Z_p$ is the function $\hat{f} : Z_2^n \to Z_p$ defined by $\hat{f}(x) = \sum_{y \in \{0,1\}^n} f(y)(-1)^{y \cdot x} \pmod p$. The convolution of two functions $f, g : Z_2^n \to Z_p$ is given by $f * g(x) = \sum_{y \in \{0,1\}^n} f(y)g(x - y)$, and its Fourier transform satisfies $\widehat{f * g}(x) = \hat{f}(x) \cdot \hat{g}(x)$. Following Kahn and Meshulam[6], we abbreviate $f * \cdots * f$ (k factors) by f^{*k}, and for an integer k and a set S set $kS = \{a_1 \oplus \cdots \oplus a_k \mid a_i \in S\}$.

Definition 2. *Let C be a circuit with inputs $\{1, x_1, x_2, \ldots, x_n\}$. For a gate g at the first level in the circuit C, the characteristic string corresponding to the gate g, denoted char(g), is an n bit binary string such that the ith bit of char(g) is equal to 1 if and only if x_i is connected to the gate g for $i = 1, 2, \ldots, n$. The span of the circuit C, which will be denoted by span(C), is defined to be the set of characteristic string corresponding to all the gates in the first level of the circuit C.* □

For a set of integers $I = \{i_1, i_2, \ldots, i_k\} \subseteq \{1, 2, \ldots, n\}$ and for a set of strings $S \subseteq \{0,1\}^n$, the set of strings $S|_I \subseteq \{0,1\}^k$ is defined by $S|_I = \{w_{i_1} w_{i_2} \cdots w_{i_k} \mid$

$w_1 w_2 \cdots w_n \in S$}. For a symmetric function f and for an integer i, $f(i)$ denotes the value of f on an input with i 1's.

Theorem 2. *Let $p \geq 3$ be a prime number and let f be a symmetric function on n variables such that $f(i) \neq f(i+2 \cdot p^k)$ for some positive integers i and k. Assume that a $(MOD_p - MOD_2)$ circuit C computes f. Then for any $I \subseteq \{1, 2, \ldots, n\}$ with $|I| = 2 \cdot p^k$, $(p-1)span(C)|_I \supseteq \{x \in \{0,1\}^{2 \cdot p^k} \mid \sharp_1(x) = p^k\}$ holds.* □

Before proceeding to the proof of Theorem 2, we show a simple fact which will be needed in the proof of Theorem 2.

Fact 3. *Let f' be a symmetric function on n' variables with $0 = f'(0) \neq f'(n') = 1$. Suppose that C is a $(MOD_p - MOD_2)$ circuit computing f'. Then there exists a $(MOD_p - MOD_2)$ circuit C' such that (i) $span(C) = span(C')$, (ii) for any gate g at the first level in C', the set of inputs of g is a subset of $\{x_1, x_2, \ldots, x_n\}$ (,i.e., the constant 1 is never connected to a MOD_2 gate in C') and (iii) every input of the top MOD_p gate of C' is the output of a MOD_2 gate (i.e., C' is a layered circuit).*

Proof (sketch). It suffices to verify that $MOD_p(MOD_2(1, I), y) = MOD_p((p-1)MOD_2(I) + 1, y)$ holds for any $I \subseteq \{x_1, x_2, \ldots, x_n\}$. In the circuit obtained by repeated transformation based on the equation, the top MOD_p gate has no constant input 1 because of the condition $f'(0) = 0$. □

Now we proceed to the proof of Theorem 2.

Proof (of Theorem 2). Let f be a symmetric function on n variables such that $f(i) \neq f(i + 2 \cdot p^k)$ for some positive integers i and k, and I be a subset of $\{1, 2, \ldots, n\}$ with size $2 \cdot p^k$. Let C be a $(MOD_p - MOD_2)$ circuit computing the function f.

When $1 = f(i) \neq f(i + 2 \cdot p^k) = 0$, let C_1 be a circuit obtained from C by replacing each input variable by its negation and let f_1 be the function computed by C_1. Note that $span(C) = span(C_1)$ and that $0 = f_1(n-i-2 \cdot p^k) \neq f_1(n-i) = 1$. Hence without loss of generality, we can assume that $0 = f(i) \neq f(i + 2 \cdot p^k) = 1$ (by putting $i := n - i - 2 \cdot p^k$ and $f := f_1$ for the case).

Restrict the circuit C by assigning the value 0 to the $n - i - 2 \cdot p^k$ input variables outside I and assigning the value 1 to the i input variables of them. The resulting circuit C' computes a symmetric function f' on the set of variables in I satisfying $0 = f'(0) \neq f'(2 \cdot p^k) = 1$, and satisfies $span(C') = span(C)|_I$. Let $n' = 2 \cdot p^k$. By Fact 3, we can assume that, for any gate at the first level in C', its inputs is a subset of $\{x_1, x_2, \ldots, x_n\}$.

Let $S = \{z_1, z_2, \ldots, z_s\}$ be the set of characteristic strings corresponding to any gate at the first level in C'. Note that S is a multiset such that if a gate g is connected to the MOD_p gate with l wires, then the characteristic strings corresponding to the gate g appears l times in S. Note that each element in S corresponds to a vector in $\text{span}(C)|_I$. The function $u : Z_2^{n'} \to Z_p$ is defined to be $u(x) = 1$ if $x = 00\cdots0$ and $u(x) = 0$ otherwise. Let $w : Z_2^{n'} \to Z_p$ denote the indicator function of the set S, i.e., $w(x) = |\{j \mid x = z_j\}|$. Set the function $g : Z_2^{n'} \to Z_p$ to $g = su - w$. For $x \in \{0,1\}^{n'}$, let H_x denote the set of vectors $\{y \in \{0,1\}^{n'} \mid y \cdot x = 1\}$. By the condition (iii) in Fact 3, the output of the circuit C' is 0 if and only if $|\{i \mid 1 \le i \le s, z_i \in H_x\}| \equiv 0 \pmod{p}$.

For any $x \in \{0,1\}^{n'}$,

$$\widehat{g}(x) = \sum_{y \in \{0,1\}^n} g(y)(-1)^{y \cdot x} = s - \sum_{i=1}^{s}(-1)^{z_i \cdot x} = \sum_{i=1}^{s}(1 - (-1)^{z_i \cdot x})$$
$$= 2|\{i \mid 1 \le i \le s, z_i \in H_x\}|.$$

Let $h = g^{*(p-1)}$. Then we have $\widehat{h}(x) = \widehat{g}(x)^{p-1}$. Thus $\widehat{h}(x) = 0$ if x satisfies the condition $|\{i \mid 1 \le i \le s, z_i \in H_x\}| \equiv 0 \pmod{p}$, and $\widehat{h}(x) = 1$ otherwise. Moreover, since C' computes the symmetric function, $\widehat{h}(x_1) = \widehat{h}(x_2)$ for any x_1 and x_2 with $\sharp_1(x_1) = \sharp_1(x_2)$.

Hence, for any fixed x with $\sharp_1(x) = p^k$,

$$h(x) = 2^{-n'}\widehat{\widehat{h}}(a) = 2^{-n'} \sum_{y \in \{0,1\}^{n'}} \widehat{h}(y)(-1)^{y \cdot x} = 2^{-n'} \sum_{i=0}^{n'} \sum_{y : \sharp_1(y)=i} \widehat{h}(y)(-1)^{y \cdot x}$$
$$= 2^{-n'} \sum_{i=0}^{n'} \sum_{y : \sharp_1(y)=i} \widehat{h}(y) \sum_{j=0}^{i}(-1)^j \binom{p^k}{j}\binom{p^k}{i-j}$$
$$= 2^{-n'}(\widehat{h}(00\cdots0) - \widehat{h}(11\cdots1)).$$

The first equality holds since

$$\widehat{\widehat{h}}(x) = \sum_{a \in \{0,1\}^{n'}} \widehat{h}(x)(-1)^{x \cdot a}$$
$$= \sum_{a \in \{0,1\}^{n'}} \Big(\sum_{y \in \{0,1\}^{n'}} h(y)(-1)^{y \cdot a} \Big)(-1)^{x \cdot a}$$
$$= \sum_{a \in \{0,1\}^{n'}} \sum_{y \in \{0,1\}^{n'}} (h(y)(-1)^{y \cdot a}(-1)^{x \cdot a})$$
$$= \sum_{y \in \{0,1\}^{n'}} \Big(\sum_{a \in \{0,1\}^{n'}} (h(y)(-1)^{(x \oplus y) \cdot a}) \Big) = 2^{n'} h(x),$$

and the last equality holds since p divides $\binom{p^k}{j}$ unless $j = 0$ or $j = p^k$. Note that for convenience $\binom{p^k}{j}$ is considered to 0 for any $j > p^k$. Since one of $\widehat{h}(00\cdots0)$ and $\widehat{h}(11\cdots1)$ is 0 and the other is not 0, the set $\{x \mid h(x) \not\equiv 0 \pmod{p}\}$, denoted $\text{supp}(h)$, contains $\{x \in \{0,1\}^n \mid \sharp_1(x) = p^k\}$.

On the other hand, $\text{supp}(h) \subseteq \{\oplus_{i=1}^{p-1} z_i \mid z_i \in S \text{ for any } 1 \le i \le p-1\}$. Therefore, $(p-1)\text{span}(C)|_I = \{\oplus_{i=1}^{p-1} z_i \mid z_i \in S \text{ for any } 1 \le i \le p-1\} \supseteq \{x \in \{0,1\}^{n'} \mid \sharp_1(x) = p^k\}$. This completes the proof of the theorem. □

5   Applications

In this section, we show some lower bounds on the size complexity of circuits with modulo gates using the techniques developed in this paper. The following theorem says that even MOD_4 function requires exponential size $((\text{MOD}_3)^l - \text{MOD}_6)$ circuits for constant l. We remark that a similar result was obtained in [5] by using the *random clustering technique*.

Theorem 4. *Let f be a symmetric function on n variables such that $f(i) \ne f(i+2\cdot 3^k)$ for some positive integers i and k. Suppose that a $(\text{MOD}_3 - \cdots - \text{MOD}_3 - \text{MOD}_6)$ circuit C of depth l computes f. Then the size of the circuit C is at least $(2^{3^k/2} - n^2)^{1/2^l}$.*

Proof. Let f be a symmetric function on n variables such that $f(i) \ne f(i+2\cdot 3^k)$ for some positive integers i and k. Let C be a depth l $(\text{MOD}_3 - \cdots - \text{MOD}_3 - \text{MOD}_6)$ circuit computing f. By Lemma 1, we obtain a depth $l+1$ $(\text{MOD}_3 - \cdots - \text{MOD}_3 - \text{MOD}_2)$ circuit C_1 computing f such that

$$\text{span}(C_1) \subseteq \text{span}(C) \cup \{x \in \{0,1\}^n \mid \sharp_1(x) \le 2\}.$$

By using Lemma 3 recursively, we obtain a $(\text{MOD}_3 - \text{MOD}_2)$ circuit C_2 computing f such that $\text{span}(C_2) \subseteq 2^l \text{span}(C_1)$. Thus,

$$\text{size of } C \ge |\text{span}(C)| \ge |\text{span}(C_1)| - n^2 \ge (|\text{span}(C_2)| - n^2)^{1/2^l}$$
$$\ge \left(\left(\frac{2\cdot 3^k}{3^k}\right)^{1/2} - n^2\right)^{1/2^l} \ge (2^{3^k/2} - n^2)^{1/2^l}.$$

The second to last inequality follows from Theorem 2. □

The following corollary is immediate from Theorem 4.

Corollary 1. *Let n be an integer such that $n = 2 \cdot 3^k$ for some integer k. Let f be a symmetric function on n variables such that $f(0) \ne f(n)$. Suppose that $((\text{MOD}_3)^l - \text{MOD}_6)$ circuit C of depth $l+1 = o(\log n)$ computes f. Then the size of the circuit C is super-polynomial in n.* □

Finally we remark that, as one of the anonymous referees has pointed out to us, a similar argument as above can be applied to obtain a super-polynomial lower bounds on the size of MOD_{2p} circuits having sublogarithmic depth (that is, $o(\log n)$ depth) that compute symmetric functions whose value for all 1's is different from all 0's.

Acknowledgments

The authors would like to thank the anonymous referees for their valuable suggestions and comments.

References

1. E. Allender, "Circuit Complexity before the Dawn of the New Millennium", *Proc. of 16th FSTTCS*, LNCS 1180, pp. 1–18, 1996.
2. R. Beigel and A. Maciel, "Upper and Lower Bounds for Some Depth-3 Circuit Classes", *Computational Complexity*, Vol. 6, No. 3, pp. 235–255, 1997.
3. H. Caussinus, "A Note on a Theorem of Barrington, Straubing and Thérien", *Information Processing Letters*, Vol. 58, No. 1, pp. 31–33, 1996.
4. V. Grolmusz, "A Degree Decreasing Lemma for (MOD$_q$ – MOD$_p$) Circuits", *Disc. Math. and Theor. Comput. Sci.*, Vol. 4, pp. 247–254, 2001. (a preliminary version appeared in *Proc. 25th ICALP*, 1998).
5. V. Grolmusz and G. Tardos, "Lower Bounds for (MOD$_p$ – MOD$_m$) Circuits", *SIAM J. Comput.* Vol. 29, No. 4, pp. 1209–1222, 2000. (a preliminary version appeared in *Proc. 39th FOCS, 1998*).
6. J. Kahn and R. Meshulam, "On Mod p Transversals", *Combinatorica*, Vol. 11, No. 1, pp. 17–22, 1991.
7. M. Krause and S. Waack, "Variation Ranks of Communication Matrices and Lower Bounds for Depth Two Circuits Having Symmetric Gates with Unbounded Fan-in", *Math. Syst. Theory*, Vol. 28, No. 6, pp. 553–564, 1995. (a preliminary version appeared in *Proc. 32nd FOCS, 1991*).
8. R. Smolensky, "Algebraic Methods in the Theory of Lower Bounds for Boolean Circuit Complexity", *Proc. 19th STOC*, pp. 77–82, 1987.
9. D. Thérien, "Circuits Constructed with MOD$_q$ Gates Cannot Compute "AND" in Sublinear Size", *Computational Complexity*, Vol. 4, pp. 383–388, 1994.
10. P. Yan and I. Parberry, "Exponential Size Lower Bounds for Some Depth Three Circuits", *Information and Computation*, Vol. 112, pp. 117–130, 1994.

XOR-Based Schemes
for Fast Parallel IP Lookups

Giancarlo Bongiovanni[1] and Paolo Penna[2,*,**]

[1] Dipartimento di Scienze dell'Informazione, Università di Roma "La Sapienza",
via Salaria 113, I-00133 Roma, Italy
bongio@dsi.uniroma1.it.
[2] Dipartimento di Informatica ed Applicazioni "R.M. Capocelli",
Università di Salerno, via S. Allende 2, I-84081 Baronissi (SA), Italy
penna@dia.unisa.it

Abstract. An IP router must forward packets at gigabit speed in order to guarantee a good QoS. Two important factors make this task a challenging problem: (i) for each packet, the longest matching prefix in the forwarding table must be computed; (ii) the routing tables contain several thousands of entries and their size grows significantly every year. Because of this, parallel routers have been developed which use several processors to forward packets. In this work, we present a novel algorithmic technique which, for the first time, exploits the parallelism of the router to also reduce the size of the routing table. Our method is scalable and requires only a minimal additional hardware. Indeed, we prove that any IP routing table T can be split into two subtables T_1 and T_2 such that: (a) $|T_1|$ can be any positive integer $k \leq |T|$ and $|T_2| \leq |T| - k$; (b) the two routing tables can be used separately by two processors so that the IP lookup on T is obtained by simply XOR-ing the IP lookup on the two tables. Our method is independent on the data structure used to implement the lookup search and it allows for a better use of the processors L2 cache. For real routers routing tables, we also show how to achieve simultaneously: (a) $|T_1|$ is roughly 7% of the original table T; (b) the lookup on table T_2 does not require the best matching prefix computation.

1 Introduction

We consider the problem of forwarding packets in an Internet router (or backbone router): the router must decide the next hop of the packets based on their destinations and on its *routing table*. With the current technology which allows to move a packet from the input interface to the output interface of a router [21,19] at gigabit speed and the availability of high speed links based on optic fibers, the bottleneck in forwarding packets is the *IP lookup* operation, that is, the task of deciding the output interface corresponding to the next hop.

---

* Research supported by the European project CRESCCO.
** Most of this work has been done while at the University of Rome "Tor Vergata", Math Department.

R. Petreschi et al. (Eds.): CIAC 2003, LNCS 2653, pp. 238–250, 2003.

In the past this operation was performed by data link Bridges [7]. Currently, Internet routers require the computation of the *longest matching prefix* of the destination address a. Indeed, in the early 1990s, because of the enormous increase of the number of endpoints, and the consequent increase of the size of the routing tables, Classless Inter-Domain Routing (CIDR) and *address aggregation* have been introduced [8]. The basic idea is to aggregate all IP addresses corresponding to endpoints whose next hop is the same: it might be the case that all machines whose IP address starts by 255.128 have output interface $I1$; therefore we only need to keep, in the routing table, a single pair prefix/output 255.128. ∗ . ∗ /$I1$. Unfortunately, not all addresses with a common prefix correspond to the same "geographical" area: there might be so called *exceptions*, like a subnet whose hosts have IP address starting by 255.128.128 and whose output interface is different, say $I2$. In this case, we have both pairs in the routing table and the rule to forward a packet with address a is the following: if a is in the set[1] 255.128. ∗ .∗, but *not* in 255.128.128.∗, then its next hop is $I1$; otherwise, if a is in the set 255.128.128.∗, then its next hop is $I2$. More in general, the correct output interface is the one of the so called *best matching prefix* BMP(a, T), that is, the longest prefix in T that is a prefix of a.

Even though other operations must be performed in order to forward a packet, the computation of the best matching prefix turns out to be the major and most computationally expensive task. Indeed, performing this task on low-cost workstations is considered a challenging problem which requires rather sophisticated *algorithmic solutions* [4,6,9,12,17,23,25]. Partially because of these difficulties, *parallel routers* have been developed which are equipped with several processors to process packets faster [21,19,16].

We first illustrate two simple algorithmic approaches to the problem and discuss why they are not feasible for IP lookup:

1. *Brute force search on the table T.* We compare each entry of T and store the longest that is a prefix of the given address a;

2. *Prefix (re-) expansion.* We write down a new table containing all possible IP addresses of length 32 and the corresponding output interfaces.

Both approaches fail for different reasons. Typically, a routing table may contain several thousands of prefixes (e.g., the MaeEast router contains about 33,000 entries [14]), which makes the first approach too slow. On the other hand, the second approach would ensure that a single memory access is enough. Unfortunately, 2^{32} is a too large number to fit in the DRAM. Also, even a table with only IP addresses corresponding to endpoints would be unfeasible: this is exactly a major reason why prefix aggregation has been introduced!

In order to obtain a good tradeoff between memory size and number of memory accesses, a data structure named *forwarding table* is constructed on the basis of the routing table T and then used for the IP lookup. For example, the forwarding table may consist of suitable hash functions. This approach, that works well in the case searching for a key a into a dictionary T (i.e., the exact matching problem), has several drawbacks when applied to the IP lookup problem:

---

[1] We consider a prefix $x*$ as the set of all possible string of length 32 whose prefix is x.

1. We do not know the length of the BMP. Therefore, we should try all possible lengths up to 32 for IPv4 [22] (128 for IPv6 [5]) and, for each length, apply a suitable hash function;

2. Even when an entry of T is a prefix of the packet address a, we are not sure that this one is the correct answer (i.e., the BMP). Indeed, the so called *exceptions* require that the above approach must be performed for all lengths even when a prefix of a is found.

1.1   Previous Solutions

The above simple solution turns out to be inefficient for performing the IP lookup fast enough to guarantee millions of packets per second [16]. More sophisticated and efficient approaches have been introduced in several works in which a suitable data structure, named *forwarding table*, is constructed from the routing table T [4,6,9,12,17,23,25]. For instance, in [25] a method ensuring $O(\log W)$ memory accesses has been presented, where W denotes the number of different prefix lengths occurring in the routing table T. This method has been improved in [23] using a technique called *controlled prefix expansion*: prefixes of certain lengths are expanded thus reducing the value W to some W'. For instance, each prefix x of length 8 is replaced by $x \cdot 0$ and $x \cdot 1$ (both new prefixes have the same output interface of x). The main result of [23] is a method to pick a suitable set of prefix lengths so that (a) the overall data structure is not too big, and (b) the value of W' is as small as possible.

Actually, many existing works pursue a similar goal of obtaining an efficient data structure whose size fits into the L2 memory cache of a processor (i.e., about 1Mb). This goal can be achieved only by considering real routing tables. For example, the solutions in [4,6,9,17] guarantee a constant number of memory accesses, while the size of the data structure is experimentally evaluated on real data; the latter affects the time efficiency of the solution.

These methods are designed to be implemented on a single processor of a router. Some routers exploit several processors by assigning different packets to different processors which perform the IP lookup operation using a suitable forwarding table. It is worth observing that:

1. All such methods suffer from the continuous growth of the routing tables [10,3,1]; if the size of the available L2 memory cache will not grow accordingly, the performance of such methods is destined to degrade;[2]

2. Other hardware-based solutions to the problem have been proposed (see [13,20]), but they do not scale, thus becoming obsolete after a short time, and/or they turn out to be too expensive;

3. The solution adopted in [21,19] (see also [16]) exploits the parallelism in a rather simple way: many packets can be processed in parallel, but the time a single packet takes to be processed depends on the above solutions, which are still the bottleneck.

---

[2] In our experiments we observed that the number of entries of a router can vary significantly from one day to the next one: for instance, the Paix router had about 87,000 entries the 1st November 2000, and about 22,000 only the day after.

Finally, the issue of efficiently updating the forwarding table is also addressed in [12,18,23]. Indeed, due to Internet routing instability [11], changes in the routing table occur every millisecond, thus requiring a very efficient method for updating the routing/forwarding table. Similar problems are considered in [15] for the task of constructing/updating the hash functions, which are a key ingredient used by several solutions.

1.2   Our Contribution

In this work, we aim in exploiting the parallelism of routers with more than one processors in order to reduce the size of the routing tables. Indeed, a very first (inefficient) idea would be to take a routing table T and split it into two tables T_1 and T_2, each containing half of the entries of T. Then, a packet is processed in *parallel* by two processors having in their memory (the forwarding table of) T_1 and T_2, respectively (see Fig. 1). The final result is then obtained by combining via *hardware* the results of the IP lookup in T_1 and T_2.

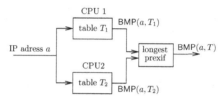

Fig. 1. A simple splitting of T into two tables T_1 and T_2 requires an additional hardware component to select the longest prefix.

The main benefit of this scheme relies in the fact that access operations on the L2 cache of the processor are much faster (up to seven times) than accesses on the DRAM memory.[3] Thus, working on smaller tables allows to obtain much more efficient data structures and to face the problem of the continuous increase of the size of the tables [10]. Notice that this will not just increase the time a single packet takes to be processed once assigned to the processors, but also the throughput of the router: while our solution uses two processors to process 1 packet in one unit of time, a "classical" solution using two processors for two packets may take 7 time units because of the size of the forwarding table.

Unfortunately, the use of the hardware for computing the final result may turn out to be unfeasible or too expensive: this circuit should take in input $\mathsf{BMP}(a, T_1)$ and $\mathsf{BMP}(a, T_2)$ and return the longest string between these two (see Fig. 1). An alternative would be to split T according to the leftmost bit: T_1 contains addresses starting by 0 (i.e., so called CLASS A addresses) and T_2 those starting by 1. This, however, does not necessarily yield an even splitting of the original table, even when real data are considered [14].

---

[3] From now on we will improperly use the term DRAM also for SRAM adopted in some routers.

The main contribution of this work is to provide a suitable way of splitting T into two tables T_1 and T_2 such that the two partial results can be combined in the simplest way: the XOR of the two sequences. This result is obtained via an efficient algorithm which, given a table T, for any positive integer $k \leq |T|$, finds a suitable subtable T_1 of size k with the property that

$$\text{LOOKUP}(a, T) = \text{LOOKUP}(a, T_1) \oplus \text{LOOKUP}(a, T \setminus T_1),$$

where $\text{LOOKUP}(a, T)$ denotes the output interface corresponding to $\text{BMP}(a, T)$, for any IP addresses a.

The construction of T_1 is rather simple and the method yields different strategies which might be used to optimize other parameters of the two resulting routing tables. These, together with the guarantee that the size is smaller than the original one, might be used to enhance the performance of the forwarding table. Additionally, our approach is *scalable* in that T can be split into more than two subtables. Therefore, our method may yield a scalable solution alternative to the simple increase of the number of processors and/or the size of their L2 memory cache. We believe that our novel technique may lead to a new family of parallel routers whose performance and costs are potentially superior to those of the current solutions [21,16,18,24].

We have tested our method with real data available at [14] for five routers: MaeEast, MaeWest, AADS, Paix and PacBell. We present a further strategy yielding the following interesting performances:

1. A very small routing table T_1 whose size is very close to 7% of $|T|$; Indeed, in all our experiments it is always smaller but in one case (the Paix router) in which it equals to: (i) 7.3% of $|T|$ when T contains over 87,000 entries, and (ii) 10.2% when $|T|$ is only about 6,500 entries.
2. A "simple" routing table $T_2 = T \setminus T_1$ with the interesting feature that *no exceptions* occur, that is, every possible IP address a has at most one matching prefix in T_2.

So, for real data, we are able to circumscribe the problem of computing the best matching prefix to a very small set of prefixes. By one hand, we can apply one of the existing methods, like controlled prefix expansion [23], to table T_1: because of the very small size we could do this much more aggressively and get a significant speed-up. By the other hand, the way table T_2 should be used opens new research directions in that, up to our knowledge, the IP lookup problem with the restriction that no exceptions occur has never been considered before. Observe that, table T_2 can be further split into subtables *without* using our method, since at most one of them contains a matching prefix.

Finally, we consider the issue of *updating* the routing/forwarding table, which any feasible solution for the IP lookup must take into account. We show that updates can be performed without introducing a significant overhead. Additionally, for the strategy presented in Sect. 3, all type of updates can be done with a constant number of operations, while keeping the structure optimality.

Roadmap. We describe our method and the main analytic results in Sect. 2. In Sect. 3 we present our experimental results on real routing tables. In Sect. 4 we conclude and describe the main open problems.

2   The General Method

In this section we describe our approach to obtain two subtables from a routing table T so that the computation of $\text{LOOKUP}(a, T)$ can be performed in parallel with a minimal amount of additional hardware: the XOR of the two partial results.

Throughout the paper we make use of an equivalent representation of a routing table by means of trees. Let us consider a routing table $T = \{(s_1, o_1), \ldots, (s_n, o_n)\}$, where each pair (s_i, o_i) represents a prefix/output pair. Given two binary strings s_1 and s_2, we denote by $s_1 \prec s_2$ the fact that s_1 is a prefix of s_2. We can represent T as a *forest* (S, E) where the set of vertices is $S = \{s_1, \ldots, s_n\}$ and for any two $s_1, s_2 \in S$, $(s_1, s_2) \in E$ if and only if (i) $s_1 \prec s_2$, and (ii) no $s \in S$ exists such that $s_1 \prec s \prec s_2$. Finally, to every vertex s_i, we attach a label o_i according to the corresponding output interface.

To simplify the presentation, we assume that T always contains the empty string ϵ, thus making (S, E) a tree rooted at ϵ. Observe that, this tree is not directly used to perform IP lookups. So, it will not be stored in the memory cache which will contain the forwarding tables derived from the subtables.

Our method consists of two phases which we describe below.

The Split Phase. Given a routing table T, let T^{up} denote *any* subtree of T having the same root. Also, for any node $u \in T$, let $l(u)$ denote its old label (i.e., its output interface in T) and let $l'(u) = l_1(u)/l_2(u)$ denote a pair of new labels. Intuitively, $l_i(u)$ represents the label of u in subtable T_i, $i = 1, 2$. We assign the new labels as follows (see Fig. 2):

- For any $u \in T^{up}$, $l'(u) = l(u)/\overline{0}$, where $\overline{0}$ denotes the bit sequence $(0, \ldots, 0)$;
- For any $v \in T \setminus T^{up}$, $l'(v) = l(x)/(l(x) \oplus l(v))$, where x is the lowest ancestor of v in T^{up}.

Let T' (respectively, T'') be the routing table obtained from T by replacing, for each $u \in t$, the label $l(u)$ with the label $l_1(u)$ (respectively, $l_2(u)$). It clearly holds that $l_1(u) \oplus l_2(u) = l(u)$. Hence,

$$\text{LOOKUP}(a, T') \oplus \text{LOOKUP}(a, T'') = \text{LOOKUP}(a, T).$$

The Compact Phase. The main idea behind the way we assign the new labels is the following (see Fig. 2):

1. All nodes in T^{up} have the second label equal to $\overline{0}$;
2. $T \setminus T^{up}$ contains upward paths where the first label each node are all the same.

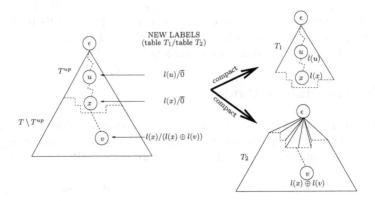

Fig. 2. An overview of our method: The subtree T^{up} corresponds to table T_1 after a compact operation is performed; similarly, $T \setminus T^{up}$ yields the table T_2.

Because of this, T_1 and T_2 contain redundant information and some entries (vertices) can be removed as follows. Given a table T, let COMPACT(T) denote the table obtained by repeatedly performing the following transformation: for every node u with a child v having the same label, remove v and connect u to all the children of v. Then, the following result holds:

Lemma 1. *Let* $T_1 = \text{COMPACT}(T')$ *and* $T_2 = \text{COMPACT}(T'')$. *Then,* $|T_1| = |T^{up}|$ *and* $|T_2| = |T| - |T^{up}|$.

Proof. We will show that no node in T^{up}, other than ϵ, will occur in T_2; similarly, no node in $T \setminus T_1$ will occur in T_1. Indeed, every node $u \in T^{up}$ have label equal to $\bar{0}$ in T'' (see Fig. 2). Since also ϵ has label $\bar{0}$, COMPACT(T'') $= T_2$ will not contain any such u. Similarly, any node $v \in T \setminus T^{up}$ has its label in T' equal to some $l(x)$, where x is the lowest ancestor of v in T^{up} (see Fig. 2). Therefore, all nodes in the path from x to v have label the same label $l(x)$ and thus will not occur in COMPACT(T') $= T_1$. This completes the proof.

Lemma 2. *For any table T and for any address a, it holds that* LOOKUP(a, T) $=$ LOOKUP(a, COMPACT(T)).

Proof. Let $u_a = \text{BMP}(a, T)$ and $v_a = \text{BMP}(a, \text{COMPACT}(T))$. If $u_a = v_a$, then the lemma clearly follows. Otherwise, we observe that, in constructing COMPACT(T), we have removed from T the node u_a and all of its ancestors up to v_a. This implies $l(u_a) = l(v_a)$, i.e., LOOKUP(a, T) $=$ LOOKUP(a, COMPACT(T)).

We have thus proved the following result:

Theorem 1. *For any routing table T and for any integer $1 \leq k \leq |T|$, there exist two routing tables T_1 and T_2 such that: (i) $|T_1| \leq k$ and $|T_2| \leq |T| - k$, and (ii) for any address a,* LOOKUP(a, T) $=$ LOOKUP(a, T_1) $\oplus$ LOOKUP(a, T_2).

The above theorem guarantees that any table T can be divided into two tables T_1 and T_2 of size roughly $|T|/2$. By applying the above construction iteratively, the result generalizes to more than two subtables:

Corollary 1. *For any routing table T and for any integers $k_1, k_2, \ldots, k_l$, there exist $l + 1$ routing tables $T_1, T_2, \ldots, T_{l+1}$ such that: (i) $|T_i| \leq k_i$, for $1 \leq i \leq l$; (ii) $|T_{l+1}| \leq |T| - k$, where $k = k_1 + k_2 + \cdots + k_l$; (iii) for any address a,* LOOKUP$(a, T) = \bigoplus_{i=1}^{l+1}$ LOOKUP(a, T_i).

Finally, we observe that the running time required for the construction of the two subtables depends on two factors: (a) the time needed to construct the tree corresponding to T; (b) the time required to compute T^{up}, given that tree. While the latter depends on the strategy we adopt for T^{up} (see also Sect. 3), the first step can be always performed efficiently. Indeed, by simply extending the partial order '$\prec$', a simple sorting algorithm yields the nodes of the tree in the same order as if we perform a BFS on the tree. Thus, the following result holds:

Theorem 2. *Let $t(|T|)$ denote the time needed for computing T^{up}, given the tree corresponding to a routing table T. Then, the subtables T_1 and T_2 can be constructed in $O(|T| \log |T| + t(|T|))$ time.*

Also notice that, if we want to obtain two subtables of roughly the same size, then a simple visit (BFS or DFS) suffices, thus allowing to construct the subtables in $O(|T| \log |T|)$ time. The same efficiency can also be achieved for a rather different strategy which we describe in Sect. 3.

2.1 Updates

In this section we show that our method does not yield an overhead in the process of updating the forwarding table. We consider two types of updates: (a) label changes, and (b) entry insertion/deletion. In particular, we assume that we have already computed the position, inside the tree T, of the newly added node or of the node to update.

Label Changes. Consider the situation in which the label of a prefix $p \in T$ changes from $l(u)$ to $l'(u)$. We distinguish three cases according to the left tree in Fig. 2:

- p is an internal node of T^{up}, i.e., $p = v$ in Fig. 2. This is the easy case, since it suffices to update the first label from $l(u)$ to $l(u)'$.
- p is a leaf of T^{up}, i.e., $p = x$ in Fig. 2. As above we have to change the first label from $l(u)$ to $l'(u)$; however, in this case p may be an ancestor of some other nodes in $T \setminus T^{up}$. This would require an update of all descendant nodes of p in T_2. We instead propose a simpler approach: move p from T^{up} into $T \setminus T^{up}$, so that the parent of p becomes a leaf of T^{up}. Even though this "lazy update" approach would make T_2 larger, this will happen only after several such updates. Therefore, we could periodically rebuild T^{up} so to maintain the two trees of roughly the same size.

- p is a node in $T \setminus T^{up}$, i.e., $p = v$ in Fig. 2. This is another simple case, since we only have to change the second label from $l(u) \oplus l(x)$ to $l'(u) \oplus l(x)$.

It worth observing that in the first and third case, one label change in T translates into one label change in either T_1 or T_2. In case p is on the "frontier" between T^{up} and $T \setminus T^{up}$, we can choose between performing one insertion/deletion (see next paragraph) or a certain number of updates in T_2. The choice of which to perform depends on several parameters, including how efficiently these operations are performed in the forwarding table.

Insertion/Deletion. Consider the situation in which a new node p must be inserted as child of some existing node of T. Again, we distinguish the following cases (see Fig. 2):

- p is child of an internal node $u \in T^{up}$. The new node can be simply inserted in its appropriate position with label pair $(l(p), \overline{0})$. As u is not a leaf of T^{up}, this will not affect any other node in $T \setminus T^{up}$; additionally, even if p may be inserted as parent of some node in T^{up} (a previously child of u), no further change is needed.
- p is a child a leaf node $x \in T^{up}$. We can deal with this case by first inserting p with the same label of x and then making a change of such label.
- p is a child of a node $v \in T \setminus T^{up}$. Another simple case since we only have to set the first label equal to $l(x)$ (see Fig. 2).

As for deletion of a node p, we only have to consider the case p is a leaf of T^{up}: we can simulate this by considering a label change of p so that its new label equals the label of its parent in T.

Finally, we mention that every label change could also imply some node deletion whenever the labels of two adjacent nodes become equal. This requires only a constant amount of time and keeps the two subtables "simplified", without computing COMPACT(T') and/or COMPACT(T'') from scratch.

3   Experimental Results

These experiments have been performed on real routing tables of five routers: Mae-East, Mae-West, AADS, Paix and PacBell. (Data available at [14].) In particular, we first observe that the tree T of the original table is a *shallow* tree, that is, its depth is always at most 6 (including the dummy node ϵ corresponding to the empty string). More importantly, the table *contains many leaf nodes*, i.e., entries that have no suffix. Based on this, we have tested the following strategy:

- The tree T^{up} contains all *non leaf* nodes of T.

The idea is that of obtaining a table T_2 with *no exceptions* and a table T_1 of size significantly smaller than $|T|$. Clearly, the smaller the size of T_1 the better is:

- The small size of T_1 (w.r.t. the size of T) basically resolves the issue of the memory size of the structure for the IP lookup in T_1;
- The particular structure of T_2 (i.e., no exceptions) may simplify significantly the problem and yield a data structure of smaller size (w.r.t. those solving the BMP problem).

It turns out that in all our experiments the size of T^{up} (and thus T_1) is always roughly 7% of $|T|$. Indeed, the only case in which it is smaller than 7% is for the Paix router of 00/10/01. Interestingly, the routing table of this router, for this day, has *over* 87,000 entries, thus showing that our method is "robust" to size fluctuations (compare the same router of other days in Table 1).

We also emphasize that, very similar results have been obtained over both a period of one week (see Table 1) and over a sample consisting of snapshots of the same for several months (see also [2]).

These two things together give a strong evidence that this method guarantees the same performance over a long period of time (see also. Table 2).

Table 1. Percentage of leaf nodes over one week (leaves/total entries, percentage).

Day (yy/mm/dd)	Mae-East	Mae-West	AADS	Paix	PacBell
00/10/01	22462/24018 93.5%	30195/32259 93.6%	27112/28820 94%	80812/87125 92.7%	34266/36313 94.3%
00/10/02	22380/23932 93.5%	30124/32178 93.6%	27066/28755 94%	21325/22887 93%	34446/36511 94.3%
00/10/03	22361/23922 93.4%	30038/32094 93.5%	27016/28730 94%	80776/87100 92.7%	34505/36557 94.3%
00/10/04	22426/23991 93.4%	30170/32239 93.5%	27121/28832 94%	81025/87372 92.7%	34315/36387 94.3%
00/10/05	22276/23820 93.5%	30249/32320 93.5%	27200/28912 94%	81030/87374 92.7%	39460/42142 93.6%
00/10/06	22252/23800 93.4%	30620/32701 93.6%	27945/29763 93.8%	81283/87638 92.7%	34465/36535 94.3%
00/10/07	22323/23876 93.4%	30414/32488 93.6%	27942/29672 94.1%	81179/87542 92.7%	34240/36308 94.3%
00/10/08	22339/23902 93.4%	7655/8140 94%	28000/29734 94.1%	5939/6536 90.8%	7824/8275 94.5%

Table 2. More results on the Mae-West for March 2002.

Day	9th	10th	11th	12th	13th	14th	15th
Leaves	27654	27670	27660	27697	27575	27527	27620
Total entries	29635	29648	29633	29686	29542	29485	29585
Percentage	93.3%	93.3%	93.3%	93.2%	93.3%	93.3%	93.3%

Justification. It is worth observing that the high percentage of leaf nodes does *not* directly derives from the fact that the table T contains very few exceptions (i.e. suffixes) which is what we want to achieve in subtable T_2 (actually, we want no exceptions at all). Indeed, in the following table we compare the number of entries of a given height in the tree [4] vs the number of entries whose subtree has a given height:

Height	0	1	2	3	4	5	6
of entry	1	16917	6041	928	123	7	1
of subtree	22462	1362	162	27	3	1	1

Notice that, the percentage of nodes which are suffixes (i.e., height bigger than 1) is roughly 29.5% of the total entries; for the same table, the number of non-leaf nodes is 6.5% only. This implies that, in practice, our strategy performs much better than the intuitive method of collecting the exceptions into a table.

4   Conclusion, Future Work and Open Problems

We have introduced a general scheme which allows to split a routing table into two (or more) routing tables T_1 and T_2 which can be used in parallel without introducing a significant hardware overhead. The method yields a family of possible ways to construct T_1: basically, all possible subtrees T^{up} as in Fig. 2.

This will allow for a lot of flexibility. In particular, it might be interesting to investigate whether, for real data, it is possible to optimize other parameters. For instance, the worst-case time complexity of some solutions for the IP lookup [25,23] depends on the number of different lengths occurring in the table. Is it possible to obtain two tables of roughly the same size and such that the set of prefix lengths is also spread between them?

Do strategies which split T into more than two tables have significant advantages in practice?

Finally, the main problem left open is that of designing and efficient forwarding table for the case of routing tables with no exceptions. Does any of the existing solutions get simpler or more efficient because of this?

Acknowledgments

We are grateful to Andrea Clementi, Pilu Crescenzi and Giorgio Gambosi for several useful discussions. We also thank Pilu for providing us with part of the software used in [4] which is also used here to extract the information from the routing tables available at [14]. Our acknowledgments also go to Corrado Bellucci for implementing the strategy described in Sect. 3 and for performing some preliminary experiments.

---

[4] The height of an entry in the tree corresponds to the number of prefixes of such an entry.

References

1. S. Bellovin, R. Bush, T.G. Griffin, and J. Rexford. *Slowing routing table growth by filtering based on address allocation policies.* http://www.research.att.com/~jrex/, June 2001.
2. G. Bongiovanni and P. Penna. XOR-based schemes for fast parallel IP lookups. Technical report, University of Salerno, 2003. Electronically available at http://www.dia.unisa.it/~penna.
3. T. Bu, L. Gao, and D. Towsley. On Routing Table Growth . In *Proceedings of Globe Internet*, 2002.
4. P. Crescenzi, L. Dardini, and R. Grossi. IP address lookup made fast and simple. In *Proc. 7th Annual European Symposium on Algorithms*, volume 1643 of *LNCS*, 1999.
5. S. Deering and R. Hinden. *Internet protocol, version 6 (IPv6).* RFC 1883, 1995.
6. M. Degernark, A. Brodnik, S. Carlesson, and S. Pink. Small forwarding tables for fast routing lookups. *ACM Computer Communication Review*, 27(4):3–14, 1997.
7. DIGITAL, http://www.networks.europe.digital.com/html/products_guide/hp-swch3.html. *GIGAswitch/FDDI networking switch*, 1995.
8. V. Fuller, T. Li, J. Yu, and K. Varadhan. *Classless Inter-Domain Routing (CIDR): and address assignment and aggregation strategy.* RFC 1519, September 1993.
9. N. Huang, S. Zhao, and J. Pan C. Su. A Fast IP Routing Lookup Scheme for Gigabit Switching Routers. In *IEEE INFOCOM*, 2002.
10. G. Huston. Analyzing the Internet's BGP Routing Table. *The Internet Protocol Journal*, 4(1), 2001.
11. C. Labovitz, G.R. Malan, and F. Jahanian. Origins of Internet Routing Instability. In *IEEE INFOCOM*, 1999.
12. B. Lampson, V. Srinivasan, and G. Varghese. IP Lookups using Multi-way and Multicolumn Search. In *INFOCOM*, 1998.
13. A. McAuley, P. Tsuchiya, and D. Wilson. *Fast multilevel hierarchical routing table using content-adressable memory.* US Patent Serial Number 034444, 1995.
14. MERIT, ftp://ftp.merit.edu/ipma/routing_table. *IPMA statistices*, 2002.
15. M. Mitzenmacher and A. Broder. Using Multiple Hash Functions to Improve IP Lookups . In *IEEE INFOCOM*, 2001.
16. P. Newman, G. Minshall, T. Lyon, and L. Huston. IP Switching and Gigabit Routers. *IEEE Communications Magazine*, January 1997.
17. S. Nilsson and G. Karlsson. Fast address look-up for internet routers. *Proc. of ALEX*, pages 42–50, February 1998.
18. D. Pao, C. Liu, A. Wu, L. Yeung, and K.S. Chan. Efficient Hardware Architecture for Fast IP Adress Lookup. In *IEEE INFOCOM*, 2002.
19. C. Partridge, P. Carvey, E. Burgess, I. Castineyra, T. Clarke, L. Graham, M. Hathaway, P. Herman, A. King, S. Kohalmi, T. Ma, J. Mcallen, T. Mendez, W.C. Miller, R. Pettyjohn, J. Rokosz, J. Seeger, M. Sollins, S. Storch, B. Tober, G.D. Troxel, and S. Winterble. A 50-Gb/s IP router. *IEEE/ACM Transactions on Networking*, 6(3):237–247, 1998.
20. T.-B. Pei and C. Zukowski. Putting routing tables into silicon. *IEEE Network*, January 1992.
21. Pluris Inc., White Paper, http://www.pluris.com. *Pluris Massively Parallel Routing.*
22. J. Postel. *J. Internet protocol.* RFC 791, 1981.

23. V. Srinivasan and G. Varghese. Faster IP Lookups using Controlled Prefix Expansion. In *Proc. of ACM SIGMETRICS (also in ACM TOCS 99)*, pages 1–10, September 1998.
24. D.E. Taylor, J.W. Lockwood, T.S. Sroull, J.S. Turner, and D.B. Parlour. Scalable IP Lookup for Programmable Routers. In *IEEE INFOCOM*, 2002.
25. M. Waldvogel, G. Varghese, J. Turner, and B. Plattner. Scalable High Speed IP Routing Lookups. In *Proc. of ACM SIGCOMM*, pages 25–36, September 1997.

The Impact of Network Structure on the Stability of Greedy Protocols

Dimitrios Koukopoulos[1], Marios Mavronicolas[2],
Sotiris Nikoletseas[1], and Paul Spirakis[1,*]

[1] Department of Computer Engineering & Informatics, University of Patras
and Computer Technology Institute (CTI),
Riga Fereou 61, P. O. Box 1122, 261 10 Patras, Greece
`koukopou,nikole,spirakis@cti.gr`

[2] Department of Computer Science, University of Cyprus, 1678 Nicosia, Cyprus
`mavronic@ucy.ac.cy`

Abstract. A *packet-switching* network is *stable* if the number of packets in the network remains *bounded* at all times. A very natural question that arises in the context of stability and instability properties of such networks is how network structure precisely affects these properties. In this work, we embark on a systematic study of this question in the context of *Adversarial Queueing Theory,* which assumes that packets are adversarially injected into the network. We consider *size, diameter,* maximum *vertex degree,* minimum number of *disjoint paths* that cover all edges of the network, and *network subgraphs* as crucial structural parameters of the network, and we present a comprehensive collection of structural results, in the form of bounds on both stability and instability thresholds for various greedy protocols:

- We present a novel, yet simple and natural, construction of a network parameterized by its size on which certain compositions of *universally stable,* greedy protocols are unstable for low rates. The closeness of the drop to 0.5 is proportional to the increase in size.
- It is now natural to ask how unstable networks with *small* (constant) size be. We show that size of 22 suffices to drop the instability threshold for the FIFO protocol down to 0.704. This results is the current state-of-the-art trade-off between size and instability threshold.
- The diameter, maximum vertex degree and minimum number of edge-disjoint paths play a significant role in an improved analysis of stability threshold for the FIFO protocol. The results of our analysis reveal that a calibration of these parameters may be a valuable asset for the design of networks with as high as possible stability threshold.
- How much can network subgraphs that are forbidden for stability affect the instability threshold? Through improved combinatorial constructions of networks and executions, we improve the state-of-the-art instability threshold induced by certain known forbidden subgraphs on networks running a certain greedy protocol.

* This work has been partially supported by IST Program the IST Program of the E.U. under contract numbers IST-1999-14186 (ALCOM-FT) and IST-2001-33116 (FLAGS).

R. Petreschi et al. (Eds.): CIAC 2003, LNCS 2653, pp. 251–263, 2003.

Our results shed more light and contribute significantly to a finer understanding of the impact of structural parameters on stability and instability properties of networks.

1   Introduction

Motivation-Framework. *Objectives.* A lot of research has been done in the
field of packet-switched communication networks for the specification of their
behavior. In such networks, packets arrive dynamically at the nodes and they
are routed in discrete time steps across the edges. In this work, we embark on
a study of the impact structural network properties have on the correctness
and performance properties of networks. We study here *greedy* protocols as our
test-bed. In some cases, we consider networks in which different switches can
use different greedy protocols. This is motivated by the *heterogeneity* of modern
large-scale networks such as the Internet.
Framework of Adversarial Queueing Theory. We focus on a basic adversarial
model for packet arrival and path determination that has been recently introduced in a pioneering work by Borodin *et al.* [3]. It was developed as a robust
counterpart to classical Queueing theory [4] that replaces stochastic by worst
case assumptions. The underlying goal is to determine whether it is feasible
to prove stability results even when packets are injected by an *adversary*. At
each time step, the adversary may inject a set of packets into some nodes. For
each packet, the adversary specifies a simple path that the packet must traverse;
when the packet arrives to its destination, it is absorbed by the system. When
more than one packets wish to cross a queue at a given time step, a *contention-
resolution* protocol is employed to resolve the conflict. A crucial parameter of
the adversary is its *injection rate* r, where $0 < r < 1$. Among the packets that
the adversary injects in any time interval I, at most $\lceil r|I| \rceil$ can have paths that
require any particular edge. We say that a packet p *requires* an edge e at time t
if the edge e lies on the path from its position to its destination at time t.
Stability. Stability requires that the number of packets in the system remains
bounded at all times. We say that a protocol P is *stable* [3] on a network $\mathcal{G}$
against an adversary $\mathcal{A}$ of rate r if there is a constant C (which may depend on
$\mathcal{G}$ and $\mathcal{A}$) such that the number of packets in the system is bounded at all times
by C. We say that *a protocol* P *is universally stable* [3] if it is stable against
every adversary of rate less than 1 and on every network. We also say that *a
network* $\mathcal{G}$ *is universally stable* [3] if every greedy protocol is stable against every
adversary of rate less than 1 on $\mathcal{G}$.
Greedy Protocols. We consider six *greedy* protocols– ones that always advance
a packet across a queue (but one packet at each discrete time step) whenever
there resides at least one packet in the queue (see Table 1).
Network Structure. Important parameters of it are: (a) the used protocols, (b)
graph parameters such as minimum degree, diameter, size, (c) forbidden subgraphs for stability and (d) the subclasses of parameterized families of networks.

Table 1. Greedy protocols considered in this paper (**US** stands for universally stable).

Protocol name	Which packet it advances:	US
Shortest-in-System (**SIS**)	The most recently injected packet	√
Longest-in-System (**LIS**)	The least recently injected packet	√
Furthest-to-Go (**FTG**)	The furthest packet from its destination	√
Nearest-to-Source (**NTS**)	The nearest packet to its origin	√
First-In-First-Out (**FIFO**)	The earliest arrived packet at the queue	**X**
Nearest-To-Go-Using-LIS (**NTG-U-LIS**)	The nearest packet to its destination or the least recently injected packet for tie-breaking	**X**

Contribution. How does the network structure precisely affect stability? In this work, we present a comprehensive collection of structural results in the form of bounds, on both stability and instability thresholds.

- We present an innovative *parameterized* adversarial construction for estimating *instability* thresholds in heterogeneous networks. Our parameterized approach considers that each execution (phase) of the adversarial construction consists of distinguished time periods (rounds) whose number depends on the parameterized network topology. We apply our construction in instances of a parameterized network family and prove that when the network size parameter k tends to infinity then the instability threshold for the compositions of LIS-SIS, LIS-NTS and LIS-FTG fast converges to 0.5.
- We present a general analysis showing that any network $\mathcal{G}$ has an upper bound on injection rate for FIFO stability that depends only on the minimum number of edge-disjoint paths that cover $\mathcal{G}$, the maximum in-degree, and the maximum directed path length of the network. This result improves the previous known upper bound for FIFO stability of [5] for all networks. Furthermore, for several networks our stability bound is better than the one estimated in [8] such as the network $\mathcal{U}_1$ in Figure 1.
- We demonstrate an ad-hoc FIFO network that uses only 22 queues and it is unstable for any $r \geq 0.704$. The corresponding parameterized network [8] for $r = 0.704$ needs at least 361 queues. Thus, we show that ad-hoc constructions may beat the parametric ones with respect to the network size.
- In the model of *non-simple paths* (paths do not contain repeated edges), we study two simple graphs ($\mathcal{U}_2$ and $\mathcal{U}_3$ in Figures 1, 4) that have been shown in [2] to be forbidden subgraphs for universal stability. Note that $\mathcal{U}_3$ is an extension of $\mathcal{U}_1$ (Figures 1) for $n = 0$, $m = 1$ and $d = 2$. For these graphs we show instability for lower rates than those in [2] via a different construction.

Related Work. *Adversarial Queueing Theory* was developed by Borodin *et al.* [3] as a more realistic model that replaces traditional stochastic assumptions in Queueing Theory by more robust, worst-case ones. Subsequently, adversarial queueing theory, and corresponding stability and instability issues, received a lot of interest and attention (see, e.g., [1,2,5,6,7,9]). The universal stability of SIS, LIS, NTS and FTG protocols was established by Andrews *et al.* [1]. The subfield

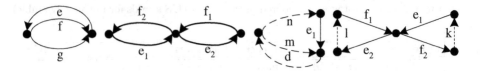

Fig. 1. Networks $\mathcal{U}_1$, $\mathcal{U}_2$ and their extensions $\Gamma(\mathcal{U}_1)$ and $\Gamma(\mathcal{U}_2)$ [2, Lemma 7]

of study of the stability properties of compositions of universally stable protocols has been opened recently by *Koukopoulos et al.* [6,7] where lower bounds of 0.683 and 0.519 on the instability threshold of the composition pairs LIS-SIS, LIS-NTS and LIS-FTG were respectively presented.

The subfield of proving stability thresholds for greedy protocols on every network was first initiated by Diaz *et al.* [5] showing an upper bound on injection rate for the stability of FIFO in networks with a finite number of queues that is based on network parameters. In an alternative work, Lotker *et al.* [8] proved that any greedy protocol can be stable in any network if the injection rate of the adversary is upper bounded by $1/(d+1)$, where d is the maximum path length that can be followed by any packet. Also, they proved that for a specific class of greedy protocols, time-priority protocols the stability threshold becomes $1/d$.

The instability of FIFO for *small-size* networks (in the model of adversarial queueing theory) was first established by Andrews *et al.* [1, Theorem 2.10] for injection rate $r \geq 0.85$. Lower bounds of 0.8357 and 0.749 on FIFO instability were presented by Diaz *et al.* [5, Theorem 3] and Koukopoulos *et al.* [6, Theorem 5.1]. An alternative approach for studying FIFO instability is based on parameterized constructions for networks with *unbounded size*. Using this approach, Lotker *et al.* [8] proved an instability threshold of $\frac{1}{2} + \epsilon$ for FIFO; the network size is a function of r that goes to infinity very fast as r goes down to 0.5.

In [2, Lemma 7], a characterization for directed network graphs (digraphs) universal stability is given when the packets follow non-simple paths (paths do not contain repeated edges). According to this characterization a digraph is universally stable if and only if it does not contain as subgraph any of the extensions of $\mathcal{U}_1$ ($\Gamma(\mathcal{U}_1)$) or $\mathcal{U}_2$ ($\Gamma(\mathcal{U}_2)$) where the parameters n, m, d, l, k represent numbers of consecutive edges with $l, k, n \geq 0$ and $m, d > 0$ (see Figure 1). These graphs have been shown to have instability thresholds of 0.84089 for a certain greedy protocol.

2   The Model

The adversarial queueing model considers a communication network that is modelled by a directed graph $\mathcal{G} = (V, E)$, where $|V| = n, |E| = m$. Each node $u \in V$ represents a communication switch, and each edge $e \in E$ represents a link between two switches. In each node, there is a buffer (queue) associated with each outgoing link. Buffers store packets that are injected into the network with a route, which is a simple directed path in $\mathcal{G}$. When a packet is injected, it is placed in the buffer of the first link on its route.

In order to formalize the behavior of a network under the adversarial queueing model, we use the notions of *system* and *system configuration*. A triple of the form $\langle \mathcal{G}, \mathcal{A}, \mathsf{P} \rangle$ where $\mathcal{G}$ is a network, $\mathcal{A}$ is an adversary and P is the used protocol on the network queues is called a system. The execution of the system proceeds in global time steps numbered $0, 1, \ldots$. Each time-step is divided into two sub-steps. In the first sub-step, one packet is sent from each non-empty buffer over its corresponding link. In the second sub-step, packets are received by the nodes at the other end of the links; they are absorbed (eliminated) if that node is their destination, and otherwise they are placed in the buffer of the next link on their respective routes. In addition, new packets are injected in the second sub-step. Furthermore, the configuration C^t of a system $\langle \mathcal{G}, \mathcal{A}, \mathsf{P} \rangle$ in every time step t is a collection of sets $\{ S_e^t : e \epsilon \mathcal{G} \}$, such that S_e^t is the set of packets waiting in the queue of the edge e at the end of step t. The time evolution of the system is a sequence of such configurations $C^1, C^2, \ldots$ such that the load restriction imposed by the Adversarial Queueing Theory [3] is satisfied.

In the adversarial constructions we study here for proving instability, we assume that there is a sufficiently large number of packets s_0 in the initial system configuration. This will imply instability results for networks with an *empty* initial configuration, as established by Andrews *et al.* [1, Lemma 2.9]. Also, for simplicity, and in a way similar to that in [1], we omit floors and ceilings and sometimes count time steps and packets roughly. This only results to loosing small additive constants while we gain in clarity.

3   Stability in Heterogeneous Networks

Before proceeding to the adversary constructions we give two basic definitions.

Definition 3.1 *We denote X_i the set of packets that are injected into the system in the ith round of a phase. These packet sets are characterized as "investing" flows because they will remain in the system till the beginning of the next phase.*

Definition 3.2 *We denote $S_{i,j}$ the jth set of packets the adversary injects into the system in the ith round of a phase. These packet sets are characterized as "short intermediate flows" because they are injected on judiciously chosen paths of the network for blocking investing flows.*

A Parameterized Network Family. We provide here a parameterized family of heterogeneous networks $\mathcal{N}_k$. The motivation that led us to such a parameterization in the network topology is *two-fold*: (a) the existence of many parallel queues in the network allows the adversary to *simultaneously* inject several short intermediate flows that block the investing flows in the system, without violating the rule of the restricted adversarial model, (b) such a parameterized network topology construction, enables *a parameterized analysis* of the system configuration evolution into distinguished rounds whose number depends on the parameterized network topology. In LIS-FTG composition, the parameterization, besides

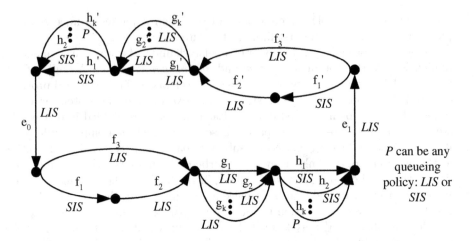

Fig. 2. A network $\mathcal{N}_k$ that uses LIS-SIS protocols

the parallel edges, includes additional chains of queues for the exploitation of FTG in blocking investing flows.

A Parameterized Adversarial Construction. In order for our adversarial construction to work, we split the time into phases. In each phase we study the evolution of the system configuration by considering distinguished time rounds. For each phase, we inductively show that the number of packets in the system increases. Applying repeatedly this inductive argument we show instability.

Theorem 3.1 *Let $r > 0.5$. There is a network $\mathcal{N}_k$ where k is a parameter linear to the number of network queues and an adversary $\mathcal{A}$ of rate r, such that the system $\langle \mathcal{N}_k, \mathcal{A}, \Pr \rangle$ is unstable if $\Pr$ is a composition of LIS protocol with any protocol of a) SIS, b) NTS and c) FTG.*

Sketch of proof. Part a) This proof is based on the preservation of all the investing flows injected during a phase into the system. We consider an instance of the parameterized network family (network $\mathcal{N}_k$ in Figure 2). All the queues use the LIS protocol except the queues $f_1, f_1', h_1, \ldots, h_{k-1}, h_1', \ldots, h_{k-1}'$ that use the SIS protocol. Moreover, the edges h_k, h_k' use protocol P, that can be any of LIS or SIS because there is no packet conflict in these queues.

Inductive Hypothesis: At the beginning of phase j, there are s_j packets that are queued in $g_1', h_1', \ldots, h_{k-1}'$ requiring to traverse the edges e_0, f_1, f_2, g_1, h_1.

Induction Step: At the beginning of phase $j+1$ there will be more than s_j packets (s_{j+1} packets) that will be queued in $g_1, h_1, \ldots, h_{k-1}$ requiring to traverse the edges $e_1, f_1', f_2', g_1', h_1'$.

We construct an adversary $\mathcal{A}$ such that the induction step holds. Proving that the induction step holds, we ensure that the inductive hypothesis will hold at the beginning of phase $j+1$ for the symmetric edges with an increased value of s_j, $s_{j+1} > s_j$. From the inductive hypothesis, initially, there are s_j packets (called $S - flow$) in the queues $g_1', h_1', \ldots, h_{k-1}'$ requiring to traverse the edges

e_0, f_1, f_2, g_1, h_1. In order to prove that the induction step works it is assumed that there is a large enough number s_j of packets in the initial system configuration. Phase j consists of $l = k + 1$ rounds with $l \geq 3$, that is $k \geq 2$. The sequence of injections is as follows:

Round 1: It lasts s_j steps. The adversary injects a set X_1 of $|X_1| = rs_j$ packets in e_0 wanting to traverse the edges $e_0, f_3, g_1, h_1, e_1, f_1', f_2', g_1', h_1'$ and a set $S_{1,1}$ of $|S_{1,1}| = rs_j$ packets in f_1 that require to traverse only the edge f_1.
Evolution of the system configuration. X_1 packets are blocked by the $S - flow$ packets in e_0. $S - flow$ packets are delayed by $S_{1,1}$ packets in f_1. Thus, at the end of this round a set Y of $|Y| = rs_j$ packets of S remain in queue f_1.

Round 2: It lasts rs_j steps. The adversary injects a set X_2 of $|X_2| = r^2 s_j$ packets in e_0 requiring to traverse the edges $e_0, f_3, g_1, h_2, e_1, f_1', f_2', g_1', h_1'$. Also, it injects a set $S_{2,1}$ of $|S_{2,1}| = r^2 s_j$ packets in f_2 wanting to traverse f_2, g_2, h_1.
Evolution of the system configuration. X_1 and $S_{2,1}$ packets are blocked in g_1 and f_2 correspondingly by Y packets. X_2 packets are blocked in e_0 by X_1 packets.

Since the number of rounds depends on the network topology (i.e. $l = k+1$), we next analyze an intermediate round t, $3 \leq t < l$.

Round t (intermediate round): It lasts $r^{t-1} s_j$ steps. The adversary injects $t - 1$ short intermediate flows $S_{t,1}, \ldots, S_{t,t-1}$ of $|S_{t,1}| = \ldots = |S_{t,t-1}| = r^t s_j$ packets. Flow $S_{t,j}$ with $1 \leq j \leq t - 1$ and $j \neq 2$ is injected in g_{j+1} wanting to traverse the edges g_{j+1}, h_j. Flow $S_{t,2}$ is injected in queue f_2 wanting to traverse the edges f_2, g_3, h_2. In addition, an investing flow X_t of $|X_t| = r^t s_j$ packets is injected in f_3 wanting to traverse the edges $f_3, g_1, h_t, e_1, f_1', f_2', g_1', h_1'$.
Evolution of the system configuration. $S_{t,j}$ packets with $3 \leq j \leq t-1$ are blocked in g_{j+1} by $S_{t-1,j-1}$ packets, while flows $S_{t,1}$ and $S_{t,2}$ are blocked in g_2 and f_2 correspondingly by $S_{t-1,1}$ packets. X_t packets are blocked in g_1 by X_{t-1} packets that were injected in round $t - 1$. At the end of round t, there is a number of $t - 3$ different cases for the queues where $X_1, \ldots, X_{t-1}$ are queued depending on their position at the beginning of the round and the injection rate r. In case i $(1 \leq i \leq t - 3)$ we have: at the beginning of round t, a portion or all X_i packets along with $X_{i+1}, \ldots, X_{t-1}$ are queued in g_1, while the rest X_i packets are queued in h_i and all the $X_1, \ldots, X_{i-1}$ packets are queued in $h_1, \ldots, h_{i-1}$ correspondingly. Then, at the end of round t, two cases can happen. In both cases, $X_1, \ldots, X_{i-1}$ packets are queued in $h_1, \ldots, h_{i-1}$ correspondingly. In the first one, a portion or all X_i packets in g_1 are queued with the rest X_i packets in h_i and $X_{i+1}, \ldots, X_{t-1}$ packets remain in g_1. In the second one, all the X_i packets in g_1 are queued with the rest X_i packets in h_i, a portion of X_{i+1} packets is queued in h_{i+1}, while the rest X_{i+1} packets along with $X_{i+2}, \ldots, X_{t-1}$ remain in g_1. Note that, in all possible system configurations at the end of round t, *the investing flows $X_1, \ldots, X_t$ remain into the system.*

Round l: It lasts $r^{l-1} s_j$ steps. The adversary injects an investing flow X_l of $|X_l| = r^l s_j$ packets in f_3 wanting to traverse the edges $f_3, g_1, h_l, e_1, f_1', f_2', g_1', h_1'$.
Evolution of the system configuration. X_l packets are blocked in g_1 by X_{l-1} that was injected in the system at round $l-1$. $X_1, \ldots, X_{l-1}$ packets are blocked in the system by flows $S_{l-1,1}, \ldots, S_{l-1,l-2}$. Therefore, at the end of round l, the number

of packets that are queued in $g_1, h_1, \ldots, h_{l-2} = h_{k-1}$ requiring to traverse the edges $e_1, f_1', f_2', g_1', h_1'$ is $s_{j+1} = |X_1| + \ldots + |X_l|$.

In order to have instability, we must have $s_{j+1} > s_j$. This holds for $r^{k+2} - 2r + 1 < 0$. This argument can be repeated for an infinite number of phases ensuring the instability of the system $\langle N_k, \mathcal{A}, \mathsf{LIS} - \mathsf{SIS} \rangle$. Also, $k \to \infty \implies r^{k+2} \to 0$, because $0 < r < 1$. Thus, for instability it suffices $-2r + 1 < 0$, i.e. $r > 0.5$.

Parts b, c) The adversarial constructions that are used in *Parts a, b, c*. One difference is the use of NTS and FTG protocols in the networks of *Parts b, c* correspondingly where SIS is used in the network of *Part a*. The topology of the used networks in *Parts a, b, c* is similar. The only difference is the use of additional paths in the network of *Part b* that start at queues that use FTG. These paths have sufficient lengths, such that the short intermediate packet flows have priority over the investing packet flows when they conflict in queues that use FTG. On the other hand, the injection of short intermediate flows in *Part b* with the same paths as in *Part a* is enough to guarantee their priority over investing flows when they conflict in queues that use NTS. As in *Part a*, it is proved that for any $r > 0.5$ the system $\langle N_k, \mathcal{A}, \mathsf{Pr} \rangle$ is unstable if Pr is a composition of LIS NTS or FTG. □

Notice that our method converges very fast to 0.5 for small values of the parameter k that depends on the network size. This can be shown easily if in the inequality $r^{k+2} - 2r + 1 < 0$ the parameters r, k are replaced by appropriate values. Therefore, for $k = 7$ the instability threshold is 0.501 and the number of network queues is 36 in the case of LIS-SIS and LIS-NTS (given by $8 + 4k$), while it is 102 in the case of LIS-FTG (given by $14 + 4k + 10(k - 1)$).

4   Structural Conditions for FIFO Stability

We denote "old" a packet that was injected in previous time periods than the current one. The earliest time step in a time period, at which all the old packets in the system have been served is denoted by M. Denote $j(\mathcal{G})$ the minimum number of edge-disjoint paths that cover the network $\mathcal{G}$. Thus, consider the disjoint paths $\Pi_1, \Pi_2, \ldots, \Pi_{j(\mathcal{G})}$. The number of packets in the path Π_j, where $1 \leq j \leq j(\mathcal{G})$, at time step M will be denoted as $s(\Pi_j)$. Furthermore, the maximum in-degree, and the maximum directed path length of the network $\mathcal{G}$ will be denoted as $\alpha(\mathcal{G})$ and $d(\mathcal{G})$ respectively. Notice that if $\alpha(\mathcal{G}) = 1$ then we have a tree or a ring, that is known to be universally stable [1], so we assume $\alpha(\mathcal{G}) > 1$. We show:

Theorem 4.1 *Let* $r_{\mathcal{G}}^2 \Sigma_{i=0}^{d(\mathcal{G})-1} (\alpha(\mathcal{G}) + r)^i = \frac{1}{j(\mathcal{G})}$. *Then for any network* $\mathcal{G}$, *and any adversary with* $r \leq r_{\mathcal{G}}$ *the system* $\langle \mathcal{G}, \mathcal{A}, \mathsf{FIFO} \rangle$ *is stable.*

Proof. Let us denote the queues of $\mathcal{G}$ as $Q_1, Q_2, \ldots, Q_m$ and their loads at time $t \geq 0$ as $q_1(t), q_2(t), \ldots, q_m(t)$. Let $P(0) = \Sigma q_i(0)$ be the initial load. We will construct an infinite sequence of consecutive distinguished time periods, t_i, at which $P(t_i) \leq P(0)$ thus keeping the network stable. The fact that we are using a FIFO protocol implies that after a certain time all the old packets will leave the system. We will compute a bound to this time.

Let's now, consider the worst case of an old packet being last in a queue Q_j at time 0 and targeted with the largest simple path in the network. Rename the queues in this simple path as $Q_j \equiv Q_{j_0}, \ldots = Q_{j_{d(\mathcal{G})-1}}$. Note that at time $M_1 = q_{j_0}$ all packets of this queue will have been served. Thus these packets have passed to the next queues in the path. Moreover, they can be delayed by at most rM_1 new injections. Furthermore, the size of any Q_{j_i} is bounded above by $(\alpha(\mathcal{G}) + r)M_1$. We repeat the same procedure, each time considering the last queue in the path that still contains old packets. After $d(\mathcal{G}) - 2$ additional steps $(M_2, M_3, \ldots, M_{d(\mathcal{G})-1})$ all the old packets would disappear or being in $Q_{j_{d(\mathcal{G})}}$. Define $P(t) = \max_{i=0}^m \{q_i(t)\}$. Working in the previous way, an absolute bound for the delay of the last old packet in Q_j is $M = M_1 + \ldots + M_{d(\mathcal{G})-1}$, where for every $0 < i < d(\mathcal{G})$, we have $M_1 \leq q(\Sigma_{j<i} M_j)$, with $M_0 = 0$. Moreover, during a period of $q(t)$ steps starting at time t, we have $P(t + P(t)) \leq (\alpha(\mathcal{G}) + r)P(t)$. Solving the recurrence, the total time is $M \leq \Sigma_{i=0}^{d(\mathcal{G})-1}(\alpha(\mathcal{G}) + r)^i P(0)$

At time step M all the old packets have been absorbed and only the injected packets in the time period $[0 \ldots M]$ will remain in the system. Because $j(\mathcal{G})$ is the minimum number of edge-disjoint paths in the network, during this period in the worst case at most $j(\mathcal{G})rM$ packets will be injected in the network. Therefore the total number of packets in the network at time step M is at most $P(M) \leq j(\mathcal{G})rM$. At time step M, $s(\Pi_j)$ packets exist in each disjoint path Π_j from the definitions. Note that the minimum number of packets in a disjoint path Π_j at time step M $(min\{s(\Pi_j)\})$ is significantly bigger comparing to the number of network edges. This allows us to assume that when a disjoint path Π_j has $s(\Pi_j)$ packets, then in each time step of a time period of $s(\Pi_j)$ time steps, r packets arrive into the path and one packet leaves it.

Assume now $s = \min\{s(\Pi_j)\}$. The change of the number of packets in the disjoint path Π_j in absolute values, Δ_{Π_j}, at $M + s$ time step will be $\Delta_{\Pi_j} = \Sigma_0^{\min\{s(\Pi_j)\}} |r - 1| = |r - 1| \min\{s(\Pi_j)\} \leq |r - 1| s(\Pi_j)$. Thus, the total change of the system configuration will be $\Sigma_{\Pi_j} \Delta_{\Pi_j} \leq \Sigma_{\Pi_j} |r - 1| s(\Pi_j) = |r - 1| \Sigma_{\Pi_j} s(\Pi_j)$. But, $P(M) = \Sigma_{\Pi_j} s(\Pi_j)$. Thus, $\Sigma_{\Pi_j} \Delta_{\Pi_j} \leq |r - 1| P(M)$ is at most the change of the system configuration for a time period with $s = min\{s(\Pi_j)\}$ steps. Consider now, the consecutive time intervals with duration: $s, rs, r^2 s, \ldots, r^k s$, where k is such that $r^k s \geq 1$ and $r^{k+1} s < 1$. The same argument as in the case of s time steps can be used for $r^i s$ time steps. For each of these time intervals the change of the system configuration will be at most $r^i(r - 1)P(M)$. Let t_1 be the time at which $r^k s$ finishes. The packets in network $\mathcal{G}$ at time t_1 are all new. Thus, the number of packets in the system at time t_1 is at most $P(t_1) \leq P(M) + (r - 1)P(M) + r(r - 1)P(M) + \ldots + r^{k-1}(r - 1)P(M) = r^k P(M)$.

For stability, we need $P(t_1) \leq P(0)$. Thus, we must choose an r such that $r^k P(M) \leq P(0)$. But, $P(M) \leq j(\mathcal{G})rM \leq j(\mathcal{G})r\Sigma_{i=0}^{d(\mathcal{G})-1}(\alpha(\mathcal{G}) + r)^i P(0)$. Thus, $r^k j(\mathcal{G})r\Sigma_{i=0}^{d(\mathcal{G})-1}(\alpha(\mathcal{G})+r)^i P(0) \leq P(0)$. For $k = 1$ this equation takes its smallest value $r^2 \Sigma_{i=0}^{d(\mathcal{G})-1}(\alpha(\mathcal{G}) + r)^i \leq \frac{1}{j(\mathcal{G})}$. This is equivalent to find in the real interval $(0, 1)$, the root $r_{\mathcal{G}}$ of the polynomial $-r^2 j(\mathcal{G})(\alpha(\mathcal{G})+r)^{d(\mathcal{G})} + r^2 j(\mathcal{G}) + \alpha(\mathcal{G}) + r - 1$. By the Bolzano Theorem, this polynomial has a root $r_{\mathcal{G}}$ in $(0, 1)$. □

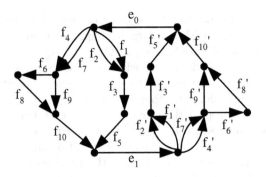

Fig. 3. The Network $\mathcal{G}$

In order to compare this bound with the bound of [5] we make the following analysis: Let m be the number of queues in any network $\mathcal{G}$. We get by definition of the minimum number of edge-disjoint paths $j(\mathcal{G})$ that $\frac{1}{j(\mathcal{G})} \geq \frac{1}{m}$ (1). The bound of [5] is the biggest r satisfying $\frac{2-r}{1-r}rm\sum_{i=0}^{d(\mathcal{G})-1}(\alpha(\mathcal{G})+r)^i \leq 1$ (2). Let us call it r_1. Our new bound is the longest value of r satisfying $r^2j(\mathcal{G})\sum_{i=0}^{d(\mathcal{G})-1}(\alpha(\mathcal{G})+r)^i \leq 1$ (3). Let us call it r_2. Thus, r_1 satisfies equation (2) as an equality. The same holds for r_2 and equation (3). Let $f(r) = \sum_{i=0}^{d(\mathcal{G})-1}(\alpha(\mathcal{G})+r)^i$. Note that $f(r)$ is monotone and increasing with $r > 0$. But, $\frac{2-r_1}{1-r_1}r_1mf(r_1) = 1 \implies \frac{1}{2m} \leq r_1f(r_1) \leq \frac{1}{m}$ (4) because $\forall r \in (0,1)$ it holds that $1 \leq \frac{2-r}{1-r} \leq 2$. Since $r_2 \leq 1$, it holds $r_2f(r_2) > r_2^2f(r_2^2) = \frac{1}{j(\mathcal{G})}$. Thus, from (1), (4) we take $r_2^2f(r_2^2) > r_1f(r_1) \implies r_2^2 > r_1$ and this implies $r_2 > r_1$ since $r_1 \in (0,1)$.

Lemma 4.2 *In all networks $\mathcal{G}$, we have $\sqrt{r_1} < r_2$ (thus, $r_2 > r_1$).*

The upper bound $1/d(\mathcal{G})$ was obtained in [8] for FIFO stability. From this and the previous lemma we conclude to the following theorem,

Theorem 4.3 *Let $r^* = \max\{r_2, \frac{1}{d(\mathcal{G})}\}$. Then for every $\mathcal{G}$, and any adversary with $r \leq r^*$ the system $\langle \mathcal{G}, \mathcal{A}, \text{FIFO}\rangle$ is stable.*

To illustrate the strength and applicability of our analytical techniques towards the threshold of $1/d(\mathcal{G})$ for FIFO stability in [8], we apply them to a simple network with three queues (network $\mathcal{U}_1$ in Figure 1). The upper bound for this is $1/3$ in [8], while in our case is:

Corollary 4.4 *Let $0 < r < 0.339$. Then, for network $\mathcal{U}_1$ and any adversary $\mathcal{A}$ with injection rate r the system $\langle \mathcal{U}_1, \mathcal{A}, \text{FIFO}\rangle$ is stable.*

5   Instability of Small-Size FIFO Networks

Theorem 5.1 *Let $r \geq 0.704$. There is a network G and an adversary $\mathcal{A}$ of rate r such that the system $\langle \mathcal{G}, \mathcal{A}, \text{FIFO}\rangle$ is unstable.*

Sketch of proof. The main ideas that are hidden behind the adversarial construction we use to prove this theorem are the following: (i) We split the time

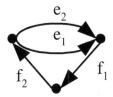

Fig. 4. Network U_3

into phases. In each phase we study the evolution of the system configuration by considering corresponding time rounds. For each phase, we inductively show that the number of packets in the system increases. This inductive argument can be applied repeatedly, thus showing instability. (ii) We use an inductive hypothesis with two parts. The first part specifies the position of the initial packets at the beginning of a phase (at the beginning of phase j, s_j packets are queued in s $e_1, f_3', f_4', f_5', f_6', f_8'$ requiring to traverse the edges e_0, f_1, f_3, f_5) and that their number is smaller than the number of packets in the corresponding subset of queues that will serve as initial packets at the beginning of the next phase (at the beginning of phase $j + 1$, s_{j+1} packets will be queued in $e_0, f_3, f_5, f_4, f_6, f_8$ requiring to traverse the edges e_1, f_1', f_3', f_5'). This part of inductive hypothesis holds for $r \geq 0.704$. The second part guarantees that the initial packets in each phase will traverse their path as a continuous flow. This holds for $r \geq 0.609$. Clearly, $r \geq \max\{0.704, 0.609\} = 0.704$ suffices for instability of the network $\mathcal{G}$. (iii) We achieve further delay of packets initially residing in the system by exploiting multiple "parallel" paths of the network topology. (iv) In order to create instability, we heavily exploit the *fair mixing* property of FIFO stating that if two packet sets arrive at the same queue simultaneously will mix according to the initial proportions of their sizes. □

6   Unstable Subgraphs

Consider the networks U_2 and U_3 (see Figures 1, 4) that use NTG-U-LIS protocol.

Theorem 6.1 *Let $r \geq 0.794$. There is a network U_i and an adversary $\mathcal{A}$ of rate r, such that the system $\langle U_i, \mathcal{A}, \mathsf{NTG} - \mathsf{U} - \mathsf{LIS} \rangle$ is unstable where U_i is the network a) U_2, b) U_3.*

Sketch of proof. We assume that there is a large enough number of packets s_j in the initial system configuration. Furthermore, we consider that the time is split into phases, each one of which consists of three distinguished time rounds with durations s_j, rs_j and $r^2 s_j$ for a phase j respectively. The proof is based on induction on the number of phases. In *Part a* the *inductive argument* states that if at the beginning of a phase j, there are s_j packets in the queues e_1, e_2 (the same queues in *Part b*) requiring to traverse the edges e_1, f_2 and e_2, f_1, f_2 correspondingly (the edge f_1 in *Part b*), then at the beginning of phase $j + 1$ there will be more than s_j packets in the queues e_1, e_2 (the same queues in *Part*

b) requiring to traverse the edges e_1, f_2 and e_2, f_1, f_2 correspondingly (the edge f_1 in *Part b*). In *Parts a,b* the basic idea behind the adversarial construction is the injection of two types of packet sets during a phase. One packet type is used for the reproduction of the inductive argument. The adversary tries to keep as many of these packets can in the queues e_1, e_2 of the network at the end of each phase. The packets of the other type are injected on judiciously chosen paths to keep as many of the packets of the other type in the system. In *Part a* at the end of round 3, there is a remaining portion of packets ($r^3 s_j$) in e_1 (e_2 in *Part b*) that require to traverse the edges e_1, f_2 (e_2, f_1 in *Part b*) and $r^3 s_j$ packets in e_2 (e_1 in *Part b*) requiring to traverse the edges e_2, f_1, f_2 (e_1, f_1 in *Part b*). Therefore, the number of packets in e_1, e_2 requiring to traverse the edges e_1, f_2 and e_2, f_1, f_2 (the edge f_1 in *Part b*) is $s_{j+1} = 2r^3 s_j$. For instability it suffices $s_{j+1} > s_j$, i.e. $r \geq 0.794$. This argument can be repeated for an infinite number of phases showing that the number of packets in the system increases forever. $\square$

7 Conclusions

Note that the technique used to get an instability threshold of 0.5 for certain compositions of universally stable protocols might produce lower bounds, if one finds a network allowing more investing flows to stay in the network. The technique that gets the upper bound on FIFO stability is based on a fundamental FIFO property, namely that in any FIFO network, old packets *exit the network* after some bounded time (by their size and the network structure). We feel that a refinement of such a technique may answer the fundamental open question of whether FIFO is stable below a certain *fixed* rate.

References

1. M. Andrews, B. Awerbuch, A. Fernandez, J. Kleinberg, T. Leighton, and Z. Liu, "Universal Stability Results for Greedy Contention-Resolution Protocols," *Journal of the ACM,* Vol. 48, No. 1, pp. 39–69, January 2001.

2. C. Alvarez, M. Blesa, and M. Serna, "Universal stability of undirected graphs in the adversarial queueing model," *Proceedings of the 14th Annual ACM Symposium on Parallel Algorithms and Architectures,* pp. 183–197, August 2002.

3. A. Borodin, J. Kleinberg, P. Raghavan, M. Sudan and D. Williamson, "Adversarial Queueing Theory," *Journal of the ACM,* Vol. 48, No. 1, pp. 13–38, January 2001.

4. H. Chen and D. D. Yao, *Fundamentals of Queueing Networks,* Springer, 2000.

5. J. Diaz, D. Koukopoulos, S. Nikoletseas, M. Serna, P. Spirakis and D. Thilikos, "Stability and Non-Stability of the FIFO Protocol," *Proceedings of the 13th Annual ACM Symposium on Parallel Algorithms and Architectures,* pp. 48–52, 2001.

6. D. Koukopoulos, M. Mavronicolas, S. Nikoletseas and P. Spirakis, "On the Stability of Compositions of Universally Stable, Greedy, Contention-Resolution Protocols," *Proceedings of the 16th International Symposium on DIStributed Computing,* LNCS 2508, pp. 88–102, October 2002.

7. D. Koukopoulos, S. Nikoletseas, and P. Spirakis, "Stability Issues in Heterogeneous and FIFO Networks under the Adversarial Queueing Model," Invited Keynote Address, *Proceedings of the 8th International Conference on High Performance Computing 2001,* pp. 3–14, December 2001.

8. Z. Lotker, B. Patt-Shamir and A. Rosen, "New Stability Results for Adversarial Queuing," *Proceedings of the 14th Annual ACM Symposium on Parallel Algorithms and Architectures,* pp. 192–199, August 2002.

9. P. Tsaparas, *Stability in Adversarial Queueing Theory,* M.Sc. Thesis, Computer Sc. Dept., Univ. of Toronto, 1997.

Improving Customer Proximity
to Railway Stations*

Evangelos Kranakis[1,**], Paolo Penna[2], Konrad Schlude[2], David Scot Taylor[3],
and Peter Widmayer[2]

[1] School of Computer Science, Carleton University, Canada
kranakis@cs.carleton.ca
[2] Institute for Theoretical Computer Science,
ETH Zentrum, CH-8092 Zürich, Switzerland
{lastname}@inf.ethz.ch
[3] Department of Computer Science, San Jose State University,
San Jose, CA 95192, USA
taylor@cs.sjsu.edu

Abstract. We consider problems of (new) station placement along (existing) railway tracks, so as to increase the number of users. We prove that, in spite of the NP-hardness for the general version, some interesting cases can be solved exactly by a suitable dynamic programming approach. For variants in which we also take into account existing connections between cities and railway tracks (streets, buses, etc.) we instead show some hardness results.

1 Models and Problems

There are many instances when public or private sector bodies are faced with making decisions on how to allocate facilities optimally. Such problems with mathematically quantifiable optimization constraints have been studied extensively in the scientific literature (e.g., see the book [3]). Recently the European Union has been encouraging the privatization of railway assets in various EU countries in order to improve system efficiency as well as customer satisfaction. In this paper we approach one such problem by studying how customer proximity can affect the railway station location. More specifically, given a set of settlements and an existing track, one wishes to build a set of new stations such that (some of) the settlements can easily access those stations and, thus, use the railway. This gives a gain in terms of (potentially) new users, but it also turns into a cost for the old ones (for instance, a new station results into a delay for those trains travelling on the track). Let us consider the following problem:

---

* Work partially supported by the Swiss Federal Office for Education and Science under the Human Potential Programme of the European Union under contract no. HPRN-CT-1999-00104 (AMORE).
** Research of E. Kranakis was supported in part by NSERC (Natural Sciences and Engineering Research Council of Canada) and MITACS (Mathematics of Information Technology and Complex Systems) grants.

R. Petreschi et al. (Eds.): CIAC 2003, LNCS 2653, pp. 264–276, 2003.

Input: A set of $P = \{p_1, \ldots, p_n\}$ of settlements (i.e. points) on the Euclidean plane, each of them with an associated *demand* d_i, and an existing railway, that is, a set of straight-line segments forming a connected polygonal and whose endpoints represent existing stations.
Solution: A set of new stations along the track.

Given a solution to this problem, we have a *gain* and a *cost* function due to the new stations. The cost of building a new station, in general, depends on the position we are placing it. In the sequel, we describe some possible definitions for the gain function. All such definitions are distance-based, that is, the gain due to the new stations depends on how far a settlement is from its closest new station. We will first assume that the distance is the Euclidean one (although, some of the results can be extended to other metrics).

Single radius. We first consider the following (simplified) scenario. A certain settlement p_i is far away from every existing station. So, for the people living there it is not worth to use the railway. If we build a new station which is close enough to p_i (let us say at distance less than R) then the railway transportation becomes "competitive" with respect to other transportations and *all* the people in p_i (let their number be d_i) will use this new station. We then have the following model: a settlement p_i uses a (newly built) station if and only if (a) this station is at distance less than or equal to some *radius* R and (b) no existing station was at distance less than R.

Notice that we can assume w.l.o.g. that no settlement in P is currently "covered" by the existing stations. Hence, the gain of a set S of new stations is the sum of the demands d_i of those p_i that are covered by the radius of some $s \in S$. Formally,

$$\sum_{i=1}^{n} d_i \cdot cover(S, p_i),$$

where $cover(S, p_i)$ equals 1 if there exists an $s \in S$ at distance less than or equal to R from p_i, and it equals 0 otherwise.

Distance based costs. Notice that the single radius model is, in some cases, too unrealistic since it assumes that a station at distance $R = 500m$, for instance, is accessible, while a station at distance $R' = 550m$ is not. A more realistic model should take into account the fact that the closer a station is the more (potential) customers from a settlement are expected. For instance, we could say that the expected number of users from p_i is $d_i/(\delta + 1)$, where δ is the distance of p_i to the closest station. More generally, given a *monotone* (decreasing) function $\alpha(\cdot)$, the gain of a set of new stations can be expressed as $\sum_{i=1}^{n} d_i \cdot \alpha(\delta(p_i, S))$, where $\delta(p_i, S)$ is the distance between p_i and the closest station in S.

Multiple radii. This setting is somewhat in between the two previous ones. Indeed, it can be used to approximate any distance based cost function with a fixed set of radii. Roughly speaking, these radii result from a "discretization" of an arbitrary function $\alpha(\cdot)$. For instance, the function $\alpha(\delta) = 1/(\delta + 1)$ can be approximated by two (or more) radii. Clearly, the more radii we consider, the better we approximate $\alpha(\cdot)$.

Two optimization problems for single radius. We now focus on the single radius model and we assume the cost of building a new station to be constant. For this version, one can envision the following two optimization problems:

- MIN NUMBER OF STATIONS (MIN STATION): minimize the number of stations needed to cover all the settlements.
- BUDGET CONSTRAINED MAX GAIN (MAX GAIN): given an integer k, with $1 \leq k \leq n$, find the placement of k stations that maximizes the gain.

We first observe that the second problem can be easily used to solve the first one: one just has to try all the k from 1 up to the smallest one for which the gain is the biggest possible, that is, we cover all the settlements. On the other hand, the other way round does not necessarily work. The limitation on the number of new stations seems to complicate things: with only k new stations at hand, we may not be able to cover all the settlements. In this case, our task is to find the best subset of settlements that can be covered with k stations only.

1.1 Previous Work

Our model is inspired by [6]. In that paper, the authors consider two different variants. The first one corresponds to what we here (re-) named single radius model (*accessibility model* in [6]). For this model the authors proved that, when a line set $\mathcal{L}$ (i.e., tracks) and a set S of *integer* points (i.e., settlements) are given, finding the best placements for k stations is NP-hard. The second model, named *travel time model*, takes into account the saved travel time over all the travellers. In contrast with the accessibility model, here it is assumed that all people in a settlement use the closest station and there is no a priori limit on the number of new stations we are allowed to build. As also mentioned above, there is a clear trade-off between the saved travel time due to a new station closer to some settlements, and the increased travel time due to the fact that trains must stop several times. Minimizing the saved travel time is also NP-hard [6].

The *geometric disk cover* problem [8] (related to the single radius model) asks to cover a set of points in the plane with the minimum number of disks of unit radius. This and other variants, in which the possible locations for the center of the disks is given in the input, admit a polynomial-time approximation scheme [8,4], which easily applies also to the MIN STATION problem. As for more general distance based functions, we observe that placing the minimum number of stations to maximize the gain is a restriction of *uncapacitated facility location with metric spaces* [5]. The latter problem (and hence our problem(s)) admits constant-ratio approximation algorithms [5,17]. Moreover, the case in which the *service cost* is the Euclidean distance (in our model this corresponds to choose $\alpha = 1/\delta$) has a polynomial-time approximation scheme [1] (in short PTAS). The same paper also gives (with the same technique) a PTAS for the k-median problem, while constant-factor approximation algorithms for the metric version are presented in [2,10]. Notice that the MAX GAIN problem is a special case of the k-median one (while in the facility location problem there is no such restriction for the number of new facilities to open). Noticeably, the Euclidean version of the k-median problem is already NP-hard [14].

The main difference between the above (more general) problems and our problem(s) concerns the restriction on the possible locations for the new stations. Indeed, in practice it is quite unlikely that the radius associated with a settlement crosses several tracks far apart from one another: if somebody walks to a station, then the distance he/she can cover is relatively small; if he/she goes by car, then probably driving for more than, let us say, 10 minutes to reach such a station would already make the railway transportation not that convenient (with respect to simply driving to the destination). This is confirmed by the data from the *Deutsche Bahn AG* used for the experiments in [6] and from the *Swiss Federal Railways SBB* [11].

Therefore, in many cases the whole network can be broken down into simpler smaller components. These components are nothing but single segments (i.e. parts of a track) and the solution of one segment does not affect the others. Actually, as already observed in [6], if every radius intersects the railway network in at most one interval, then the MIN STATION problem is polynomially solvable. Also, the MAX GAIN problem can be formulated as (uncapacitated) k-facility location problem with unimodular matrices [16], which is solvable in polynomial time [13, Chapt. 3.1] (see also [15] for more efficient methods). In [9] the case of facilities and customers located both on a line at n given points and cost service function satisfying the unimodal property (a generalization of Euclidean case [7]) an efficient dynamic programming approach is given. This result can be also used to obtain efficient algorithms for the single track versions in which we do no have a single radius per settlement (namely monotone cost functions, which include the multiple radii case). In the same paper, the authors also proved the NP-hardness of generalization of the unimodal case (namely, bimodal functions).

Finally, in [16] several variant problems have been studied, including non-continuous versions in which the possible locations of the stations is not given a priori.

1.2   Our Results

In this work we focus on the MAX GAIN problem in the single radius model. In particular, we aim in finding efficient exact algorithms for interesting cases that *do not* satisfy the unimodal property assumption [6].

To this aim, in Sect. 2 we present a novel dynamic programming approach for the single straight-line track. This restriction is solvable (within a better time complexity) using the results in [7,18]: Indeed, the result in [7] implies an $O(n^2)$-time exatct algorithm, which can be further improved to $O(kn \log n)$ by using orthogonal range queries [18]. However, we use the ideas contained in our dynamic programming approach to solve more complex situations where the results of [7,9] do not apply and do not yield exact polynomial-time algorithms. The natural extension of the single straight-line track is the case in which we have two parallel tracks and radii may intersect *both* of them. As we discuss at the beginning of Sect. 3, this apparently simple version of the problem already contains some complicating factors that make a natural extension of the dynamic approach fail. However, we are still able to modify our dynamic programming to exactly solve the following two versions:

- There is a minimum distance between consecutive stations we want to place; (Sect. 3.1)
- All settlements lie in between the two tracks. (Sect. 3.2)

The first variant is motivated by the practical consideration that putting two stations very close to each other has the only negative effect of delaying trains. The second case makes a non-trivial use of geometric properties of the radii generated by the settlements and might be of theoretical interest towards a characterization of those instances that admit polynomial-time exact algorithms. In both cases, the techniques used in [7,9] do not apply.

In Sect. 4 we show how an exact algorithm for the single track problem can be also used to exactly solve a problem of simultaneously building a new straight-line track and new stations on it: in this case we also have to decide where the new track should lie.

Finally, in Sect. 5, we go back to non Euclidean cases (motivated by the existence of streets/buses connecting settlements to the tracks) and show that, even with a single track, this version of the MAX GAIN problem is NP-hard. Moreover, the corresponding MIN STATION problem is hard to approximate within a factor $c \log n$, for some $c > 0$. This highlights the role played by the "geometry" in our solutions and indicates the need of some assumptions on the geometry of the streets to obtain exact (in some cases even approximate) solutions in polynomial time.

Due to lack of space some proofs are only sketched or omitted in this version. These proofs can be found in the extended version of this work [12].

2   Dynamic Programming for One Straight-Line Track

In this section we describe our exact algorithm for the problem restricted to single radius and constant cost per new station for the case of only a single track. The main ideas of this algorithm will be used in the sequel to solve the two parallel track versions.

We first observe that a circle around a settlement p_i with radius R intersects the track in an interval I_i. By construction, this interval is the only region that can contain a new station serving p_i. Moreover, if we place a station in the intersection of two (or more) intervals I_i, I_j, then this station will cover *all* the corresponding settlements. So, the MAX GAIN problem translates into the following one:

- MAX GAIN 1 TRACK: Given a collection of (weighted) intervals[1] on a line L and an integer k, find k points on L that maximize the sum of the weights of the intervals containing at least one of such points.

Notice that there is only a finite set of points on the line that must be taken into account to optimally place the k stations: the endpoints of the intervals. Hence, a simple brute-force approach yields an algorithm whose complexity is $O(n^k)$, where n is the number of intervals. However, a more efficient approach can be used to have a running time polynomial in both n and k. Consider a set of k stations s_1, s_2, up to s_k from left to right. Then, we hope that the set of intervals that contain *only* the station s_k can be computed by looking (only) at the position of s_{k-1} (i.e. it is independent from what the solution on the left of s_{k-1} is). Intuitively, this will allow us to break the instance into two *independent* subproblems: on the "left" of s_{k-1} (there we have to use at most $k-1$ stations) and on the right of it (where only one station must be located). This will be accomplished by a suitable dynamic programming approach. Before describing this, we first prove formally the above statement.

Lemma 1. *Given a collection of intervals on a line L and given three points i_1, i_2 and i_3 on L, with $i_1 < i_2 < i_3$, it holds that* contain$(i_2, i_3) \subseteq$ contain(i_1, i_3), *where* contain(i, j) *denotes the set of intervals containing j and not containing i.*

Intuitively, if we are given a set of stations in which the rightmost one is in position i and we add a new station in position j on the right of i, the gain due to the new station is given by

$$\mathsf{gain}(i, j) = \sum_{I \in \mathsf{contain}(i,j)} w(I), \tag{1}$$

where $w(I)$ is the weight associated to the interval I. We are now in a position to describe the dynamic programming algorithm to solve the problem. Given a set of intervals, let opt$(i; k)$ denote the minimum cost among the solutions that use k stations with the rightmost one in position i. Then, the following lemma states how opt$(i; k)$ can be computed if we have already computed this value for all the possible $i' < i$ and $k' < k$.

Lemma 2. *For any i and for any integer $k \geq 2$, the following condition holds:*

$$\mathsf{opt}(i; k) = \max_{i' < i} \{\mathsf{opt}(i'; k-1) + \mathsf{gain}(i', i)\}, \tag{2}$$

where gain$(\cdot, \cdot)$ *is defined as in Eq. 1.*

The above lemma allows us to efficiently compute an optimal solution. Indeed, we first observe that we need to compute tables gain$(\cdot, \cdot)$ and opt$(\cdot; \cdot)$ only for at most $2n$ values: the endpoints of the intervals. Hence in the following the point (or position) i will denote the ith endpoint of this set from left to right.

---

[1] The weight of interval I_i equals the demand d_i.

Lemma 3. *The table* gain(i, j) *can be computed for all* $1 \leq i < j \leq 2n$ *within* $O(n^2)$ *time.*

We are now in a position to prove the main result.

Theorem 1. *The* MAX GAIN 1 TRACK *problem can be solved within* $O(k \cdot n^2)$ *time, where* n *is the number of settlements.*

3   Two Parallel Tracks

We now consider the following extension of MAX GAIN 1 TRACK: instead of one track, we are given *two parallel tracks* (segments) and we have to place k stations on them. Then, the natural extension of the dynamic programming algorithm for the single track problem is as follows. We consider two parameters i^T and i^B which are the positions of the rightmost station on the top and on the bottom track, respectively. Then, opt$(i^T, i^B; k)$ is defined accordingly. Let us observe that every settlement p_i turns into a *pair* of weighted intervals I_i^T and I_i^B (with one of the two or both possibly empty). Moreover, such two intervals *cannot be considered separately*: if a station j^T intersects I_i^T and a station j^B intersects I_i^B, then the gain due to such two stations is not the sum of the weights of I_i^T and I_i^B. Indeed, those two stations are satisfying the same settlement p_i. Because of this, we have to measure the contribution of a station as (the sum of the weights of) the pairs (I_i^T, I_i^B) such that I_i^T and I_i^B do not contain any other station. So, we define contain(i^T, i^B, j^T) and contain(i^T, i^B, j^B) accordingly. Thence, we would like to extend Lemma 1 and prove that, for any $i_1^T \leq i^T < j^T$ and $i_2^B \leq i^B$,

$$\text{contain}(i_1^T, i_2^B, j^T) \subseteq \text{contain}(i^T, i^B, j^T).$$

Unfortunately, *the above statement is false.* (Fig. 1 shows a counterexample: $p_j \in \text{contain}(i^T, i^B, j^T)$ but $p_j \notin \text{contain}(i^T, i_2^B, j^T)$.)

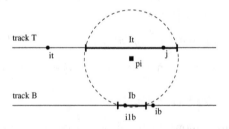

Fig. 1. A counterexample to the extension of dynamic programming algorithm to the case of two (parallel) tracks.

3.1   Minimum Distance between the Stations

The above example suggests a (reasonable) restriction of the problem for which, instead, the dynamic programming approach works. Assume we require the distance between any two consecutive stations in the solution to be at least $2R$. Then, in the example of Fig. 1, the station i_2^B would be on the left of I_j^B. We can prove that this is always the case.

We start with some useful observations. For any two points i and j, let $d_x(i, j)$ denote the distance between the respective projections on the x-axis. Then, the following fact holds:

Fact 2 *For any $I = (I^T, I^B)$ and for any two points i and j on the tracks, if $d_x(i, j) > 2R$ then not both i and j intersect I.*

Lemma 4. *For any $i_1^T \leq i^T < j^T$, and $i_2^B \leq i^B$, such that $d(i_1^T, i_1^T) > R$ and $d(i_2^B, i^B) > R$, the following holds:*

$$\text{contain}(i_1^T, i^B, j^T) \cup \text{contain}(i^T, i_2^B, j^T) \subseteq \text{contain}(i^T, i^B, j^T). \qquad (3)$$

Moreover, the same holds by considering some $j^B > i^B$ in place of j^T.

The above lemma easily implies that the dynamic programming algorithm for the single track can be extended for two parallel tracks if we impose this restriction on the minimum distance between two consecutive stations. Let us consider the following problem restriction:

> MAX GAIN r-ST: We require any solution $S = \{s_1^T, \ldots, s_l^T, s_1^B, \ldots, s_m^B\}$ to satisfy $d(s_i^T, s_{i+1}^T) \geq 2R/r$ and $d(s_j^B, s_{j+1}^B) \geq 2R/r$, for any $1 \leq i \leq l - 1$ and $1 \leq j \leq m - 1$.

Also MAX GAIN r-ST admits a polynomial-time exact algorithm:

Theorem 3. *The MAX GAIN r-ST problem can be solved within $O(n^{2r+2})$-time.*

3.2   Settlements in between the Tracks

We now consider the following problem:
MAX GAIN INNER: We consider instances in which all the settlements are located in between the two parallel tracks.

The remaining of this section is devoted to the proof of the following result:

Theorem 4. *The MAX GAIN INNER problem can be solved within $O(n^6)$ time, where n is the number of settlements.*

We will provide a polynomial-time algorithm for the MAX GAIN INNER as follows: (i) we first restrict to solutions that have a particular structure and show that the optimum can be found in polynomial time via (a variant of) our dynamic programming used for MAX GAIN r-ST; (ii) then, we show that every instance of MAX GAIN INNER has an optimal solution with the same structure. We begin with some definitions (see Fig. 2):

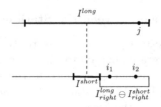

Fig. 2. A non essential solution.

Definition 1. *For a pair* $I = (I^T, I^B)$, *we denote by* I^{long} *(resp., I^{short}) the longer (resp., shorter) between* I^T *and* I^B. *Moreover, for every interval* I^*, I^*_{right} *denotes its right half. We say that an interval is* long *if it is the longer in its pair.*

Definition 2 (Essential solution). *Let* $I^{long}_{right} \ominus I^{short}_{right}$ *denote the interval obtained by the projection of* I^{long}_{right} *on the other track minus* I^{short}_{right}. *A station placement is* essential *if, for every pair* $I = (I^{long}, I^{short})$, *if* I^{long}_{right} *contains a station, then no two stations fall in* $I^{long}_{right} \ominus I^{short}_{right}$.

We first show that optimal essential solutions are computable in polynomial time.

Lemma 5. *The optimal essential solution can be computed in* $O(n^6)$ *time.*

Proof. For each track we consider the *two rightmost* stations, provided that their position satisfies Def. 2. Let $i^T = (i_1^T, i_2^T)$, and $i^B = (i_1^B, i_2^B)$ denote such stations. Also let $i'^T = (i_1', i_2^T)$ where $i_1' \leq i_1^T$. (Similarly, we define i''^T with respect to i_2^T.) Def. 2 easily implies the following fact:

$$\mathsf{contain}(i'^T, i^B, j^T) \cup \mathsf{contain}(i^T, i''^B, j^T) \subseteq \mathsf{contain}(i^T, i^B, j^T), \qquad (4)$$

where $\mathsf{contain}(i, j)$ denotes those intervals intersecting j and not intersecting any of the stations in i. Notice that the above inclusion is similar to that of Lemma 4, and it also holds by considering some $j^B > i^B$ in place of j^T. This guarantees that the two rightmost stations are all we need to compute the contribution of a new station j^T (or j^B) when added to a partial essential solution. Therefore we can define a function $\mathsf{gain}(i^T, i^B, j)$ and solve the problem in a similar way of MAX GAIN r-ST for $r = 2$.

The next technical lemma will be used to prove the optimality of the essential solutions vs. more general ones.

Lemma 6. *Let* $I = (I^T, I^B)$ *and* $J = (J^T, J^B)$ *be two pairs of intervals such that* I^{long} *and* J^{long} *lie on the same track. Also, let* $J^{short} \subseteq I^{long}_{right} \ominus I^{short}_{right}$. *Then,* $I^{long}_{right} \subseteq J^{long}$.

Lemma 7. *Every instance of* MAX GAIN INNER *has an optimal solution which is also essential.*

Proof. We show that any solution not satisfying Def. 2 can be transformed into an essential one that covers the same set of intervals. Let I be an interval pair for which Def. 2 is violated and let i_1 and $i_2 > i_1$ be two stations both in $I^{long} \ominus I^{short}$. We will show that moving i_1 on the right endpoint i' of I^{short} yields a (essential) solution that covers the same set of interval pairs. Let J be an interval pair covered by i_1 but not covered by i_2 nor by i'. If $i_1 \in J^{short}$, then we can apply Lemma 6 and conclude that J^{long} must contain the station in I^{long}_{right}. Otherwise, it must be the case that J^{long} is on the same track of I^{short}. It is then easy to see that Lemma 6 implies that J^{long} covers I^{short}_{right} (simply consider the interval J' obtained by exchanging J^{short} with J^{long}). Therefore, we have $i_2 \in I^{short}_{right} \subseteq J^{long}$. In both cases, the interval J is still covered in the transformed solution.

By putting together Lemmata 5 and 7 we obtain Theorem 4.

4   Building Tracks and Stations

Here we consider the following problem: we are allowed to build a new track and to place new stations on it. (Again, we are given a maximum budget for the new track and stations.) Rather than solving the whole realistic problem, we aim at showing some interesting cases in which an efficient algorithm for the MAX GAIN problem immediately translates into an efficient algorithm for this track-station placement problem.

In the sequel we show how to optimally decide the location of a new track: we are allowed to build a straight line track (no matter how long) and to place at most k stations on it so as to maximize the gain (defined in the usual way). We denote this problem as MAX TRACK GAIN.

The main idea is to show that, given n settlements, there are at most $O(n^4)$ lines to be considered. One among those gives an optimal solution. In particular, given a straight line track, what really matters is the underlying weighted interval graph. The latter changes whenever (i) the line crosses the border of some settlement radius or (ii) it crosses the intersection of two radius borders. We thus obtain the following result (see [12] for the details):

Theorem 5. *Let $t(n)$ be the running time of an algorithm for the* MAX GAIN *problem. Then the* MAX TRACK GAIN *problem can be solved within $O(n^4 t(n))$ time.*

Notice that, if one of the endpoints of the line is given in input, then the running time is $O(n^2 t(n))$. Indeed, we only have to consider lines passing through some of the $O(n^2)$ points defined above.

5   Existing Streets: Even a Single Track Is Hard

We consider the following generalizations of MIN STATION/MAX GAIN: every settlement is connected to the track by means of a certain number of streets. Moreover, whenever two streets leading to different settlements intersect, one street passes over the other by using a bridge. So, people traveling on one street cannot switch to another one[2]. We prove that this variant of MIN STATION, denoted as GENERALIZED MIN STATION, is hard to approximate within $c \ln n$, for some $c > 0$. Although this problem looks quite unnatural, it gives us a strong indication: even if we only want to have good (i.e. constant ratio) approximate solutions, we have to take into account some "geometry" of the streets, in particular how they intersect. Since this result also implies the NP-hardness of GENERALIZED MAX GAIN, similar geometric properties should be considered (at least) in deriving polynomial-time exact algorithms.

Our hardness proof is an adaptation of the NP-hardness proof given in [7] for the k-facility location problem restricted to bimodal matrices. For the details we refer the reader to [12].

Theorem 6. *It is* NP-*hard to approximate* GENERALIZED MIN STATION *within* $c \ln n$, *for some* $c > 0$. *Moreover, the* GENERALIZED MAX GAIN *problem is* NP-*hard, even when all settlements have the same demand.*

Finally, our reduction implies the same hardness results even if we assume the existence of a (common) street $\bar{l}$ parallel to the track: people from p_i first take some street from p_i leading to $\bar{l}$ and then walk (i.e. move along the track) up to some distance R.

6   Open Problems

We mention here some problems whose solution does not seem to be a straight-forward consequence of our results:

- Consider non-Euclidean distances (e.g., existing streets connecting settlements to the track might be considered). For the simple case of one street leading to the track per settlement, the problem can be solved in the same way: every p_i still corresponds to one weighted interval on the track; the length of such an interval depends on the "cost" of travelling through that street. On the other hand, if we make no assumption on the way settlements are connected to the track, then the problem becomes NP-hard, or even worse if we look at the extension of MIN STATION (see Sect. 5).
- Consider the variant of the MAX TRACK GAIN problem in which the new track(s) must connect *two* cities: Which is the best placement of a new track between, let us say, Konstanz and Zürich if we consider polylines, or even 1-bend segments?

---

[2] Or at least we can assume the majority of the people not able to jump from a bridge.

- The complexity of some of the above extensions is not polynomial if the budget value B or the number t of parallel tracks is not bounded. Are those problems NP-hard for some values of B and t?
- As a cost function of a set of new stations, consider how much a train is slowed down, with respect to the previous configuration without such stations. Clearly, the speed of the train will depend on the relative position of the new stations and, in general, cannot be expressed as the sum of costs b_i of building a new station in position i.

Acknowledgments

We are grateful to Anita Schöbel and Dorothea Wagner for many useful discussions on the model and on the issue of unimodular matrices, and for providing us with a copy of [6,16]. We thank Arie Tamir for pointing out a connection between our problem on the single track and [7], and the improvement in [18]. We also wish to thank Jörg Jermann for showing us the system developed for SBB [11].

References

1. S. Arora, P. Raghavan, and S. Rao. Polynomial Time Approximation Schemes for Euclidean k-medians and related problems. In *Proc. of the 30th ACM STOC*, pages 106–113, 1998.
2. M. Charikar, S. Guha, É. Tardos, and D.B. Shmoys. A constant-factor approximation algorithm for the k-median problem. In *Proc of the 31st ACM STOC*, pages 1–10, 1999.
3. M.S. Daskin. *Network and Discrete Location: Models Algorithms, and Applications*. Wiley-Interscience series in Discrete Mathematics and Optimization. John Wiley and Sons, Inc, 1995.
4. M. Galota, C. Glasser, S. Reith, and H. Vollmer. A Polynomial-Time Approximation Scheme for Base Station Positioning in UMTS Networks. In *Proc. of the 5th International Workshop on Discrete Algorithms and Methods for Mobile Computing and Communications, DIALM*, pages 52–59, 2001.
5. S. Gupta and S. Kuller. Greedy strikes back: improved facility location algorithms. In *Proc. of 9th ACM-SIAM Symposium on Descrete Algorithms, SODA*, pages 649–657, 1998.
6. H.W. Hamacher, A. Liebers, A. Schöbel, D. Wagner, and F. Wagner. Locating new stops in a railway network. In *Proc. of the ATMOS workshop, Algorithmic Methods and Models for Optimization of Railways*, Electronic Notes in Theoretical Computer Science, pages 15–25. Elsevier Science, 2001.
7. R. Hassin and A. Tamir. Improved complexity bounds for location problems on the real line. *Operations Research Letters*, 10:395–402, 1991.
8. D.S. Hochbaum and W. Maass. Fast approximation algorithms for a nonconvex covering problem. *Journal of Algorithms*, 8:305–323, 1987.
9. V.N. Hsu, T.J. Lowe, and A. Tamir. Structured p-facility location problems on the line solvable in polynomial time. *Operations Research Letters*, 21:159–164, 1997.
10. K. Jain and V. V. Vazirani. Primal-dual approximation algorithms for metric facility location and k-median problems. In *IEEE FOCS*, pages 2–13, 1999.

11. J. Jermann. Private communication. A decision support tool for placing new stations in the SBB railway network.

12. E. Kranakis, P. Penna, K. Schlude, D.S. Taylor, and P. Widmayer. Improving customer proximity to railway stations. Technical Report 371, Institute for Theoretical Computer Science, ETH Zurich, http://www.inf.ethz.ch, 2002.

13. G.L. Nemhauser and L.A. Wolsey. *Integer and Combinatorial Optimization*. Wiley, 1988.

14. C. H. Papadimitriou. Worst-case and probabilistic analysis of a geometric location problem. *SIAM J. Comput.*, 10, 1981.

15. A. Schöbel. Set covering problems with consecutive ones property. Technical report, Universität Keiserslautern, 2001.

16. A. Schöbel, H.W. Hamacher, A. Liebers, and D. Wagner. The continuous stop location problem in public transportation networks. *Submitted to international journal*, 2002.

17. D.B. Shmoys, É. Tardos, and K. Aardal. Approximation algorithms for facility location problems. In *Proc. of the 29th ACM STOC*, pages 265–274, 1997.

18. A. Tamir. Personal comunication. July 2002.

Differential Approximation
for Some Routing Problems

Cristina Bazgan[1], Refael Hassin[2], and Jérôme Monnot[1]

[1] Université Paris Dauphine, LAMSADE, 75016 Paris, France
{bazgan,monnot}@lamsade.dauphine.fr
[2] Dep. of Statistics and Operations Research,
Tel-Aviv University, Tel-Aviv 69978, Israel
hassin@post.tau.ac.il

Abstract. We study vehicle routing problems with constraints on the distance traveled by each vehicle or on the number of vehicles. The objective is to minimize the total distance traveled by vehicles. We design constant differential approximation algorithms for some of these problems. In particular we obtain differential bounds: $\frac{1}{2}$ for METRIC 3VRP, $\frac{3}{5}$ for METRIC 4VRP, $\frac{2}{3}$ for METRIC kVRP with $k \geq 5$, $\frac{1}{2}$ for the non-metric case for any $k \geq 3$, and $\frac{1}{3}$ for CONSTRAINED VRP. We prove also that MIN-SUM EkTSP is $\frac{2}{3}$ differential approximable and has no differential approximation scheme, unless $\mathbf{P} = \mathbf{NP}$.

Keywords: differential ratio, approximation algorithm, VRP, TSP

1 Introduction

Vehicle routing problems that involve the periodic collection and delivery of goods and services as mail delivery or trash collection are of great practical importance. Simple variants of these real problems can be modeled naturally with graphs. Unfortunately even simple variants of vehicle routing problems are **NP**-hard. In this paper we consider approximation algorithms, and measure their efficiencies in two ways. One is the *standard* measure giving the ratio $\frac{apx}{opt}$, where *opt* and *apx* are the values of an optimal and approximate solution, respectively. The other measure is the *differential* measure, that compares the worst ratio of, on the one hand, the difference between the cost of the solution generated by the algorithm and the worst cost, and on the other hand, the difference between the optimal cost and the worst cost. Formally, the differential measure gives the ratio $\alpha = \frac{wor-apx}{wor-opt}$, where *wor* is the value of the optimal solution for the complementary problem. In [11], the measure $1 - \alpha$ is considered and it is called there z-approximation. Justification for this measure can be found for example in [1,5,18,11,14].

The main subject of this paper is differential approximation of the routing problems. In these problems n *customers* have to be served by *vehicles* of limited capacity from a common *depot*. A solution consists of a set of routes, where each starts at the depot and returns there after visiting a subset of customers, such

R. Petreschi et al. (Eds.): CIAC 2003, LNCS 2653, pp. 277–288, 2003.
© Springer-Verlag Berlin Heidelberg 2003

that each customer is visited exactly once. We refer to a problem as a VEHICLE
ROUTING PROBLEM (VRP) if there is a constraint on the (possibly weighted)
number of customers visited by a vehicle. This constraint reflects the assumption
that the vehicle has a finite capacity and that it *collects* from the customers (or
distributes among them) a commodity. The goal is to find a solution such that
the total length of the routes is as small as possible. In other cases, the vehicle
is just supposed to *visit* the customers, for example, in order to serve them. In
such cases we refer to the problem as a TSP problem.

The problems that are considered here generalize the (undirected) TRAVEL-
ING SALESMAN PROBLEM (TSP). Differential approximation algorithms for the
TSP are given by Hassin and Khuller [11] and Monnot [14]. We will sometimes
use these algorithms to generate approximations for the problems of this paper.
However, we note an important difference. In the TSP, adding a constant k to
all of the edge length does not affect the set of optimal solutions or the value of
the differential ratio. The reason is that every solution contains exactly n edges
and therefore every solution value increases by exactly the same value, namely
$n \cdot k$. In particular, this means that for the purpose of designing algorithms with
bounded differential ratio, it doesn't matter whether d is a metric or not (it can
be made a metric by adding a suitable constant to the edge lengths). In contrast,
in some of problems dealt with here, the number of edges used by a solution is
not the same for every solution and therefore it may turn out, as we will see,
that in some cases the metric version is easier to approximate.

It is easy to see that 2VRP is polynomial time solvable. For $k \geq 3$, MET-
RIC kVRP was proved **NP**-hard by Haimovich and Rinnooy Kan [8]. In [9],
Haimovich, Rinnooy Kan and Stougie gave a $\frac{5}{2} - \frac{3}{2k}$ standard approximation
for Metric kVRP. We study for the first time the differential approximability of
kVRP. More exactly we give a $\frac{1}{2}$ differential approximation for the non-metric
case for any $k \geq 3$. We improve it to $\frac{3}{5}$ for METRIC 4VRP and to $\frac{2}{3}$ for MET-
RIC kVRP with $k \geq 5$. An approximation lower bound of $\frac{2219}{2220}$ is given here for
METRIC nVRP with length 1 and 2 using a lower bound of TSP(1,2) [6].

MIN-SUM EkTSP is a generalization of TSP where we search to cover the
customers by *exactly* k vehicles such that the total length is minimum. Bellmore
and Hong [3] showed that when we search for a solution with *at most* k cycles
then MIN-SUM kTSP is equivalent to TSP on an extended graph. As for MIN
TSP, MIN-SUM EkTSP is differential equivalent to METRIC MIN-SUM kTSP
or MAX-SUM EkTSP. We show in this paper that METRIC MIN-SUM EkTSP
is $\frac{2}{3}$ differential approximable and it has no differential approximation scheme
unless **P** = **NP**.

The paper is organized as follows: In section 2 we give the necessary definitions. In section 3 we give a constant differential approximation algorithm for
GENERAL kVRP, and a better constant differential approximation for the metric case. In section 4 the main result is a constant differential approximation for
CONSTRAINED VRP. In section 5 we show that MIN-SUM EkTSP is constant
differential approximable and has no differential approximation scheme.

2   Terminology

We first recall a few definitions about differential approximability. Given an instance x of an optimization problem and a feasible solution y of x, we denote by $val(x, y)$ the value of the solution y, by $opt(x)$ the value of an optimal solution of x, and by $wor(x)$ the value of a worst solution of x. The *differential approximation ratio* of y is defined as $\delta(x, y) = \frac{|val(x,y)-wor(x)|}{|opt(x)-wor(x)|}$. This ratio measures how the value of an approximate solution $val(x, y)$ is located in the interval between $opt(x)$ and $wor(x)$. More exactly it is equivalent for a minimization problem to prove $\delta(x, y) \geq \varepsilon$ and $val(x, y) \leq \varepsilon opt(x) + (1 - \varepsilon)wor(x)$.

For a function f, $f(n) < 1$, an algorithm is a *differential $f(n)$-approximation algorithm* for a problem Q if, for any instance x of Q, it returns a solution y such that $\delta(x, y) \geq f(|x|)$. We say that an optimization problem is *constant differential approximable* if, for some constant $\delta < 1$, there exists a polynomial time differential δ-approximation algorithm for it. An optimization problem has a *differential polynomial time approximation scheme* if it has a polynomial time differential $(1 - \varepsilon)$-approximation, for every constant $\varepsilon > 0$. We say that two optimization problems are *differential equivalent* (reps., *standard equivalent*) if a differential δ-approximation (reps., standard δ-approximation) algorithm for one of them implies a differential δ-approximation (reps., standard δ-approximation) algorithm for the other one.

We consider in this paper several routing problems. The problems are defined on a complete undirected graph denoted $G = (V, E)$. The vertex set V consists of a *depot vertex* 0, and *customer vertices* $\{1, \ldots n\}$. There is also a function $d : E \rightarrow R$, where $d_{i,j} \geq 0$ denotes the *length* of edge $(i, j) \in E$. In the rest of the paper we call a such graph a *complete valued graph*. We refer to the version of the problem in which d is assumed to satisfy the triangle inequality as the *metric case*. The output to the problems consists of simple cycles, $C_1, \ldots, C_p$, such that $V(C_i) \cap V(C_j) = \{0\}, \forall i \neq j$, and $\cup_{i=1}^{p} V(C_i) = V$. We call such a set of cycles a *p-tour*. We now describe the problems. For each one we specify the input, the problem's constraints, and the output.

kVRP
Input: A complete valued graph.
Constraint: $|C_j| \leq k + 1$, $j = 1, \ldots, p$.
Output: A p-tour minimizing the total length of the cycles.

CONSTRAINED VRP
Input: A complete valued graph and a metric $\ell : E \rightarrow R_+$, and $\lambda > 0$.
Constraint: $\sum_{(i,j)\in C_q} \ell_{i,j} \leq \lambda$, $q = 1, \ldots, p$.
Output: A p-tour minimizing the total length of the cycles.

kWVRP
Input: A complete valued graph and a function $w : \{1, \ldots, n\} \rightarrow R$ where w_i denotes the weight of i.
Constraint: $\sum_{i\in C_j} w_i \leq k$, $j = 1, \ldots, p$.
Output: A p-tour minimizing the total length of the cycles.

MIN-SUM EkTSP
Input: A complete valued graph.
Constraint: $p = k$.
Output: A p-tour minimizing the total length of the cycles.

For a problem Q, we denote by Q(1, 2) the version of Q where lengths are 1 and 2. We will use in this paper the following problems:

MIN TSP PATH(1,2) is the variant of MIN TSP(1,2) problem where in place of a tour we ask for a Hamiltonian path of minimum length. We can prove that MIN TSP PATH(1,2) where the graph induced by edges of weight 1 is Hamiltonian and cubic has no differential approximation scheme, unless $\mathbf{P} = \mathbf{NP}$. The result follows since MIN TSP PATH(1,2) on these instances has no standard approximation scheme, unless $\mathbf{P} = \mathbf{NP}$, Bazgan [2].

PARTITIONING INTO PATHS OF LENGTH k (kPP): Given a graph $G = (V, E)$ with $|V| = (k + 1)q$, is there a partition of V into q disjoint sets $V_1, \ldots, V_q$ of $k + 1$ vertices each, so that each subgraph induced by V_i has an Hamiltonian path? 2PP have been proved **NP**-complete in [7] whereas, more generally, the **NP**-completeness of kPP is proved in Kirkpatrick and Hell [12] as a special case of G-PARTITION PROBLEM.

A *binary 2-matching* (also called *2-factor* or *cycle cover*) is a subgraph in which each vertex of V has a degree of exactly 2. A *minimum binary 2-matching* is one with minimum total edge weight. Hartvigsen [10] has shown how to compute a minimum binary 2-matching in $O(n^3)$ time. More generally, a *binary f-matching*, where f is a vector of size $n + 1$, is a subgraph in which each vertex i of V has a degree of exactly f_i. A *minimum f-matching* is one with minimum total edge weight and is computable in polynomial time, Cook et al. [4].

3 kVRP

nVRP is standard equivalent with TSP. So, when the distance is not metric and using the result of Sahni and Gonzalez [17] we deduce that nVRP is not $2^{p(n)}$ standard approximable for any polynomial p, unless $\mathbf{P} = \mathbf{NP}$. In fact for any $k \geq 5$ constant the problem is as hard to approximate as nVRP.

3.1   General kVRP

When d is a metric, the reduction of TSP to nVRP is straightforward, and it easily follows that computing *opt* is **NP**-hard. However, there is no such reduction between the corresponding maximization problems MAX TSP and MAX nVRP leading to the conclusion that computing *wor* is also **NP**-hard.

Proposition 1. *Computing a worst solution for kVRP is* **NP**-*hard for any $k \geq 3$ even if the distance function takes only two values.*

Proof. We use a reduction from PARTITIONING INTO PATHS OF LENGTH k (kPP). Let $G = (V, E)$ with $V = \{1, \ldots, (k+1)q\}$ be an instance of kPP. We construct an instance of kVRP by the following way: we add the depot vertex 0 and set $d_e = 3$ if $e \in E$ and $d_e = 1$ otherwise. It is easy to verify that the answer to kPP is positive if and only if $wor \geq q(3k + 2)$.

In the following we give a $\frac{1}{2}$ differential approximation for non-metric kVRP. We apply ideas similar to those in Hassin and Khuller [11].

We first compute a lower bound LB. Then we generate a feasible solution for G with value $apx = LB + \delta_1$. Next, we generate another feasible solution of value $bad = LB + \delta_2$ where $\delta_2 \geq \delta_1$. This proves that the approximate solution with value apx is an α-differential approximation where

$$\alpha = \frac{wor - apx}{wor - opt} \geq \frac{bad - apx}{bad - opt} \geq \frac{\delta_2 - \delta_1}{bad - LB} = \frac{\delta_2 - \delta_1}{\delta_2}, \tag{1}$$

since for a minimization problem $wor \geq bad \geq apx \geq opt \geq LB$. To generate LB we replace 0 by a complete graph with a set V_0 of $2n$ vertices and zero length edges. The distance between a vertex of V_0 and a vertex i of $V \setminus V_0$ is the same as the distance between 0 and i. Denote the resulting graph by G'. Compute in G' a minimum weight binary 2-matching M.

Lemma 1. *Let LB denote the weight of M. Then $opt_{VRP} \geq LB$.*

Theorem 1. kVRP *is $\frac{1}{2}$ differential approximable.*

Proof. We transform each cycle of M into a cycle in G, replacing the vertices of V_0 by a single occurrence of the depot vertex 0. We work on each cycle of M separately. For each cycle we describe solutions sol_1, sol_2 and sol_3 such that $\delta(sol_1) + \delta(sol_2) = \delta(sol_3)$. We define apx to be the VRP solution obtained by concatenating the shortest of sol_1 and sol_2 for every cycle, and we define bad similarly but using sol_3. We obtain therefore that $\delta(bad) \geq 2\delta(apx)$, and the theorem is proved by (1). In this proof, sol_3 will be always described by the cycles $(0, 1, 0), \ldots, (0, m, 0)$.

First, consider a cycle that does not contain the depot, and w.l.o.g. denote its vertices by $(1, \ldots, m)$. The construction depends on the parity of m.

Suppose that m is even ($m \neq 2$ since the 2-matching is binary). Let sol_1 consist of the cycles $(0, 1, 2, 0), \ldots, (0, m - 1, m, 0)$. Let sol_2 consist of the cycles $(0, m, 1, 0), \ldots, (0, m - 2, m - 1, 0)$ (See Figure 1 for $m = 6$).

Suppose now that m is odd. We modify sol_1 by choosing $(0, m - 2, m - 1, m)$ as the last cycle (see Figure 2). We modify sol_2 by using $(0, m - 1, 0)$ as the last cycle.

Consider a cycle $(0, 1, \ldots, m, 0)$ in M. If $m \leq k$ we don't change it. Otherwise:

If $m \equiv 0 \bmod 3$ then let sol_1 consist of the cycles $(0, 1, 0), (0, 2, 3, 0), \ldots, (0, m - 2, 0), (0, m - 1, m, 0)$. Let sol_2 consist of the cycles $(0, m, 1, 2, 0), \ldots, (0, m - 3, m - 2, m - 1, 0)$.

If $m \equiv 1 \bmod 3$ then let sol_1 consist of the cycles $(0, 1, 0), (0, 2, 3, 0), \ldots, (0, m - 3, 0), (0, m - 2, m - 1, m, 0)$. Let sol_2 consist of the cycles $(0, m, 1, 2, 0), \ldots, (0, m - 4, m - 3, m - 2, 0)$, and $(0, m - 1, 0)$.

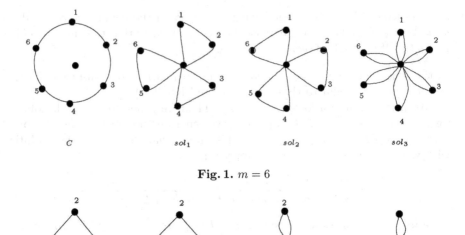

Fig. 1. $m = 6$

Fig. 2. $m = 3$

If $m \equiv 2 \mod 3$ then let sol_1 consist of the cycles $(0, 1, 0),(0, 2, 3, 0),\ldots,(0, m-3, m-2, 0),(0, m-1, m, 0)$. Let sol_2 consist of the cycles $(0, m, 1, 2, 0),\ldots,(0, m-5, m-4, m-3, 0)$, and $(0, m-2, m-1, 0)$.

Let $\delta(sol_i)$ denote the added cost of sol_i with respect to the length of C. Since M was computed to have a minimum cost, $\delta(sol_i) \geq 0$ and we have $\delta(sol_3) = \delta(sol_1) + \delta(sol_2)$ which complete the proof.

3.2 METRIC kVRP

When d is a metric, computing a worst solution becomes easy since any feasible cycle can be broken into two cycle without decrease the weight and then:

Lemma 2. $wor_{VRP} = 2 \sum_{i=1}^{n} d_{0,i}$

In Theorem 1 we have shown that kVRP is $\frac{1}{2}$ differential approximable. We now show that in the metric case, the same bound can be achieved by a simpler algorithm: we compute a minimum weight perfect matching M on the subgraph induced by $\{1,\ldots,n\}$, if n is even, or by $\{0, 1,\ldots,n\}$ if n is odd. We link each endpoint different of 0 of M to the depot. It is easy to see that $opt_{VRP} \geq 2d(M)$ by walking around an optimum solution for kVRP and by shortcut it in order to obtain a Hamiltonian cycle. Using Lemma 2 and the construction of the approximate solution, we obtain: $apx = d(M)+\sum_{i=1}^{n} d_{0,i} \leq \frac{1}{2} opt_{VRP}+\frac{1}{2} wor_{VRP}$ proving that the result is a $\frac{1}{2}$ differential approximation.

In Haimovich et al. [9], a $\frac{5}{2} - \frac{3}{2k}$ standard approximation for METRIC kVRP is obtained by reduction to METRIC TSP and using Christofides' algorithm. We proceed similarly for the differential case.

Theorem 2. METRIC kVRP *is* $\delta \cdot \frac{k-1}{k}$ *differential approximable, where* δ *is the differential approximation ratio for* METRIC TSP.

Proof. Our algorithm modifies the *Optimal Tour Partitioning* heuristic of Haimovich, Rinnooy Kan and Stougie [9]: first construct a tour T on V using the differential approximation algorithm for TSP of value $val(T)$. W.l.o.g., assume that this tour is described by the sequence $(0, 1, \ldots, n, 0)$. We produce k solutions sol_i for $i = 1, \ldots, k$ and we select the best solution. The first cycle of sol_i is formed by the sequence $(0, 1, \ldots, i, 0)$ and then each other cycle (except eventually the last) of sol_i has exactly k consecutive vertices (for instance, the second cycle is $(0, i+1, \ldots, i+k, 0)$) and finally, the last cycle is formed by the unvisited vertices (connecting n to the depot 0). Denote by apx_i for $i = 1, \ldots, k$ the values of the k solutions and by apx the value of the best one.

Consider $sol_1, \ldots, sol_k$; each edge of $T \setminus \{(0, 1), (0, n)\}$ appear exactly $(k-1)$ times and each edge $(0, j)$ for $j \neq 1, n$ appears exactly twice. Finally, the edges $(0, 1)$ and $(0, n)$ appears exactly $(k+1)$ times. Using Lemma 2, we deduce: $apx \leq \frac{1}{k}\sum_{i=1}^{k} apx_i \leq \frac{(k-1)}{k}val(T) + \frac{2}{k}wor_{VRP}$. By hypothesis, T satisfies: $val(T) \leq (1 - \delta)wor_{TSP} + \delta opt_{TSP}$ and since it is possible to construct from an optimum solution of VRP a solution of TSP with smaller a value (using the triangle inequality), it follows that $opt_{TSP} \leq opt_{VRP}$. Also, by connecting the depot twice with each customer, we can construct from a solution of TSP a solution of VRP with greater value, and therefore $wor_{TSP} \leq wor_{VRP}$.

Using the previous inequalities we obtain that

$$apx \leq \delta\frac{k-1}{k}opt_{VRP} + \left(1 - \delta\frac{k-1}{k}\right)wor_{VRP}.$$

Since the best known differential approximation algorithm for TSP is $\frac{2}{3}$ [11,14] then the algorithm of Theorem 2 is an $\frac{2}{3} \cdot \frac{k-1}{k}$ differential approximation algorithm for metric kVRP. So, for $k \geq 5$, we obtain a ratio strictly better that the bound produced by the Theorem 1 or by the previous algorithm build on a matching. Now, we will improve all the previous bounds for $k > 3$ since we give a $\frac{3}{5}$ (resp., $\frac{2}{3}$) differential approximation for METRIC 4VRP (resp., METRIC kVRP for $k \geq 5$).

Theorem 3. METRIC kVRP *is* $min\{\frac{2}{3}, \frac{k-1}{k+1}\}$ *differential approximable.*

Proof. Our algorithm works as follow: we compute a minimum weight binary 2-matching $M = (C_1, \ldots, C_q)$ on the subgraph induced by $\{1, \ldots, n\}$. Then, for each cycle of it, we produce several solutions and we take the best one among them. These different solutions depend on the size m_i of the cycle C_i of M. Without loss of generality, assume that this cycle is described by the sequence $(1, \ldots, m_i, 1)$ with $m_i \geq 3$. Let $wor_i = 2\sum_{j=1}^{m_i} d_{0,j}$.

If $m_i \leq k$, then we produce m_i solutions $sol_1, \ldots, sol_{m_i}$ where sol_j is obtained just by deleting the edge $(j, j + 1)$ (mod m_i) and by connecting j and $j + 1$ to the depot. Since $wor_i \geq d(C_i)$ by the triangle inequality and $m_i \geq 3$ we have: $apx_i = \min_j d(sol_j) \leq \frac{m_i-1}{m_i}d(C_i) + \frac{1}{m_i}wor_i \leq \frac{2}{3}d(C_i) + \frac{1}{3}wor_i$.

Now, assume that $m_i = kp + r$ with $p \geq 1$ and $0 \leq r \leq k - 1$. We still produce m_i solutions $sol_1, \ldots, sol_{m_i}$ but in this case in a different way. In order to build the solution sol_j, we delete from C_i the edge $(j - 1, j)$ and the edges $(j - 1 + r + k\ell, j + r + k\ell)$ for $\ell = 1, \ldots, p$ (the indices are taken mod m_i) and we add for each endpoint of these paths, the link with the depot. Note that when $r = 1$, the first path is the isolated vertex j and then in this case, we add once again the edge $(0, j)$.

If $r = 0$, then we deduce: $apx_i \leq \frac{p(k-1)}{kp}d(C_i) + \frac{p}{kp}wor_i \leq \frac{2}{3}d(C_i) + \frac{1}{3}wor_i$.

If $r \geq 1$, then we obtain: $apx_i \leq \frac{p(k-1)+r-1}{kp+r}d(C_i) + \frac{p+1}{kp+r}wor_i \leq \frac{k-1}{k+1}d(C_i) + \frac{2}{k+1}wor_i$.

Finally, since $d(M) \leq opt_{VRP}$, by taking the minimum between $\frac{2}{3}$ and $\frac{k-1}{k+1}$ and by summing over i the previous inequalities we obtain the expected result.

Since nVRP and TSP are standard equivalent by using the result of Papadimitriou and Yannakakis [16] we deduce immediately that nVRP(1,2) has no standard approximation scheme. Also TSP(1,2) has no differential approximation scheme, Monnot et al. [15] but we cannot deduce immediately that nVRP(1,2) has no differential approximation scheme since wor_{nVRP} and wor_{TSP} can be very far. However, we prove in the following a lower bound for the approximation of nVRP(1,2).

Theorem 4. METRIC nVRP(1, 2) *is not* $\frac{2219}{2220} - \varepsilon$ *differential approximable, for any constant* ε, *unless* **P = NP**.

Proof. Since $wor_{nVRP} \leq 4n \leq 4opt_{nVRP}$, a δ differential approximation for nVRP(1, 2) gives a $\delta + 4(1 - \delta)$ standard approximation for nVRP(1, 2). Using the negative result given in Engebretsen and Karpinski[6]: TSP(1,2) is not $\frac{741}{740} - \varepsilon$ standard approximable, then we obtain the expected result.

4   CONSTRAINED VRP

We assume now that each edge is associated with a weight ℓ (where $\ell_{i,j} \geq 0$ denotes the cost/time of traversing the edge (i, j)) satisfying the triangle inequality, and the solution must satisfy that the total weight on each cycle does not exceed λ. Note that if we do not assume that ℓ is a metric then even deciding whether the problem has any feasible solution is **NP**-complete.

Theorem 5. *Deciding the feasibility of* CONSTRAINED VRP *is NP-complete.*

Proof. In order to prove the NP-hardness, we reduce HAMILTONIAN $s - t$ PATH problem to CONSTRAINED VRP. From a graph $G = (V, E)$ with $V = \{1, \ldots, n\}$, we construct a graph G' instance of CONSTRAINED VRP by adding a depot vertex 0. We define the function l as follows: $\ell_{0,s} = \ell_{0,t} = 1$, $\ell_{0,i} = \lambda$, for $i \neq s, t$, $\ell_{i,j} = 1$ if $(i, j) \in E$ and $\ell_{i,j} = \lambda$ if $i, j \in \{1, \ldots, n\}$ and $(i, j) \notin E$. Trivially there is a feasible solution for G' only if $\lambda \geq n + 1$. It is easy to see that CONSTRAINED VRP has a feasible solution iff G contains a Hamiltonian path between s and t.

Therefore, we assume that ℓ satisfies the triangle inequality, and to ensure feasibility we also assume that $2\ell_{0,i} \leq \lambda$ for $i = 1, \ldots, n$. Using similar ideas as those used for kVRP we can prove:

Theorem 6. CONSTRAINED VRP *is $\frac{1}{3}$ differential approximable.*

Proof. We start with a binary 2-matching as described in Lemma 1 except that the initial graph is not a complete undirected graph K_{n+1} but a partial graph G' of it built by deleting the edges (i, j) for $i \neq 0$ and $j \neq 0$ such that $\ell_{0,i} + \ell_{i,j} + \ell_{j,0} > \lambda$. It is easy to observe that M is still a lower bound of an optimal solution of CONSTRAINED VRP. As previously, we work on each cycle of M separately.

First, consider a cycle of M that does not contain the depot, and w.l.o.g. denote its vertices by $(1, \ldots, m)$. The construction depends on the parity of m. If m is even, we produce two same solutions sol_1 and sol_2 of Theorem 1, i.e., $(0, 1, 2, 0), (0, 3, 4, 0), \ldots, (0, m-1, m, 0)$ and $(0, m, 1, 0), (0, 2, 3, 0), (0, 4, 5, 0), \ldots, (0, m - 2, m - 1, 0)$. Now, if m is odd, we produce m solutions $sol_1, \ldots, sol_m$ where the solution sol_i for $i = 1, \ldots, m$ consist of the cycles $(0, i, 0), (0, i + 1, i + 2, 0), (0, i + 3, i + 4, 0), \ldots, (0, i - 2, i - 1, 0)$ (the indices are taken mod m).

Second, consider a cycle $(0, 1, \ldots, m, 0)$ in M. If $\sum_{i=1}^{m} w_i \leq \lambda$ we don't change it. Otherwise, we have $m \geq 3$ and we produce two solutions sol_1 and sol_2 depending on the parity of m. Let us consider the two cases:

Suppose that m is even. Let sol_1 consist of the cycles $(0, 1, 2, 0), (0, 3, 4, 0), \ldots, (0, m - 1, m, 0)$. Let sol_2 consist of the cycles $(0, 1, 0), (0, 2, 3, 0), \ldots, (0, m - 2, m - 1, 0), (0, m, 0)$. Now, suppose that m is odd. Let sol_1 consist of the cycles $(0, 1, 0), (0, 2, 3, 0), \ldots, (0, m - 1, m, 0)$. Let sol_2 consist of the cycles $(0, 1, 2, 0), (0, 3, 4, 0), \ldots, (0, m - 2, m - 1, 0), (0, m, 0)$.

In all cases the solution $bad = (0, 1, 0), \ldots, (0, n, 0)$ satisfies $\min_i \delta(sol_i) \leq \frac{2}{3} \delta(bad)$ (this relation holds with equality when each cycle has exactly three vertices and does not contain the depot, since in this case $\delta(sol_1) + \delta(sol_2) + \delta(sol_3) = 2\delta(bad)$). Therefore, by (1), the best of the solutions sol_i is a $\frac{1}{3}$ differential approximation. $\blacksquare$

In Haimovich et al. [9], the authors consider two versions of kVRP with additional constraint on the length of each cycles. In the first problem that we will call here WEIGHTED kVRP (kWVRP), each customer has a weight and we want to find a solution such that the total customer weight on each cycle does not exceed k. In the second, called in [9] MIN METRIC DISTANCE, we want to find a solution such that the total distance on each cycle does not exceed a given bound λ. For each of these two problems, we give a differential reduction preserving approximation scheme from CONSTRAINED VRP.

Lemma 3. *A δ differential approximation solution for CONSTRAINED VRP (respectively, metric case) is also a δ differential approximation for kWVRP (respectively, metric case).*

Proof. Let $G = (V, E)$ with d and w be an instance of kWVRP. We construct an instance of CONSTRAINED VRP as follows. First we set $\lambda = k$. The graph and the function d are the same whereas the function ℓ is defined by: $\ell_{i,j} = \frac{w_i + w_j}{2}$ where we assume that $w_0 = 0$. This function satisfies the triangle inequality. Moreover, let C be a cycle linking the depot to a subset of customers. We have $\sum_{q \in C} w_q \leq k$ iff $\sum_{(i,j) \in C} \ell_{i,j} \leq \lambda$.

Corollary 1. kWVRP *is* $\frac{1}{3}$ *differential approximable.*

MIN METRIC DISTANCE is a particular case of METRIC CONSTRAINED VRP where the function ℓ is exactly the function d. Thus, from Theorem 6 we deduce the corollary:

Corollary 2. MIN METRIC DISTANCE *is* $\frac{1}{3}$ *differential approximable.*

5   MIN-SUM EkTSP

It is easy to see that MIN-SUM EkTSP is differential equivalent to METRIC MIN-SUM EkTSP since the number of edges in every solution is the same (like in the TSP case). Hence, we add a constant to all the edge lengths and achieve the triangle inequality without affecting the best and worst solutions.

Theorem 7. METRIC MIN-SUM EkTSP *is* $\frac{2}{3}$ *differential approximable,* $\forall k \geq 1$.

Proof. Add to every edge incident with the depot a parallel copy. Compute a minimum binary f-matching M on G where $f(0) = 2k$ and $f(v) = 2$ for $v \in V \setminus \{0\}$. Compute by using an algorithm of Hassin and Khuller [11] or Monnot [14] a solution C' for TSP on the subgraph G' of G induced by $V' = V \setminus (\cup_{i=1}^{k-1} V(C_i)) \cup \{0\}$, where $C_1, \ldots, C_k$ are the cycles of M containing the depot 0. The approximate solution sol is composed of C' and the cycles $C_1, \ldots, C_{k-1}$. See Figure 3. Since M is an minimum binary f-matching M on G then $M' = M \setminus (\cup_{i=1}^{k-1} C_i)$ is an optimum binary 2-matching on G'. We denote by $r = \sum_{i=1}^{k-1} d(C_i)$; the TSP algorithm gives a solution satisfying $val \leq \frac{2}{3} d(M') + \frac{1}{3} wor_{TSP}(G')$. Since $wor_{kTSP}(G) \geq wor_{TSP}(G') + r$ and $opt_{kTSP}(G) \geq d(M)$, we deduce that the value of sol satisfies: $apx = val + r \leq \frac{2}{3}[d(M') + r] + \frac{1}{3}[wor_{TSP}(G') + r] \leq \frac{2}{3} opt_{kTSP}(G) + \frac{1}{3} wor_{kTSP}(G)$

Theorem 8. *Unless* $\mathbf{P} = \mathbf{NP}$, MIN-SUM EkTSP(1,2) *has no standard and differential approximation scheme.*

Proof. We reduce MIN TSP PATH (1,2) on Hamiltonian cubic graphs to MIN-SUM E2TSP. From a graph $G = (V, E)$ on n vertices, we construct a graph G' instance of MIN-SUM E2TSP. G' consists of two copies of G and a vertex 0 (the depot). Within a copy, the edges have the same distance as in G; $d_{0,i} = 1$, for each vertex i in one of the two copies; $d_{i,j} = 2$ if i and j are vertices in different copies. We have $opt(G') = 2opt(G) + 4$ and $wor(G') = 2wor(G) + 4$. Given a solution S of G' with two cycles, we can transform it in another one S' that contains exactly two cycles $(0, P_1, 0)$, $(0, P_2, 0)$, each of these two paths are contained in a copy of

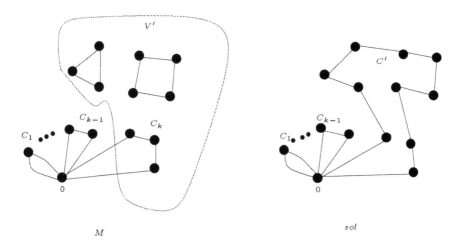

Fig. 3. M and sol

G and with a better value. The idea for doing this is to remove the edges between the two copies and to replace them by the missing edges in the two copies. We consider as solution for G the path with the smallest value among the two. So, $val = \min\{val(P_1), val(P_2)\} \leq \frac{val(P_1)+val(P_2)}{2} = \frac{val(S')-4}{2} \leq \frac{val(S)-4}{2}$. Since $opt(G) = \frac{opt(G')}{2} - 2$ and $wor(G) = \frac{wor(G')}{2} - 2$ then a δ differential approximation of MIN-SUM E2TSP gives a δ differential approximation for MIN TSP PATH (1,2) on Hamiltonian and cubic graphs and the conclusion follows.

References

1. G. Ausiello, A. D'Atri and M. Protasi, "Structure preserving reductions among convex optimization problems," *Journal of Computing and System Sciences* **21** (1980) 136-153.
2. C. Bazgan, *Approximation of optimization problems and total function of NP*, Ph.D. Thesis (in French), Université Paris Sud (1998).
3. M. Bellmore and S. Hong, "Transformation of Multi-salesmen Problem to the Standard Traveling Salesman Problem," *Journal of the Association for Computing Machinery* **21** (1974) 500-504.
4. W.J. Cook, W.H. Cunningham, W.R. Pulleyblank, and A. Schrijver *Combinatorial Optimization* John Wiley & Sons Inc New York 1998 (Chapter 5.5).
5. M. Demange and V. Paschos, "On an approximation measure founded on the links between optimization and polynomial approximation theory," *Theoretical Computer Science* **158** (1996) 117-141.
6. L. Engebretsen and M. Karpinski, "Approximation hardness of TSP with bounded metrics," http://www.nada.kth.se/~enge/papers/BoundedTSP.pdf
7. M. R. Garey and D. S. Johnson, "Computers and intractability. A guide to the theory of NP-completeness," *Freeman, C.A. San Francisco* (1979).

8. M. Haimovich and A. H. G. Rinnooy Kan, "Bounds and heuristics for capacitated routing problems," *Mathematics of Operations Research* **10** (1985) 527-542.

9. M. Haimovich, A. H. G. Rinnooy Kan and L. Stougie, " Analysis of Heuristics for Vehicle Routing Problems," in Vehicle Routing Methods and Studies, *Golden, Assad editors, Elsevier* (1988) 47-61.

10. D. Hartvigsen, *Extensions of Matching Theory.* Ph.D. Thesis, Carnegie-Mellon University (1984).

11. R. Hassin and S. Khuller, "z-approximations," *Journal of Algorithms* **41** (2001) 429-442.

12. D. G. Kirkpatrick and P. Hell, "On the completeness of a generalized matching problem," *Proc. of the 10th ACM Symposium on Theory and Computing* (1978) 240-245.

13. C-L. Li, D. Simchi-Levi and M. Desrochers, "On the distance constrained vehicle routing problem," *Operations Research* **40** 4(1992) 790-799.

14. J. Monnot, "Differential approximation results for the traveling salesman and related problems," *Information Processing Letters* **82** (2002) 229-235.

15. J. Monnot, V. Th. Paschos and S. Toulouse, " Differential Approximation Results for the Traveling Salesman Problem with Distances 1 and 2," *Proc. FCT* (2001) 275-286.

16. C. Papadimitriou and M. Yannakakis, "The traveling salesman problem with distances one and two," *Mathematics of Operations Research* **18(1)** (1993) 1-11.

17. S. Sahni and T. Gonzalez, "P-complete approximation problems" *Journal of the Association for Computing Machinery* **23** (1976) 555-565.

18. E. Zemel, "Measuring the quality of approximate solution to zero-one programming problems," *Mathematical Operations Research* 6(1981) 319-332.

Author Index